International Handl
Technology Educat

INTERNATIONAL TECHNOLOGY EDUCATION STUDIES

Volume 7

Scope
Technology Education has gone through a lot of changes in the past decades. It has developed from a craft oriented school subject to a learning area in which the meaning of technology as an important part of our contemporary culture is explored, both by the learning of theoretical concepts and through practical activities. This development has been accompanied by educational research. The output of research studies is published mostly as articles in scholarly Technology Education and Science Education journals. There is a need, however, for more than that. The field still lacks an international book series that is entirely dedicated to Technology Education. *The International Technology Education Studies* aim at providing the opportunity to publish more extensive texts than in journal articles, or to publish coherent collections of articles/chapters that focus on a certain theme. In this book series monographs and edited volumes will be published. The books will be peer reviewed in order to assure the quality of the texts.

International Handbook of Primary Technology Education

Reviewing the Past Twenty Years

Edited by

Clare Benson and Julie Lunt
Birmingham City University, UK

SENSE PUBLISHERS
ROTTERDAM/BOSTON/TAIPEI

A C.I.P. record for this book is available from the Library of Congress.

ISBN: 978-94-6091-544-4 (paperback)
ISBN: 978-94-6091-545-1 (hardback)
ISBN: 978-94-6091-546-8 (e-book)

Published by: Sense Publishers,
P.O. Box 21858,
3001 AW Rotterdam,
The Netherlands
www.sensepublishers.com

Printed on acid-free paper

TABLE OF CONTENTS

Part B: Issues within Design and Technology Education

PREFACE

The *International Handbook for Primary Technology Education* has its roots in the biennial international primary Design and Technology conference initiated in 1997 and hosted by the Centre for Research in Primary Technology (CRIPT) at Birmingham City University, England. The first conference was held seven years after the subject was introduced into the National Curriculum for all primary schools in England and Wales in 1990 and it was at this time that other countries worldwide were also introducing and developing the subject. It became, therefore, more urgent for a forum to be created through which research and curriculum development could be shared and disseminated, discussed and evaluated. The Proceedings from the Conferences have become one of the main publications that focus on primary developments (children aged 3–13 years) and this Handbook is based mainly on papers that have been delivered over the past 13 years. During this time, the subject has gathered pace around the world and it is now a compulsory subject in many countries in all continents. It has been interesting to watch its development, sometimes cyclical in nature; there have been periods of time when the subject has been high on the agenda of Governments and then declined as changing political power has affected educational issues. The development of the subject in different countries is outlined in Part A of the book; whilst in Part B many of the issues that arise from learning and teaching in Design and Technology are highlighted.

It is clearly evident from a study of different countries that the subject can be called Technology or Design and Technology. In some countries Technology is more aligned with the use of computers; although in others this would be Information and Communication Technology (ICT). In addition, Technology can be more focused on technical knowledge such as in Germany. However in this Handbook, it is usually the case that the underlying principles are the same whichever word is used. The common concepts relate to the designing and making of a product. These include consideration of user and purpose; functionality; and values and technical issues which need to be taken into account as the product is being created. The use of the word 'Technology' in the title reflects the fact that this term is the one most commonly used worldwide.

It is interesting that the growth of the subject has been worldwide and it is worth considering reasons for this. Apart from the political pressures, it is a subject that provides young people with unique opportunities to develop a range of process skills such as critical and creative thinking skills in addition to their practical skills, through undertaking authentic tasks, in which they can see a real purpose. The subject offers opportunities to link learning in a number of different disciplines. Through a variety of activities, the young people are able to take risks and develop skills that will be important to them both in their work and everyday lives.

We would like to give our thanks to all those who have contributed to this Handbook and for those children, teachers, teacher educators, and researchers who have supported the authors in countries worldwide. We hope that it will prove a valuable resource for all who use it and make a significant contribution to the on-going research in Primary Technology. Our final thanks go to Netta Pickett whose eagle eye and tenacity have proved invaluable in the formatting and final checks of the manuscript.

Clare Benson and Julie Lunt
Editors

PART A:
COUNTRY STUDIES

CLARE BENSON

1. TWENTY YEARS OF PRIMARY DESIGN AND TECHNOLOGY IN ENGLAND

Winners and Losers[1]

INTRODUCTION

It hardly seems possible that twenty years has passed since the first National Curriculum for primary schools in England (DES/WO, 1990) was introduced. It was felt that all children had an entitlement to experience a similar curriculum, although schools were free to expand the skills and content and to deliver the curriculum in ways that they felt were most appropriate for their children.

Included in this document was a new subject – that of Design and Technology. It appeared that it might be a 'loser' almost immediately from the way in which it was introduced to schools. Despite this inauspicious start, there has been so much to commend it - a vehicle through which children can enjoy and value education, develop lifelong skills, and engage with the designed and made world around them. Teachers who understand, and have confidence, to deliver the subject have certainly identified themselves as winners and have developed their teaching skills in so many ways.

This chapter provides a commentary on the development and implementation of Design and Technology in primary schools since 1990. Evidence has been gathered from the many documents, articles and books that have been published, from Inspection reports – Office for Standards in Education (OFSTED), from data gathered from primary teachers undertaking one day and extended primary continuing professional development (CPD) courses and from personal observations working in schools throughout England. 'Winners' and 'losers' during the implementation and development of this subject will be identified, and ways forward offered, as England waits once again on the brink of a possible radical change to primary education.

THE NEW NATIONAL CURRICULUM FOR DESIGN AND TECHNOLOGY

In the years before the National Curriculum was introduced, pupils in primary education (aged 3–11 years) had opportunities to do craft, woodwork, sewing and cooking. In science they mainly studied the natural world relating to flora and fauna, and areas such as mechanisms, forces, structures, and energy were largely ignored. Implementation of Design and Technology was to start from September 1990 in all state schools in England and Wales for children aged 5–16 years. (Wales and

C. Benson and J. Lunt (eds.), International Handbook of Primary Technology Education: Reviewing the Past Twenty Years. 3–11.

England shared the same curriculum until 1995 when Wales developed its own curriculum. Scotland and Northern Ireland had different curricula and have continued their own development of the subject. Scotland had until recently National Guidance rather than a National Curriculum, and both countries incorporate Technology within broader areas of experience).

The content of Design and Technology was developed by a working group (the Parkes Review, 1988), composed mainly from business and industry and those with secondary or higher education backgrounds. Primary views were certainly not strongly represented. Had primary educators been involved, the final document would almost certainly have been one, that when planning, primary teachers could relate to, and have used successfully. Instead it was overloaded with content and used vocabulary that was unfamiliar to primary teachers. For example, children were to be given experiences related to artefacts, systems, and environments – words that were used with little explanation. In addition, Design and Technology was joined with Information Technology in one document entitled Technology in the National Curriculum (DES/WO, 1990). It was in two parts: Design and Technology capability and Information Technology capability. If there had been confusion and misconceptions before, this only served to add to the problem. Primary teachers, who mostly had little knowledge and expertise in the area, immediately thought that the subject focused on computers (data gathered from teacher courses in 1990/91/92).

As implementation got underway, perhaps one of the most disappointing developments was the way in which the subject was delivered. Primary teachers saw the process as linear - to be worked through, rather than areas that could be covered in innovative and flexible ways. Children were given a design and make brief such as design and make a bag; they then might look at some bags; choose fabric that was given to them, make it, and evaluate the product. This often resulted in 30 identical bags. There was very little opportunity to design, to investigate a range of fabrics and fasteners, and to identify a user and purpose. This was hardly surprising. With little previous knowledge, and little continuing professional development (CPD), teachers looked for strategies that would help them; working through the four Attainment Targets in the National Curriculum (1990) gave them a structure, albeit a rigid one. Little, if any, support for implementation was organised nationally; few resources were available, partly due to the short time scale between publication of the National Curriculum and its implementation; CPD was almost non-existent in Design and Technology, partly due to the very small number of people who could provide relevant courses; and, perhaps most importantly, there was confusion as to the nature of the subject. It would be fair to say that many primary teachers concentrated on implementing the core subjects (English, Maths and Science) as the pressure of testing and target-setting grew (Ive, 1997, 1999; HMI, 1991).

Certainly this period could have been one of 'winners'. Design and Technology can offer children so many exciting, relevant opportunities for learning in a range of contexts. Through Design and Technology children can develop their process skills such as critical evaluation and creative thinking, as well as practical skills. Teachers were able to put the interests of the children at the centre of learning and to make appropriate cross curricular links. However, whilst it can be argued that there were no absolute 'losers' during this period, clearer documentation and teacher support would have resulted in greater success during this initial implementation period (Ive, 1997).

'GROWING' THE SUBJECT

1990–1995

It might be imagined that after such a confusing and pressured start, teachers would be given time to gain an understanding of the new Order, to plan appropriate schemes of work, find suitable resources, trial activities, and evaluate and change practice as they grew in confidence. This was not to be. Almost immediately, concerns were raised about the way in which the subject was developing and the document Technology in the National Curriculum - Getting it Right (Smithers & Robinson, 1992), prepared for the Engineering Council, talked about 'Mickey Mouse' activities with egg boxes and suggested that the subject had become generalised problem solving without a specific knowledge base. There was much debate as to the specific knowledge base that was part of the subject, some arguing that it needed to be very limited, whilst others promoted the notion that it should draw on a range of subjects, including Science, Art, Mathematics, IT (now ICT), Home Economics, and Business Studies.

Primary teachers started to create some exciting projects, but as more subjects came on stream in 1991, pressures of time and lack of understanding hindered these developments. New draft orders were published in 1992, and again in 1994 before the new final Order was published (SCCA, 1995). Thus, it seems hardly surprising that progress was slow during 1990–1995, and that teachers began to lose sight of the nature and importance of the subject. However, the Schools' Inspectors report '*Subject and Standards 1994–5*' (HMI, 1996) indicated that pupils were almost always enthusiastic about the subject and found the work enjoyable and interesting. Primary standards and standards in teaching (satisfactory and above) in English, Mathematics, Science, and Design and Technology were all graded at 80% with the exception of Key Stage 2 (7–11 years) teaching; this was graded at 75%. The format of reports changed after this, thus making comparisons in future years impossible.

1995–2000

The new National Curriculum in 1995 (DFE, 1995) clarified the nature of the subject through the statement:

> Pupils should be taught to develop their Design and Technology capability through combining designing and making skills with knowledge and understanding in order to design and make products. (p. 58)

The content of the 1995 document was slimmer than previous ones; the language of the document was more easily accessible for primary teachers; the attainment targets were reduced to two: designing and making; and the holistic nature of Design and Technology was emphasised through the level descriptions.

One of the major positive changes for primary teachers was the introduction of the section relating to opportunities through which pupils could develop their capability. Three opportunities were identified: assignments in which pupils design and make products (DMAs); focused practical tasks in which they develop and practise particular skills and knowledge (FPTs); and activities in which they investigate, disassemble, and evaluate simple products (IDEAs). This offered a clear structure for teaching, and whilst it did lead to linear teaching programmes, it promoted the idea that teaching pupils skills and knowledge was essential for good quality Design and Technology. Ive (1997) confirmed that HMI identified that this structure had been responsible for raising standards, certainly in primary Design and Technology.

Primary teachers were faced with major changes in 1998. The government, faced with poor standards in literacy and numeracy, introduced two new initiatives – one for literacy (Department for Education and Employment (DfEE), 1998) and one for numeracy (DfEE, 1999) – and whilst these Strategies were not mandatory, head teachers felt pressured into implementing them. There were time implications now as schools tried to cover the new, suggested content in the Strategies, and foundation subjects such as Design and Technology were once again under threat of being marginalised. The government, through the document *Maintaining breadth and balance* (QCA, 1998a), did suggest that schools should keep a balance in their curricula, but the threat of poor inspection results did much to negate the content of this document.

The publication of the national exemplar Schemes of Work (QCA, 1998b) for Key Stage 1 (5–7 years) and Key Stage 2 (7–11 years) did go some way, particularly at primary level, to offset the damage to the development of Design and Technology. The schemes of work were created to give schools possible ways of delivering the curriculum. They were only guidance and they were for schools to use as a tool against which they could evaluate, monitor, and plan new units of work. It was never the intention that these schemes should be followed in a step by step approach. However, problems arose when teachers kept rigidly to the suggested activities and did not develop the units to meet the needs of their own pupils and to fit in with the school's individual curriculum. Initially the Schemes therefore did much to help raise standards and, if used appropriately, continued to do so. However, where schools made no adaptation to the national Scheme, it could be seen to constrain the development of creativity and thinking skills.

The five year moratorium on National Curriculum change drew to an end and during 1998–9 fresh debates took place with regard to proposals for the new 2000 Curriculum (DfEE/QCA, 1999b). Perhaps the most significant change in Design and Technology was the inclusion of an importance statement at the beginning of the programmes of study. There was still some confusion as to the nature and importance of the subject and it was felt to be crucial to have a clear statement that set out the rationale for the subject, together with key elements relating to the nature of the subject. The user, purpose, and function of a product were emphasised and this is something that needs to be focused on throughout all phases of education, but particularly in the primary school. The content of the new document was slimmed down again and the two Attainment Targets became one to emphasise the holistic nature of Design and Technology and the importance of assessing Design and Technology as a whole, not as a series of 'can do' statements.

At the same time, a new National Curriculum for children in the Foundation Stage (3–5 years) was published (DfEE/QCA, 1999a). This curriculum was divided into six areas of experience, and whilst Design and Technology capability can be developed through all the areas, the main focus is within Knowledge and Understanding of the World. Through a Department for Education and Skills (DfES) funded projected, 'Developing Designerly Thinking in the Foundation Stage', Benson (2003, 2007) identified the lack of opportunities that 3–5 year olds were offered in developing their Design and Technology capability through the activities they experienced. There was still much emphasis on the natural world, and little on the designed and made world. Children were given many activities in relation to making, but few that would develop design skills. If pupils are to develop their capability, then it is vital that they have appropriate experiences from the start of their education.

Throughout this period, Initial Teacher Education (ITE) was developing specialist Design and Technology courses for primary trainees, thus providing new entrants to the profession who were well equipped to lead the subject area throughout the primary phase. Schools who employed these new entrants were certainly 'winners' as they had teachers who were confident in their subject knowledge and understanding and were able to support whole school planning in the subject.

2000–2010

Since 2000, primary OFSTED inspection reports (e.g. OFSTED, 2001, 2002) indicate that whilst there is much good practice in Design and Technology, much remains to be done at all levels, particularly in the areas of designing and developing teacher knowledge and understanding. However, schools have not been given much time to implement the new curriculum. Primary schools had yet another initiative to contend with in 2003 when the government started a review of the whole curriculum, starting with the publication of 'Excellence and enjoyment – a primary strategy' (DfES, 2003a) and the subsequent creation of the Primary Strategy (DfES, 2003b). Literacy and numeracy were still emphasised, but the inclusion of the importance of the development of thinking skills, questioning, and problem solving did link with Design and Technology. In addition, misconceptions about the content of the Primary Strategy arose. Schools started to plan using a topic, or thematic, approach based on the perception that the Strategy promoted cross-curricular links. This has led to a growth in making, or craft, activities rather than Design and Technology and the user, purpose, and functionality of the product is often lost. Topics linked to history often provide the worst examples, with children making Viking boats from card templates and Tudor houses using straw and card boxes. The latest OFSTED Inspection report (OFSTED, 2007) covering the three years 2003–6, provides a clear overview of the state of Design and Technology, and suggests key areas that need to be addressed. These do not surprise those working to develop primary Design and Technology and include teachers' limited expertise and confidence; schools' reluctance to allow teachers to attend staff development courses due mainly to funding; and the need to improve assessment, recording, and reporting of pupils' progress.

Unfortunately the specialist courses for ITE that were so successful before 2000, started to be withdrawn for a number of reasons and only a very few remain in 2010. Firstly the Teacher Training Agency (TTA) took away the requirement for primary teachers to have a specialism; the students can be given a choice in their course between Art or Design and Technology, and History or Geography; inspections do not cover foundation subjects including Design and Technology; and costs can be reduced by cutting specialisms. Both schools and trainees are definitely losers as there are so few trainees starting their careers with the confidence to teach and lead Design and Technology in schools.

By contrast, winners during this period have been primary teachers that have attended extended Design and Technology CPD, part funded through the Teacher Development Agency (TDA), formerly the TTA, and all have to be at MA Ed level. Courses have been available throughout England and include practical as well as theoretical content. Quality of the courses has been rated at the highest level by OFSTED and identified as bringing about positive impact on standards through school inspections.

It was during 2000–2010 that research in Design and Technology grew. In 1997, as primary Design and Technology became more established, the Centre for Research in Primary Technology (CRIPT) was created at University of Central England (now Birmingham City University). Its first international biennial conference was held in the same year, and conference proceedings have provided an excellent resource for research findings covering current key issues. Almost the only large scale project for primary work has been funded by the DfES and was undertaken by the Centre for Research in Primary Technology (CRIPT) from 2003–2006 (Benson 2003; 2005; 2008). The project – Designerly thinking in the Foundation Stage – supported the development of designerly thinking through the engagement of young children with the designed and made world around them, and involved 400 Foundation Stage teachers. More recently in 2008 the National Endownment For Science, Technology and the Arts (NESTA) has funded a curriculum development project entitled Butterflies in my Tummy. The focus for this project was the links between Social and Emotional Aspects of Learning (SEAL), risk taking and designing. A range of support materials from this project are now available at www.data.org.uk and an evaluation of the project was undertaken by Benson and Lunt (2009). One of the main findings from the evidence gathered was that there are a number of advantages for bringing together designing activities and SEAL strategies to support innovation and risk-taking. These include broadening children's repertoire of techniques for designing; developing children's awareness of how their feelings can impact upon their learning and their designing activity; and developing a supportive ethos for innovation and risk-taking. However, it is important that children are clear how these activities support the particular designing and making focus of their project. An appropriate balance between paper-based tasks and three-dimensional designing is also important in maintaining children's interest and motivation.

With regard to setting a research agenda, Baynes and Johnsey (1997) set out a platform on which to build research in primary education. This was followed in 2003 by a report from Harris and Wilson (2003), funded by the DfES to review, comment on, and offer recommendations for areas that should be funded and systematically researched. Their report provides an excellent, if not complete, picture of research relating to both primary and secondary Design and Technology. They concluded, as many suspected, that there were few large scale projects; most were small scale and case studies that often made generalisation difficult, if not impossible. However, a positive development in recent years has been the growth of action research undertaken by primary teachers involved in MA courses (e.g. Benson, Lawson & Till, 2005; Benson, Lawson, Lunt & Till, 2007) and funded through the TDA.

CONCLUSIONS - WINNERS AND LOSERS

It can be argued that it is a great tribute to teachers that Design and Technology, within the primary school curriculum, has become an established and well-liked subject (Benson & Lunt, 2007) at least amongst children. Without the teachers' resilience, hard work, and commitment to the subject, and their belief in its value to the children they teach, the subject could have disappeared after one of the numerous, major changes over the years. However, there are already some indications of changes that will affect the subject. A primary review was undertaken of the whole curriculum, led by Sir Jim Rose, but the resulting curriculum proposals failed to get through Parliament before the May 2010 General election. Therefore the 2000 National Curriculum is still the legal document that schools should use. There is divided opinion as to whether this would have been a 'winner' or 'loser'. Design and Technology was included in the area of learning 'scientific and technological understanding'. Would the subject have been weakened by further misconceptions relating to the nature of the subject or would it have been strengthened as appropriate links were made with science?

During the past twenty years it can be argued that winners and losers have balanced each other, tipping the scales up and down. Certainly there is evidence that some documentation, lack of resources and CPD tipped the scales down, whilst elements of documentation, support materials, creative teaching, and children's enthusiasm for the subject tipped the scales in a positive direction.

Whatever changes take place, it is vital that the nature of the subject is always clear for those who have to implement the curriculum. In no National Curriculum document to date is there a statement of its philosophy or underlying ideals (Kelly, 2004); however the notion of subjects, knowledge and skills can be found, together with general statements such as a desire to raise standards. Any new curriculum should be based on previous research findings and have a clear philosophical standpoint. One thing is certain: it will be crucial for Design and Technology to keep abreast of technologies in the rapidly changing world, enabling the pupils who experience the subject to become:

> ...responsible citizens who make a positive contribution to society. (DfEE/ QCA, 1999b, p. 51)

The new curriculum is awaited with interest.

NOTES

[1] This is based on a paper given at the Technology Education Research Conference, Gold Coast, Australia, December 2010. *Proceedings of the Sixth Biennial International Conference on Technology Education Research.*

REFERENCES

Baynes, K., & Johnsey R. (1997). Research in preschool and primary technology. In R. Ager & C. Benson (Eds.), *Proceedings of the first international primary Design and Technology conference* (pp. 43–46). Birmingham, UK: CRIPT at University of Central England.

Benson, C. (2003). Developing designerly thinking in the Foundation Stage. In C. Benson, M. Martin & W. Till (Eds.), *Fourth international primary Design and Technology conference* (pp. 5–8). Birmingham: CRIPT at the University of Central England.

Benson, C. (2005). Developing designerly thinking in the foundation stage: Perceived impact on teachers' practice and children's learning. In C. Benson, S. Lawson, & W. Till (Eds.), *Fifth international primary Design and Technology conference* (pp. 15–19). Birmingham: CRIPT at University of Central England.

Benson, C., Lawson, S., & Till, W. (2005). *Fifth international primary technology conference: Excellence through enjoyment.* Birmingham: CRIPT at University of Central England.

Benson, C. (2007). Making the difference: Twenty years of primary design and technology. In C. Benson, S. Lawson, J. Lunt & W. Till (Eds.), *Sixth international primary technology conference* (pp. 30–33). Birmingham: CRIPT at Birmingham City University.

Benson, C., Lawson, S., Lunt, J., & Till, W. (2007). *Sixth international primary technology conference.* Birmingham: CRIPT at Birmingham City University.

Benson, C. (2008). Young children as discriminating users of products: Essential experiences for language development. In H. Middleton & M. Pavlova (Eds.), *Proceedings of the fifth biennial international conference on technology education research. Exploring technology education: Solutions in a globalised world* (Vol. 1, pp. 19–26). Brisbane: Griffith University.

Benson, C., & Lunt, J. (2007). It puts a smile on your face! What do children actually think of Design and Technology? Investigating the attitudes and perceptions of children aged 9–11 years. In J. Dakers, W. Dow & M. J. de Vries (Eds.), *PATT 18 international conference on Design and Technology educational research: Teaching and learning technological literacy in the classroom* (pp. 296–306). Glasgow: University of Glasgow.

Benson, C., & Lunt. J. (2009). Innovation and risk-taking in primary Design and Technology: Issues arising from the evaluation of the pilot phase of the curriculum development project 'Butterflies in My Tummy'. In E. Norman & D. Spendlove (Eds.), *The Design and Technology education and international research conference 2009* (pp. 37–46). Wellesbourne: Design and Technology Association.

Department of Education and Science/Welsh Office (DES/WO). (1990). *Technology in the national curriculum.* London: Her Majesty's Stationery Office.

Department for Education (DFE). (1995). *Design and technology in the national curriculum.* London: Her Majesty's Stationery Office.

Department for Education and Employment [DfEE]. (1998). *The national literacy strategy.* London: Her Majesty's Stationery Office.

DfEE. (1999). *The national numeracy strategy*. London: Her Majesty's Stationery Office.

DfEE/Qualifications and Curriculum Authority [QCA]. (1999a). *Early learning goals*. London: DfEE.

DfEE/QCA. (1999b). *The national curriculum*. London: DfEE.

Department for Education and Skills [DfES]. (2003a). *Excellence and enjoyment: A strategy for primary schools*. London: DfES.

DfES. (2003b). *The primary strategy*. London: Her Majesty's Stationery Office.

Harris, M., & Wilson, V. (2003). *Designs on the curriculum? A review of the literature on the impact of design and technology in schools in England*. London: DfES.

Her Majesty's Inspectors (HMI). (1991). *Technology in the primary school: HMI report*. London: Department of Education and Science.

Her Majesty's Chief Inspector of Schools. (1996). *Subjects and standards*. London: Her Majesty's Stationery Office.

Ive, M. (1997). Primary Design and Technology in England – Inspection evidence and 'good practice'. In R. Ager & C. Benson (Eds.), *First international primary Design and Technology conference proceedings* (pp. 1–6). Birmingham: CRIPT at the University of Central England.

Ive, M. (1999). The state of primary design and technology in England. In C. Benson & R. Ager (Eds.), *Proceedings of the second international primary Design and Technology conference* (pp. 1–5). Birmingham, UK: CRIPT at the University of Central England.

Kelly, A. V. (2004). *The national curriculum: A critical review*. London: Paul Chapman.

National Curriculum Council [NCC]. (1992). *Technology for ages 5 to 16*. London: Department for Education/Welsh Office.

Office For Standards in Education [OFSTED]. (2001). *Report 1999/00*. London: Author.

OFSTED. (1998). *OFSTED annual report: Design and Technology*. London: Her Majesty's Stationery Office.

OFSTED. (2001). *Report 1999/00*. London: Author.

OFSTED. (2002). *Report 2000/01*. London: Author.

OFSTED. (2007). *Education for a technologically advanced nation: Design and Technology in schools 2003–6*. London: Author.

Parkes, L. (1988). *National curriculum: D&T working group. Interim report*. London: DES/WO.

Qualifications and Curriculum Authority [QCA]. (1998a). *Maintaining breadth and balance*. London: QCA.

QCA. (1998b). *A scheme of work for KS 1& 2: Design and Technology*. London: SCAA.

School Curriculum and Assessment Authority [SCAA]. (1994). *Design and Technology in the national curriculum*. Draft proposals. London: Author.

School Curriculum and assessment Authority (SCAA). (1995). *Design and Technology in the national curriculum*. London: Author.

Smithers, A., & Robinson, P. (1992). *Technology in the national curriculum: Getting it right*. London: Engineering Council.

Clare Benson
Director, CRIPT
Birmingham City University
England

MARJOLAINE CHATONEY AND JACQUES GINESTIÉ

2. PRIMARY TECHNOLOGICAL EDUCATION FOR ALL IN FRANCE

A Study of the Role of Technology in the Primary School System and Teacher Training over the last Twenty Years

INTRODUCTION

Understanding the contemporary technical environment and the socio-technical investment accompanying it is essential to the development of a country and of its citizens. Technological education contributes to this in different ways and at different levels within the education system. Its integration into general education in France is recent. This chapter is organised into two parts. The first part discusses technology in the education system and takes into account, through the specificity of different structures in the French system, the place allocated to technological education and the forms it takes according to the structure within which it is taught. The second part is dedicated to teacher training in the technological domain. Training which until recently used a degree as its starting point is now part of the Degree-Masters-Doctorate study sequence, more commonly known as LMD. LMD opens up new possibilities for organising teacher training which can be started in the different phases of L and M.

THE EDUCATION SYSTEM

The Main Principles of the French Education System

The education system is linked to the history of the republic. The principles of compulsory schooling, secularism, equality and its being free of charge are at the heart of the aims of schooling.

Schooling is compulsory for children between 5–16 years of age and from all families, both French and foreign nationals. In practice, schooling actually begins at the age of 3 until the age of 18, for a variety of reasons such as higher qualification requirements or the job market. Parents may choose to teach their children themselves, or to place them either in a public or private school. The first option is rare. It requires parental agreement, as well as approval for the given teaching. Ideological impartiality and equal opportunities are compulsory.

The principle of secularism is unique. It affirms the separation of power between the church and the state. All children are welcomed irrespective of their faith and beliefs, even in confessional private establishments. Teaching is given with total respect for freedom of thought. Curricula and school books are neutral. The personnel

C. Benson and J. Lunt (eds.), International Handbook of Primary Technology Education: Reviewing the Past Twenty Years. 13–27.

are secular. Proselytising and propaganda are forbidden. To the notion of secularism is added that of political neutrality and impartiality. The last of which, in the same way as secularism, applies to staff, syllabuses and books.

The principle of equality prohibits discrimination of any kind. The education system has to guarantee equal opportunities for both sexes, as well as social and cultural equality.

Historically, the notion of schooling being free of charge preceded its being obligatory. School equipment is provided by communes or departments for public (state) schools. School manuals are provided in both public and private schools. In secondary school, costs are paid by the families.

Organisation of the Education System

The education system is organised into levels and cycles. Primary school constitutes the first level or grade. It is split into two schools, nursery school (optional) and lower primary/elementary school (compulsory). Middle school and high school comprise the second level or grade. Two courses of study are available in secondary school: a general technical diploma and a professional (practical) one. In this chapter, we will only look at the general technical option.

The first level is organised into 3 cycles, starting at nursery school and continuing in elementary school. Cycle one corresponds to the first two years of nursery school. The children are aged between three and five years. The second cycle starts in the final year of nursery school, and ends after the first two years of elementary school. Children are aged from 5–8 years old. Cycle 3 refers to the last three years of elementary school, for children aged 8–11. Primary school education is organised using a national programme (syllabus) established by the Ministry for National Education. No diploma is awarded at the end of this period. Pupils are assessed throughout the whole of primary school. First degree education is funded by municipalities (city councils) which provide the buildings and equipment needed for the school to function. Teachers are paid by the education ministry. Administrative staff are managed and paid for by the city council.

Second level education begins in an establishment known as a 'college' (middle school) and ends in another one called a 'lycée' (secondary school). Middle school lasts for four years, pupils are aged 11–15. Secondary/high school lasts for three years. Pupils are aged 15–18, sometimes 19. Approximately 83% go to a school of this kind.

Middle schools adhere to a national curriculum. It is structured in three cycles - the adaptation cycle, the central cycle and the orientation cycle. These cycles last for one year, two years and one year respectively. Pupils are evaluated by means of continuous assessment to gauge their knowledge and a final exam, with a view to obtaining the diploma known as the 'Brevet des colleges'. The budget for middle schools is allocated by the Ministry for National Education for staff costs and salaries, whereas investment and equipment are funded by departments (France is split into 96 administrative departments). At the end of middle school, pupils have two choices over which course of study to take. They can either continue their learning at a general technological secondary school, or at a general one. All these choices make a wide range of qualifications available to young people, such as the 'certificat d'aptitude professionnel' (CAP), 'le brevet d'études professionnelles' (BEP) or a 'baccalauréat general or professionnel' (Bac, BP). Study for the 'baccalauréat général' is in a 'lycée general', and the 'baccalauréat professionnel' in a 'lycée professionnel'.

General and technological high schools are structured in two cycles: the determination cycle and the terminal (final) cycle. The determination cycle allows pupils to choose either a literary, scientific or technological (industrial, tertiary or biotechnological) route at the end of the determination cycle. It lasts for one year. The final/terminal cycle lasts for two years, culminating in the baccalauréat diploma. This qualification is the recognition given for completed secondary school studies. Each study choice has a national syllabus. The budget for secondary schools' staff salaries and operations is managed by the Ministry for National Education, whereas investment and equipment is dealt with by regional authorities.

TECHNOLOGICAL EDUCATION IN GENERAL TEACHING

A Generalist Technological Education from Nursery to Secondary School

Technology has been taught in France from nursery to secondary school since 1985. The different structures (primary, middle and secondary school) give it varying forms. In primary school (pupils aged 3–11 years old) technology education is linked to scientific learning (physics-chemistry, biology, geology). In middle school (11–15 years), technology is a compulsory subject linked to sciences. In secondary school (15–17 years) technological education takes the form of an option or study choice called 'Sciences for Engineers' chosen along with others (e.g. sciences, arts, economic and social sciences). Besides the specific formats and progressive organisation of subject areas and specialities (options), all the different levels are interested in situations that allow for the production of technical objects (objects or systems).

Technological Education for Science and Technology-Based Teaching in Primary Schools

In cycle 1 (known as the 'cycle des apprentissages premiers' or early learning cycle) a subject area called 'Découvrir le monde' (Discovering the world) aims to teach pupils about the richness of the world surrounding them (objects and living things). Besides experiences already known to young children, nursery school allows a child to be curious about things by discovering some of the phenomena of life, matter and man-made objects (Bulletin Officiel de l'Education Nationale (BOEN), 2002a). The teacher makes the pupil aware of the fact that (s)he can handle and transform the objects around them; that they can be put in order and classified; and that their qualities can be distinguished in doing so.

In cycle 2 ('cycle des apprentissages fondamentaux' or fundamental learning cycle) the Discovering the World module continues. Pupils learn how to use technical objects correctly. They learn to ask themselves questions and to think about their actions. They alter, handle, observe, compare, classify and experiment with things. They go beyond their initial ideas by learning to apply them to real situations. Hence, they learn about materials that are available to them. They question themselves and develop their practical know-how. The teacher allows pupils to structure their thinking and actions through basic construction or building projects, heightening their sense of innovation and inventiveness (BOEN, 2002a). The weekly time dedicated to this domain switches between three and three and a half hours.

In cycle 3 (known as 'cycle des approfondissements' or in-depth learning) a discipline called 'Experimental Sciences and Technology' targets a more rational way of thinking about materials/matter and living things, through careful observation and analysis of phenomena which pupils are interested in. The aim is to prepare pupils for living in a society where technical objects play a major part, as well as teaching them about the benefits of science. The science and technology curriculum is heavily centred on the experimental approach. The knowledge offered is much more aptly put together, as it is the result of questions being asked when activities involving observations and changes are conducted (MEN, 2002a). Such teaching leads to discussions about major ethical problems of our times to which children are particularly sensitive (economic development, environment or health). Weekly time dedicated to this field can vary from two and a half to three hours.

In primary school, the teacher is versatile. S/he has to teach all subjects. But due to the reality of institutional demands and practices, teachers focus more upon French and mathematics. As a result, scientific and technological education is not taught much. In order to aid the development of sciences and technology, the ministry put a plan in place to revamp the teaching of science and technology call PRESTE. This project was preceded by 'la main à la pâte' operation, initiated by Georges Charpak of the 'Académie des Sciences' and 'Prix Nobel de Science'. The results are favourable: science and technology teaching rose from 3% to 25% in three years.

The Plan to Reform Science and Technology

A plan to reform science and technology was put in place by the ministry in 2002 with the aim of making primary school science and technology teaching more effective and giving it an experimental dimension (Ministère Education Nationale (MEN), 2002b). The idea behind this was to increase pupils' ability to discuss and reason, as well as to progressively implement scientific concepts.

The pedagogical approach is based on questioning and investigation, through the use of concrete processes for experimenting, complemented by documentary research if necessary. The pedagogical approach therefore creates the conditions for a genuine intellectual activity for pupils. On the one hand, it allows pupils to build their knowledge by being involved in scientific activities, whilst on the other hand encouraging pupils to discuss and express their ideas together with observations, hypotheses and conclusions. This leads them to share ideas, throw their points of view together and formulate provisional or final results, orally or by writing. Listening to and respecting other pupils' input and taking their opinions into account allows pupils to develop the ability to reason and to critique something constructively. Finally, the activity is part of a coherent process which places the emphasis on common sense and helps to create inter-disciplinary links, notably a general grasp of language and citizenship.

These learning paths target the acquisition of knowledge, and the use of analytical methods and reasoning. The advantage of the defined framework is that it clearly explains the situation and gets away from any institutional blurring of boundaries. It also reassures the people involved.

This plan to reform the teaching of science and technology does not ask teachers to find and create everything. It requires them to refer to existing documents and to define precise and realistic objectives. Teachers will be in a position to create conditions for a real learning scenario, in which pupils will be able to build lasting knowledge.

With the science reform plan, access to knowledge of technology in primary schools forms part of an investigation process, as is the case for science. This process adheres to a principle of unity and diversity in the choice of learning methods, materials and objectives. Technological exposure within this structure is difficult to find, because the transition between science and technology cannot be taken for granted (Agassi, 1997; Chatoney, 2003, Rowell, 2004). It requires effort to be made to identify knowledge, which is not easy to do (Chatoney, 2006) due to limited training time being dedicated to the teaching of science and technology for primary school teachers in their early and subsequent training.

THE TRAINING OF PRIMARY SCHOOL TEACHERS

The objectives and demands of teacher training are defined by ministerial decrees and must be adhered to by everyone with regard to the organisation of competitive entrance exams within the public sector. Teachers employed in the public sector by the state represent approximately 80% of all teachers in France. Around 18% have private teaching contracts. The state helps to run these kinds of establishments by dealing with teachers' salaries. In return, such teachers have to meet the same demands as public state teachers; primary, middle or secondary schools are at the very least, obliged to follow national curricula. For practically all teachers working in France, training carries with it the same requirements set by the Ministry for Education. Such a training course is overseen by the Instituts Universitaires de Formation des Maîtres (IUFM) (National Teacher Training Institutes) set up in 1991 with a new law for teacher training being passed[1]. These institutes are in charge of training all teachers in 1st and 2nd degree education for all disciplines, including technology.

Teacher training in France is sequential. Firstly, students wishing to become teachers go to university to obtain a degree. Once they have this, the students attend the IUFM for a 2-year training course.

In keeping with the Bologna protocol/agreement, French universities, like their European counterparts, are changing university courses to make them the same as studies in Europe and on an international level (Chatoney & Ginestié, 2006). The Licence (Degree), Master, (Masters), Doctorat (Doctorate) (LMD) system is adopted in three cycles. Each qualification now corresponds to the same period of study, that is to say three years for the degree, five years for the masters and eight years for the doctorate. The stakes are high for Europe, for universities and for students. Europe is aiming for the free movement of people in the European space. Universities are targeting a strengthening of their training and research policies in the European space. Students want their qualifications and university education to be recognised on a European level.

The IUFM, which were previously independent from universities, are like many other institutes and specialised schools in France, progressively integrated into universities. (The IUFM for Aix - Marseille was the first to be integrated, in January 2007). Teacher training in France in its current form is quite difficult to integrate into the LMD system for several reasons: the first being that prospective teachers are recruited after gaining a degree and passing national entrance exams. These exams are difficult; it is competitive and institutional demands are high. The second reason is that it is difficult to socially place the job of teacher. Certain people think that training is not needed to become a teacher, and that some socially acquired skills and good academic knowledge are sufficient to be able to teach. Professional (practical) skills in teaching are given no credence. They are believed to be innate or gained by trial and error when working with children. Others think socially acquired skills coupled with good academic knowledge are insufficient in doing the job. For them, knowing how to teach requires having working knowledge that is specific to the job. There are actions, techniques, ways of organising things, and epistemological tools that one has to know about (Ginestié et al., 2006). In this context, integrating teacher training into the LMD format hints at moving from the idea of a professional training course which is supervised by the employer (the entrance exam), to professional university training (skills).

The putting into place of the LMD structure has brought with it openings for teaching jobs prior to the two years of training at the IUFM, leading to a new idea for such training. In this chapter, we will first of all present the LMD idea, and the problem it poses for finding opportunities in teaching jobs. We will then move on to discuss using the LMD plan in the teaching of science and technology, and the new format for training primary school science and technology teachers.

The LMD Idea

The LMD system is the result of the actions of four education ministers in 1998 (French, Italian, English, German) at the Sorbonne[2]. This declaration started the process of constructing the current European space, and led to a conference on the harmonisation of studies in European universities[3] (MEN, 1998). A year later, European education ministers in Bologna (Italy) drew up the main reference points for European study courses and diplomas, and expanded the process to incorporate all European countries (MEN, 1999). Since then, the idea of a European teaching space has made progress and begun to take shape. All these efforts have allowed an agreement to be reached about the quality of training, equivalent diplomas and the free movement of students in Europe. This last point is strongly backed by a whole array of European programmes, such as Socrates, Comenius, and Leonardo. Work on this harmonising process is still ongoing.

The LMD system has been set up to facilitate the understanding of university training and research policies, whilst also respecting the wide range of offers.

This harmonisation process is not however a standardisation process for higher education; it has to allow universities to offer their own courses and diplomas, as is the case with all major universities worldwide.

The LMD has two objectives: to reassert the value of national diplomas to give students the certainty of having a qualification that is recognised in all European countries, and to construct a European higher education teaching space based on an understanding and trust between the range of national systems in Europe. This trust is founded on methods for the evaluation of the quality of training courses and diplomas (MEN, 2002c).

On a European level, the LMD system clarifies the organisation of training schemes, making them workable and understandable in the European space. It involves:

- decompartmentalisation of curricula, allowing specialist subject areas in establishments to be more clearly identified;
- more flexibility in training schemes, in order to favour progressive orientation processes and the teaching of a wide range of students;
- the putting in place of a modular teaching and European credits system (ECTS) to improve moving from training programmes to working, and also between countries and establishments.

In terms of universities and students, the LMD scheme links training and research. It also heightens competitiveness within establishments, which leads to the following structural changes:

- the putting in place of main areas of training together with research;
- availability of a wide range of study choices to cater for student diversity and developing demand for professional training;
- promoting innovation and experimentation in teaching in order to move away from the classic lecture (cours) - seminar (TD) - practical class (TP) structure;
- supporting innovative choices of multi-subject type, incorporating general knowledge and culture, modern languages, technology and professional training in an attempt to gain more subject flexibility;
- scientific involvement for training teams by conducting research to provide coherence between scientific powers;
- recognition of professional Masters' programmes to comply with local economic, social and cultural needs.

Availability of Teaching Jobs in the LMD Structure

The LMD system progressing towards masters and doctorate poses no real problem to a university within typical university courses. However, the putting in place of professional paths towards teaching jobs is more problematic. The Ministry for National Education has reservations about the IUFM's ability to train teachers from the point when teacher training will be integrated into Masters' programmes:

> It is important … to express concerns about certain projects planning to create a Masters in teaching jobs... A university – IUFM partnership risks putting the IUFM in the midst of a university diploma process, even though it should be

> focusing on professional teacher training for those having passed the entrance exam. (MEN, 2002c)

To facilitate IUFM integration, universities have agreed to contribute fully to preparing the teachers of the future in terms of degree courses offered and entrance exam preparation, explaining that:

> The integration of the IUFM preparation will be more thorough, involving degree courses... The question of a Masters being awarded to teachers at some point during their professional training will lead to new answers with the IUFM pedagogically integrated into universities. (MEN, 2002c)

With these plans, the availability of teaching jobs is not limited to the final two years of training at the IUFM.

Like others, the universities in Marseilles have integrated modules linked to teaching into their degree courses. How does the professional element of teaching appear in technological subjects in the first LMD at the University of Provence (started in 2003)? What position can IUFM take in the university integration process and with regard to Masters' courses in teaching?

To look at this question, we will present an LMD system put in place for scientific degrees as of 2003, leading to teaching jobs for life and earth sciences, physics and chemistry students.

Multiple Subject Degree Course – Training and Scientific Knowledge

The science department of the University of Provence (Aix-Marseille 1) offers students a new choice of course corresponding to the LMD system called *Multiple subject degree course – training and scientific knowledge*. This degree leads to jobs in teaching, notably primary school teachers and in the technological sector of secondary education. Where and how does this degree course fit into the LMD structure? The illustration below shows the possible career choices available to students in the science sector at the University of Provence.

Figure 1 allows us to understand the whole structure and general format of the LMD system, which in turn allows a student to plan their choices throughout their studies. When starting a degree course, a student has several possible choices of what to study. In the general degree structure there are three possible choices: Mathematics, Information Technology and Physics and Chemistry (MI-SPC), Life Sciences and Universe and Environmental Sciences (SV-SUE) and Sciences for Engineers – mechanical and electronic engineering and automatism (SPI). The scientific multiple subject degree is a third year course. It leads (as indicated by two arrows in the illustration) either to IUFM teacher training programmes, or to a professional or research masters. It is different from professional (manual) masters courses, which lead directly to jobs as technicians.

The multiple subject degree course is done by a small number of university students who are young and highly motivated by the availability of courses which fit in with the LMD philosophy. The aim is to give students a good chance of passing the primary school teaching exams and other entrance exams for state jobs which

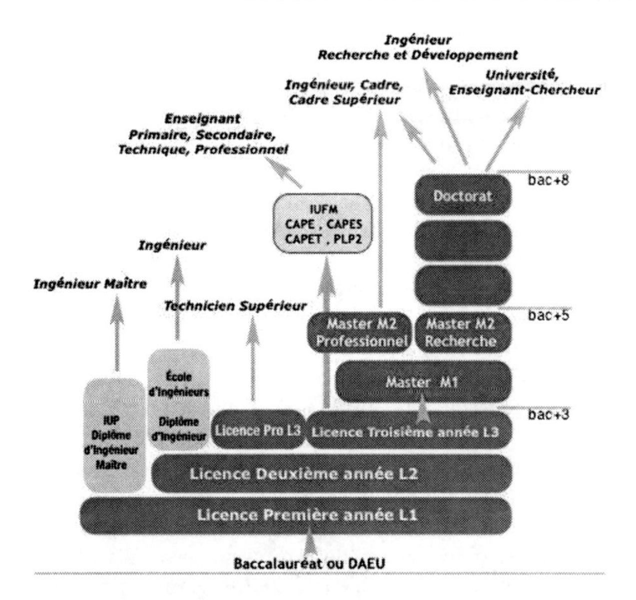

Figure 1. Possible career choices available to students in the science sector at the University of Provence.

are offered. This provides the opportunity to take exams to become either a primary school teacher or a national education advisor. It takes place prior to exam preparation given by the IUFM - preparation that is accessible via an exam which 90% to 95% of students on this course pass. With options being carefully chosen, it also allows students to go on to take socio-educational exams for state jobs at degree level.

In order to reach its multiple subject and cultural objective, this degree course includes an important scientific unit (Physics and Chemistry or Biology and Earth Sciences) which involves completing a scientific multiple subject dissertation which is presented as an exposé, lessons in Mathematics, Technology, Information & Communication Technologies (ICT), a modern foreign language, an introduction to methods of literary analysis and the writing of a story. The course also includes an introduction to the epistemology of sciences and techniques, and the possibility of doing an internship or basic constitutional law and social and political sciences.

Two modules (technology and internship organisation) are intentionally geared towards becoming a primary school teacher. These modules are formed in a partnership with the IUFM in Aix – Marseille. The technology module targets the acquisition of knowledge needed to produce technical objects/tools, and how such objects work or exist. It also aims to cover the basics of different technological approaches (structural, functional, systemic etc.) and the language or jargon relating to them. The aim is to acquire the skills required to conduct a technical project, notably in terms of anticipation, planning and organisation (dependence systems, causal chains, schemes of interaction, value analysis, etc.). It is a subject-based module. Teaching is centred upon the history of inventions and techniques, the study of social organisations for the production of objects, processes in technical projects and industrial conception, interdependence between operation, functions, structures and forms of technical objects and systems, graphs and jargon, relationships between man-object, man-machine, man-tools.... Students also receive an initiation to 3D modelling software. This teaching corresponds to what is taught in primary schools as an introduction to technology, with the help of videos, school documents, and even documents produced by pupils. Hence, students have to study in settings that are limited by pupils' understanding and the conditions or equipment. For example, students must produce a technical project outside the classroom. From this file, they produce the technical object in a practical class. In this project, the technical aid or object can only be produced using tools and materials used in primary schools. This raises students' awareness of the restrictions in primary school equipment and the importance of preparation in this kind of activity. They learn for example the majority of techniques that can be used at school, feasibility, and technical jargon. The technical dossier is limited by precise specifications. Managing the dossier complies with common reference procedures for technology teaching in France. It is a thorough process: investigation-research, operational approach, research and choice of technical solutions, technical object drawings, organising its production, and conformity. The dossier and creating the object are evaluated by means of a written statement. As far as teaching is concerned, the problem linked to graphic jargon is supported by examples from primary school (for example drawings, key, and coding). Systems for the transformation and transmission of movement are begun in a practical class about automatism adapted to primary school level. The invention history class is a practical application session adapted for 10 year-old pupils.

Such a conception of technological training naturally goes beyond the subject area for which it is intended. The module allows students to gain awareness of what technology is from a general point of view, in what ways it can be used and find viable conditions for teaching it to young pupils. This approach complements the training given by the IUFM which, due to a lack of time[4], is unable to develop these different aspects and especially contextualise them in academic practices. Spontaneously, students tend to consider that the technological initiation comes from manual education, despite the fact that such teaching disappeared in the1980s. This more thorough teaching allows students to get away from the naive perception of the expectations of primary school technological education teaching. It also allows them to become aware of the relationships between technology and other subject areas. There is of course an important epistemological element in terms of academic references and ideas regarding multiple subject courses and flexibility. These relationships are examined further in the IUFM training.

The option of doing a teaching practice in a primary school involves a preparation class for the work placement, a teaching practice in a primary or nursery school with a mid-term report, and a report at the end, concluding with a written account. The teaching practice includes approaches to a primary school and teaching through observing what the teacher does in the classroom and the school generally.

The ongoing aim is to develop the ability to analyse the teacher's actions and teaching-learning situations. Lessons about the role of the school, the teacher, the education system, syllabuses, and teaching-learning scenarios precede the work placement in the school. The system of class-internship-class-internship-dissertation progressively prepares students to observe carefully. It will be easier for them to understand the teacher's teaching choices, and to see how pupils react in a given situation. Observing how the teacher operates is also an important part of preparing a class (for example teaching preparation, thinking about teaching materials, risk prevention, and multiple subject structure) and a teacher's subsequent analysis of their own choices (for example strategy, activity accompaniments, and organisation). Of course, during the work placement, students experience school life on a daily basis. They also find out about parts of the job which are not talked about enough, such as team work, relationships with parents, lesson supervision, lunch time and other breaks, sick pupils, and school trips. This option does not have a professional (practical working) aim. It is a discovery of the working environment and the work of a teacher in general terms. Training teachers professionally is the job of the IUFM at a later date.

As we can see, both these modules created in the university and IUFM partnership help to develop student knowledge of what being a primary teacher entails. Notably, it is a question of highlighting the importance of the introduction to science and technology in the primary school, to ensure that these subjects are well taught and hopefully to encourage more pupils and students to choose studies of this kind. It is also a matter of developing the contribution made by scientific and technical culture to pupils' general knowledge.

The introduction and placing of the multiple-subject degree course in the LMD setup has been a well-documented source of controversy in academic quarters. Some people think that the decision to opt for jobs in teaching means that the degree course in question becomes a kind of cul de sac option. They go on to say that it takes the weakest students who are not good enough to do scientific studies at a higher level. In other words, being a teacher does not require exceptional academic prowess or completion of high-level university studies. This idea fits in with the one which claims that everyone is capable of teaching. Others think that general knowledge and orientation have no place among scientists. Such resistance embedded in an academic model of scientific knowledge is proof of differing opinions with regard to the role of a teacher, what the job is said to entail, and hence upon what knowledge the job is founded. People are still debating the fact that it may suffice to merely have good knowledge of a discipline in order to be capable of teaching it. Such a notion is disparaging to the work of a teacher, notably the mastery of knowledge and the skills necessary to teach a subject. The main reason for this knowledge and skills debate is that it means that teachers must perform their educational duties within the framework of a school system which forms part of a socio-cultural, socio-professional and socio-economic environment. This was in line with earlier comments in the chapter, about time being necessary to bring an end to generalisations regarding degree courses geared towards teaching jobs.

Hence, since 2002/2003, modules dedicated to knowledge of teaching and teaching jobs have been offered to students who later wish to become teachers. IUFM integration is ongoing. The IUFM in Aix-Marseille was the first one to be integrated, in January 2007. In order to do this, and purely in view of the training, the IUFM had to specify the training content, give it a university level, organise the training into modules, consider evaluations based on the European System of Transfer of Accumulation and Credits (ECTS) credits, establish a link with research, and offer new study options in line with the LMD idea.

The IUFM Masters

The IUFM has redesigned its two years of training into a Masters. This was not easy, due to confusion between recruitment by means of exams to teach in the public sector, and professional teacher training. This contradiction is clearly seen in the first year of IUFM training, the sole aim of which being to prepare students for the exams, whilst taking the format of a first year Masters. Passing the entrance exams cannot be a viable means of evaluating the elimination of a first year Masters, because it depends on the ratio of the number of positions offered and the number of candidates. IUFM training couples Masters 1 with passing the competitive entrance exams.

First year training is subject and culture based, and also involves teaching such as training for success. The school subjects' part is vast, given that primary school teachers are required to teach many subjects. They have to study for exams in French, mathematics, one of history and geography or science and technology, but also languages, sport, art and music. As well as all of this, there is also the work placement in a primary school. Such placements allow students to get an idea of the working environment. This is something which they are obliged to talk about as part of the entrance exam.

The 2nd year of training forms part of a more typical university training and working structure, the validation of which is entirely in the hands of the training institute. There are three parts to be completed in order to validate this section: work placement, teaching modules and a dissertation. The system in place satisfies the university requirement of awarding ECTS and that of an academic jury charged with awarding candidates their state school teacher status. Second year training is made up of three components, one of which is pedagogical and professional teaching. It allows teachers to learn, to link practice and theory, to analyse practical elements and to introduce common themes for all teachers. The second component regards the work placements. It is used to apply knowledge, experiment, observe, and become familiar with institutions and other educational systems. Interns do a placement where they take charge of a class at a middle school, an observatory placement in a primary or secondary school, or a supervised practical placement and a placement in a company, or a placement abroad. The third component is the completion of a dissertation about the work placement. It is used to analyse, write up, and ask a question about a specific area of teaching linked to educational research. Its aim is methodological.

Training for Technology Teaching at Primary School

In the first year, students study for the Concours de Recrutement des Professeurs des Ecoles (CRPE) teaching exam. With this in mind, students must choose to take the exam in either history and geography, or science and technology. But whichever choice they make, all students will have to answer a question to assess their knowledge of history, geography, science and technology. The remainder of the exam is dedicated to science and technology or history and geography. Due to the flexible nature of the exam, first year training has two components: a major component of 36 hours, and a minor one of 28 hours. Both target expertise in key scientific and technological concepts and their links to living things, matter, or man-made objects.

The objective of the training course is first and foremost to increase students' knowledge and skills for all subjects taught at primary school, to help them pass the exam. In order to achieve this, the chosen training structure has opted to introduce students to the process of transposing scientific and technological ideas to allow them to build their own basic knowledge of a scientific and technological nature.

Candidates are examined on the content taught at the IUFM. This content is detailed in the entrance exam syllabus (BO n° 21 du 26/05/05). The content touches upon extremely varied subject areas such as measurement, matter, energy, electricity, life forms, the earth, astronomy.

Students use a dossier to procure the main concepts of different disciplines. For the minor component, these files group together various documents (for example descriptive experimental, object drawings, processes for building, summarising, and production options) and serve as a study and analysis aid. For the major component, students compile dossiers which allow them to bring together essential scientific and technological notions and develop learning sequences with an emphasis on the investigative aspect. Students are twice put into an exam situation as a means of training.

In second year, training is centred exclusively on the teaching and professional (working) aspects. Students learn about teaching systems using existing resources, conduct evaluations, organise study, manage materials, and think about tools to help knowledge transposition such as posters, flowcharts - all of this in 15 hours. A collaborative working platform joins the link between students' private study and training.

CONCLUSION

Teacher training at the IUFM today is split into four distinct sections, the importance of which all have to be appreciated. The first time scale is that which precedes starting at the IUFM, a time to build subject knowledge and the first steps on a path leading to jobs in the field. The second part concerns preparation for the IUFM M1 exam. The third is hands-on practical or professional training, which comprises teaching, practical experience in schools and thinking about the job. This corresponds to the second year of IUFM, M2 IUFM. The fourth section is the duration of a teacher's career, a form of continual training from start to finish. Diplomas are awarded for these sections; degree, masters and doctorates for some.

Teacher training is not limited to knowing about a subject. Of course this is necessary, but it is not sufficient to be able to train a professional teacher able to transmit knowledge in different academic setups with highly heterogeneous groups of pupils. Training also requires being able to teach this knowledge and having a thorough knowledge of the educational system. This knowledge must be used in practice when the teacher is doing his/her job. The IUFM M1 and M2 training schemes attempt to find a balance between that which is practical, teaching, training time and training components. Such a balance is struck in the passing from M1 to M2 to allow every student involved in such studies to build their own professional working identity. This is a long road to take in the LMD system.

Training to teach science and technology can no longer be considered in terms of the juxtaposition between earth and life science teaching, physics, chemistry and technology, as was previously the case. The idea of a training course with integrated teaching is difficult to contemplate. It relies upon the teacher's ability to adopt different points of view about the subject they are dealing with. This raises several questions: that of the theme and its multiple subject nature, but also the handling of points of view and the notions and concepts to be constructed through these viewpoints, in other words, the question of what knowledge and epistemology is required in the training of teachers.

NOTES

[1] The orientation law of 1989 led to the disappearance of normal schools which trained primary school teachers and regional teaching centres which trained secondary teachers.

[2] The Sorbonne is a major Parisian university.

[3] Colloque de la Sorbonne, « vers l'harmonisation Européenne des cursus universitaire », Paris, 1999

[4] Total training time for science and technology is approximately 30 hours at the IUFM.

REFERENCES

Agassi, J. (1997). Thought, action and scientific technology. In M. de Vries (Ed.), *International Journal of Technology and Design Education, 7*, 49–63.

Chatoney, M. (2003). *Construction du concept de matériau dans l'enseignement des sciences et de la technologie à l'école primaire*. Thèse, Aix-Marseille: Université de Provence.

Chatoney, M. (2006). The evolution of knowledge objects at primary school: Study of material concept as taught in France. In M. de Vries (Ed.), *International Journal of Technology and Design Education, 16*, 143–161.

Chatoney, M., & Ginestié, J. (2006). *La formation universitaire et professionnelle des enseignants dans le LMD: Exemples d'intégration et de structuration des cycles universitaires dans les domaines scientifiques et technologiques*. Symposium international FORAPEVAL, Fez, Maroc.

Ginestié et al. (2006). *Teacher training, preparing young people for their future lives. An international study in technology education*. Santiago, Chile: ALFA Europe Aide Cooperation Office.

Ministère Education Nationale (MEN). (1998). *Harmoniser l'architecture Européenne d'enseignement supérieur (Déclaration de la Sorbonne)*. Retrieved from http://www.amuefr/.

MEN. (1999). *Déclaration de Bologne des ministres Européens de l'éducation*. Retrieved from http://www.amuefr/.

MEN. (2002a). *Programme de l'école primaire*. Paris: BOEN HS n 1 du 14 février.

MEN. (2000b). *Plan de rénovation des sciences*. Retrieved from http://www.eduscol.education.fr/.

MEN. (2002c). *Circulaire de mise en œuvre du schéma Licence – Master – Doctorat (LMD) du 14 novembre 2002*. Retrieved from http://www.amuefr/.

Rowell, P. (2004). Developing technological stance: Children's learning in technology education. In M. de Vries (Ed.), *International Journal of Technology and Design Education, 14*, 45–59.

Marjolaine Chatoney and Jacques Ginestié
UP-IUFM d'Aix-Marseille
UMR ADEF – GESTEPRO
France

VICKI COMPTON

3. TECHNOLOGY IN THE PRIMARY SECTOR IN NEW ZEALAND

The Journey this Far and Where to Next…

INTRODUCTION

In this chapter I provide a brief discussion of the entry of technology as an essential learning area into New Zealand's core primary school curriculum. I outline the aim of the original 1995 technology curriculum and how this was implemented within the primary sector – that is years 1–8 (approximate age 5–12). Examples of the sorts of activities students in the primary sector have been involved in will be provided alongside a discussion of the issues that teachers have faced in implementing this curriculum. I then discuss the revision of the technology curriculum that occurred during 2004–2007, particularly in terms of the re-definition of technological literacy, the development of new curriculum constructs and the implications for the primary sector in terms of implementation of the 2007 technology curriculum scheduled for 2010.

ENTER TECHNOLOGY EDUCATION

Technology has had a formal presence in New Zealand's compulsory education sector since 1995 when the original technology curriculum (Ministry of Education, 1995) was released. Previous to this, some moves to shift vocationally oriented technical education into more technology-like general education, had been undertaken in years 7–13 (approximate age 11–17) with varying levels of success (Compton, 2001; Harwood & Compton, 2007). However, within the primary sector, the only pre-cursers to technology were references made within the science (as an applied science view) or social science (as a largely technological, determinist view) curricula. However, a draft technology document was released in 1993 for consultation and this was followed by the release of the final 1995 technology curriculum two years later. Both the draft and the final technology curriculum documents presented primary teachers, particularly those of years 1–6 students (approximate age 5–10) with a significant new focus for their teaching. Since 1999, the year in which the 1995 curriculum was 'gazetted' or made mandatory, all schools in New Zealand, including those in the primary sector, have had to offer technology programmes as part of their core curriculum in order to meet official New Zealand compliance requirements. Most primary schools responded positively to this addition to the curriculum. In fact,

C. Benson and J. Lunt (eds.), International Handbook of Primary Technology Education: Reviewing the Past Twenty Years. 29–38.

in some cases technology was seen by many primary teachers as a validation of more thematically oriented teaching that had begun to be 'squeezed out' by a growing emphasis on literacy and numeracy.

The aim of *Technology in the New Zealand Curriculum* (Ministry of Education, 1995) was to support the development of technological literacy as based on the three strands:

- Technological Knowledge and Understanding
- Technological Capability
- Technology and Society

These three strands needed to be brought together in all technology programmes to ensure students were provided with opportunities to undertake technological practice (Compton & Harwood, 2003). Therefore, technological practice was seen as the vehicle through which students could develop their technological literacy (Compton & Harwood, 2006). It was argued that a strong sociological focus was needed for student learning to move away from a purely 'functional' orientation based on learning how to use technology and 'make' technological products, to one that was liberatory or critical in nature. This included learning about technology as culturally situated and providing opportunities for student empowerment through undertaking socially justifiable technological practice (Compton, 2001; Compton & Harwood, 2003; Compton & Jones, 2003; Davies Burns, 2000; Pacey, 1983; Petrina 2000).

Initial support material published did not expand upon this notion of technological literacy however, and the teaching community was largely left to work out for themselves what this might look like and how it could be supported in classrooms. In an attempt to provide guidance to teachers for developing their students' technological literacy, classroom-based research for the next few years became focused on developing better understandings of technological practice and technology pedagogical content knowledge (see for example, Compton and Harwood, 1999a; Compton & Harwood, 1999b; Compton & Harwood, 2003; Compton & Harwood, 2005; Jones & Moreland, 1998; Moreland, Jones & Northover, 2001).

IMPLEMENTATION OF THE 1995 TECHNOLOGY CURRICULUM

The strong focus in primary schools (years 1–6) on literacy and numeracy that was beginning to take hold at the time the technology curriculum was mandated, meant technology was positioned as a 'minor' focus. This meant that technology, along with science and social studies, was relegated within many primary schools to being part of what was commonly referred to as 'topic'. Topic usually covered a revolving unit focus on technology, science and social studies within a term (10 weeks) and was normally part of the afternoon programme (approximately 1.5 hours), with language and mathematics programmes running in the mornings. Although the amount of time allocated to these units varied across schools and across teachers within schools, the most common configuration was that each unit would be between 3 and 4 weeks long. Primary teachers often attempted to make links to their language programme, and in some cases, links were made between the topic units themselves.

For example, a science unit on electricity may lead into a technology unit focused on developing electronic toys. However, links between subsequent technology units were rarely made. The result of this structure for technology was that primary students from years 1–6 tended to have a series of 'one-off' learning experiences that did not focus on learning in technology in a seamless fashion as part of an overall programme. It rather focused on learning specific context knowledge and skills within the particular unit focus. This is not to say many of these learning experiences were not valuable - in fact many have provided opportunities for students to confidently and competently make a difference to their own lives, the school and/or to their local community. Examples of the types of learning experiences that years 1–6 primary students were involved in during the implementation of the 1995 technology curriculum include those outlined below.

Taonga[1]: Year 1 students were involved in developing an understanding of how identity can be symbolically represented through jewellery. They researched types of pendants and explored the requirements that a personal pendant would need to have if it was to be cast in pewter. The students went through a process of modelling design ideas in plasticine, selecting one of these to use as a former, making a mould and casting pewter using the mould. The unit was supported by the grandfather of one of the students who discussed the spiritual significance of wearing a taonga and 'blessed' the students' individual taongas before they wore them. (For details of this unit please see Compton & Harwood, 2005.)

Meeting Seating: A group of students worked alongside a practising engineer to design and develop a concrete taniwha[2] to serve as a seat that would provide a meeting space and support discussions for this group and others in the school. It was situated outside in the garden and a range of design ideas were trialled to strike a balance between functional and aesthetic attributes. (For details of this unit please see Ministry of Education, 2005a.)

Sniff, Swing, Swipe: Students from a range of primary schools and intermediate schools were involved in a project with four New Zealand zoos. Each zoo asked the students to design ideas for 'enrichment activities' for specific animals to keep them entertained and happy in a zoo enclosure. Students worked with the zoo staff and based on research and feedback, they developed prototypes for the animals to test. (For details of this work please see Ministry of Education, 2006a.)

This differed somewhat to the experiences of students in the final years of their primary schooling - years 7–8 (age approximately 11–12). Years 7–8 was, and still is, the stage in New Zealand that technology is first offered to students through a programme involving specialist technology teachers and facilities. These programmes are most commonly offered through Intermediate Schools (catering for Years 7–8 only), although some are part of Full Primary Schools (years 1–8), Middle Schools (years 7–10) or Area Schools (years 1–13). How these programmes are constructed, and the time allocated for students to attend these specialist facilities, has been variable across the country. In many cases, 'technology specialists' tend to reflect technical rather than technology education programmes. For example, it was common for students to experience traditional learning experiences focused on developing cooking/sewing/metal and woodwork skills in a contrived context.

In some cases, these were expanded to include the development of information and communication skill development. Programmes with such a focus bear little resemblance to the intent of the 1995 technology curriculum[3]. However, in those technology centres where teachers had made the transition into technology education, this had begun to change significantly, resulting in learning experiences more in keeping with the 1995 document being offered. Examples of such learning experiences are outlined below.

Hot Bread Snacks: Students identified a personal need created by their attendance at the technology centre. As they attended for full day programmes they had to bring food lunch and during the winter they liked this to be hot. The hot food was mostly pies so together the classes looked at other possibilities for quick meal-snack ideas that would be appealing and nutritious. (For details of this unit see http://www.techlink.org.nz/Case-studies/Classroom-practice/archive-2006/hotbread/index.htm)

Pataka[4]: A group of students found that their school garden was producing more vegetables than could be used during particular growing seasons. They worked alongside a community expert to develop a 'pataka' for storing these so the gardening efforts would not go to waste. (For details of this unit please see Ministry of Education, 2005b).

These brief outlines of learning opportunities across years 1–8 show examples of primary technology in New Zealand that have sought to provide students with opportunity to be involved in authentic projects. Such projects require innovative and creative solutions and often involve students collaborating with others to ensure their technological practice is informed, socially and/or culturally suitable, and successful in terms of the identified intentions. However, as indicated earlier, these excellent learning experiences were commonly 'one-off' in nature and were not part of a larger technology programme based on progressing understandings of technology per se. The 1995 curriculum did not provide the type of assessment tools to support such progression (Compton & Harwood, 2005) and therefore teachers struggled to support progression through formative interactions and summative assessment strategies based on technological practice. In light of this lack of support therefore, primary teachers tended to resort to formative and summative assessment that focused on specific knowledge and skill determined by the context (for example understanding how to make a mould out of plaster of Paris); knowledge from other areas (for example, spelling of technology related words, calculations associated with resources); or more commonly reverted to assessing essential skills or behaviour and/or attitude (for example: works well with others, completed outcome on time, enjoyed the unit) (reported on in research findings from Compton & Harwood, 1999a; Compton & Harwood, 1999b; Jones & Moreland, 1998).

The lack of guidance provided in the 1995 technology curriculum around assessment was a feature of the feedback from teachers in the New Zealand Curriculum Project Stocktake Report (Ministry of Education, 2002). In relation to technology, 33% of teachers wanted to make changes to the curriculum and of these, the most common were in terms of making the curriculum easier to understand and requests for better developed learning and assessment examples (Harlow, Jones, & Cowie, 2002). This was further supported by teacher statements that the curriculum had been least helpful… in communicating student achievement to teachers in other schools' (Harlow, Jones & Cowie, 2002, p. 164).

In addition to the Stocktake feedback, technology results from the National Certificate of Achievement (NCEA)[5] began to highlight many issues for teachers and students as well. Subsequent analysis of this data found that the nature of students' technological literacy appeared to be limited in breadth and depth and lacked the level of critical analysis required for informed decision making. It was postulated that this situation may have arisen because of a programme focus on developing students' understandings of and about technology almost exclusively within the context of their own technological practice (Compton & Harwood, 2006).

2004–2007 REVISIONS

Experiences and data from ten years of technology in New Zealand provided evidence that undertaking technological practice allowed students to collaborate with others and make a difference to their own lives and in their immediate community. It often resulted in high levels of student engagement and allowed students to take increasing ownership of their learning and feel empowered to make decisions regarding the nature of their outcomes. However, the nature of the technological literacy resulting from students undertaking technological practice alone was limited, and the current curriculum achievement objectives did not support the development of progression based programmes that would allow for seamless education in technology from years 1–13 and beyond.

This situation led to the realisation in 2004 that technological practice on its own was not enough. Research was then undertaken to identify gaps in the 1995 curriculum and how these might be addressed through revising the technology curriculum (Compton & France, 2007). This resulted in technology education being restructured around three new strands:

- Technological Practice
- Nature of Technology
- Technological Knowledge.

How each strand contributes to the 'whole' of technological literacy is outlined below.

The Technological Practice strand pulls together the three strands of the 1995 technology curriculum and provides achievement objectives that support teachers to assess technological practice undertaken by students throughout their technology education. It therefore provides teachers with more guidance to enable students to undertake their own technological practice within a particular setting and to reflect on the technological practice of others. This guidance is provided through the provision and explanation of three key components of technological practice – Planning for Practice, Brief Development and Outcome Development and Evaluation (Compton & Harwood, 2005) and their associated achievement objectives (Ministry of Education, 2006b). In keeping with the 1995 document, learning opportunities focused within this strand will allow students to continue to gain a sense of empowerment as they undertake their own technological practice to find solutions to identified needs and/or realise identified opportunities. Under the revised curriculum however, this strand also provides opportunities to embed the philosophical ideas from the nature of technology and generic technological knowledge in order to better inform student practice.

The Nature of Technology strand provides students with the ability to develop a critical understanding of technology as an interventionary force in the world, and that technological developments are inevitably influenced by and have influence on historical, social and cultural events. This strand will enable students to develop a philosophical view of technology through a focus on two identified components of the nature of technology - Characteristics of Technology and Characteristics of Technological Outcomes (Compton & France, 2007), and these are also supported with achievement objectives to guide formative and summative assessment (Ministry of Education, 2006b). Such understandings will provide opportunities for informed debate about contentious issues and the complex moral and ethical aspects that surround technological development. It will also provide an opportunity to examine the fitness for purpose of technological outcomes in the past and to make informed predictions about future technological directions at a societal and personal level. Such philosophical understandings are essential to the development of a broad and critical literacy for New Zealand students.

The Technological Knowledge strand provides students with a basis for the development of key generic concepts underpinning all technological practice and/or technological outcomes. These concepts allow students to understand evidence that is required to defend not only the feasibility of a technological outcome, but also its desirability in a wider societal sense. This strand will enable students to develop technological understandings that underpin the three identified components of technological knowledge – Technological Modelling, Technological Products and Technological Systems (Compton & France, 2007) and these are again supported with achievement objectives to guide formative and summative assessment (Ministry of Education, 2006b). The focus on functional modelling will allow students to develop an understanding of simulated environments as compared to 'real' environments, allowing them to appreciate both the power and limitations of such modelling. This should result in them undertaking more informed technological practice as they move away from a simplistic 'build and fix' approach to one based on more informed decisions making and reducing the potential for wasting resources. Gaining a better understanding of prototyping will allow students to optimise their own technological outcomes. Knowledge of materials underpinning technological products and components within technological systems will enable students to better understand how technological outcomes work, how they are constructed, and their material and/or component properties. It will also enable students to infuse their technological practice with a higher level of technological understanding and therefore support more informed decision making.

IMPLICATIONS FOR FUTURE PRIMARY PROGRAMMES

The development of these new technology component within the three revised strands, and their levelled achievement objectives (Ministry of Education, 2006b), now provides teachers in New Zealand with the opportunity to design and deliver technology programmes that allow students to move through their technology education in a seamless fashion. The achievement objectives are further supported by a series of matrices known as the 'indicators of progression'[6]. These indicators provide formative assessment tools whereby teachers can be helped to determine where their students currently are, and the next steps required in their learning. They can also provide tools to support the reporting of student achievement to allow for effective communication between teachers and across transition points - for example, when students move from one school to another. The indicators of progression for the achievement objectives of the technological practice strand have been developed through research based trialling (Compton & Harwood, 2005). However, the indicators to support the two new strands are still in draft form and are being developed further from classroom-based research currently underway (Compton & Compton, 2009).

The implications for the primary sector in terms of technology programme development can be described in terms of short and long term shifts. In the short-term, it was recommended that all primary teachers should ensure their programmes focused on progression. Given the importance of the technological practice strand as foundational learning in technology, and the support material currently available for teachers and students around technological practice (see www.techlink.org.nz), it was further recommended that most primary teachers focus only on this strand for assessing student achievement and reporting purposes between 2007 and 2009. However, they were encouraged to begin incorporating some of the ideas behind the two new strands as well in order to enhance learning opportunities within the technological practice strand.

During 2008–2010 the Ministry of Education funded a classroom-based research project (Compton & Compton, 2009) to begin to develop further support for the two new strands by the time the revised curriculum is mandated. In 2010, all teachers, including primary teachers, will be expected to begin further development of their programmes to allow for more explicit teaching of technological knowledge and the nature of technology, thereby ensuring all eight components of the 2007 technology curriculum are progressed.

Teachers in the primary sector, particularly those with no specialist technology education background (normally those teaching years 1–6), will most likely need to access additional support to aid them in the delivery of future technology learning experiences – particularly those associated with the Technological Knowledge strand. It is unrealistic to assume these generalist teachers can suddenly become expert in all this knowledge. Therefore, accessing resources such as practising technologists (through initiatives such as Futureintech – see www.futureintech.org.nz), local community experts, and other teachers will become increasingly important for this sector. It will also be important that some further support material focused on the 'big ideas' underpinning this strand and the Nature of Technology strand are provided in accessible ways to support teachers and students as they grapple with these new concepts.

CONCLUSION – WHERE TO NEXT…

In conclusion, New Zealand is seeking to set a new direction for learning in technology that builds confidently on past strengths, acknowledging past weaknesses, and is cognisant of advances in understandings - particularly about the nature of technological literacy required by students today in order to become empowered citizens of tomorrow. The greatest implication of our journey so far is the need to fully recognise the shifts required by teachers as we move from curriculum development to curriculum implementation. At such a time, it is essential to recognise that it is teachers who make up the community upon which this learning area relies to reach its potential. Further to this, it is teachers in the *primary* sector that provide the foundations for success in later years. The identification of where 'teachers are at' will be an important next focus, so as to work alongside them and their students in ways that enhance their capability as a community of professional educators. We cannot afford to repeat the mistakes of the past[7] and assume the revised national curriculum framework in technology will be translated effortlessly into school programmes and classroom experiences.

Instead we must collaborate to align pre-service and in-service teacher education, policy, research, and implementation practice, and develop research-informed material support. All these dimensions are recognised as key aspects to meet these challenges. To acknowledge the importance of these synergies, mechanisms are currently underway to ensure this work is undertaken efficiently and effectively (Dinning, 2007; Harwood, 2007; Keith, 2007). This is an exciting time for the New Zealand technology education community as a whole and we look forward to the challenges that implementing and supporting the revised curriculum will provide.

NOTES

1 A Taonga is a common New Zealand term of Maori origin, referring to a precious object with identifiable significance - the closest English equivalent is 'treasure' although this does not capture the full depth of the word particularly in terms of the level of 'significance'.

2 A taniwha is a mythical Maori creature with serpent and dragon-like qualities.

3 The reasons for this are a complex mix and include such factors as teacher background, lack of professional development, perceived community/student expectations, and a past lack of clear direction from the Ministry of Education.

4 A Pataka is the Maori name for a hut on stilts designed for the purpose of storing food.

5 New Zealand's new standards based qualification system.

6 The indicators of progression can be found at http://www.techlink.org.nz/curriculum-support/indicators/index.htm

7 It is my opinion that the 1995 technology curriculum was not adequately supported in its implementation phase.

REFERENCES

Compton, V. J. (2001). *Developments in technology education in New Zealand 1993–1995: An analysis of the reflections of key participants.* Unpublished Doctoral Thesis, New Zealand: University of Waikato.

Compton, V. J., & Compton, A. D. (2009). Technological knowledge and nature of technology: Establishing the nature of progression. In *Pupils' Attitudes Towards Technology (PATT) International Design & Technology education conference: Strengthening technology education in the school curriculum.* Delft, The Netherlands: PATT.

Compton, V. J., & France, B. (2007). Redefining technological literacy in New Zealand: From concepts to curriculum constructs. In *Pupils' Attitudes Towards Technology (PATT) International Design & Technology education conference: Teaching and Learning technological literacy in the classroom* (pp. 260–272). Glasgow: University of Glasgow.

Compton, V. J., & Harwood, C. D. (1999a). *TEALS Research Project: Starting points and future directions.* 2nd TENZ Conference (pp. 14–16). Auckland, New Zealand: TENZ.

Compton, V. J., & Harwood, C. D. (1999b). *TEALS Research Project: From directions to classroom practice.* 30th Australasian Science Education Research Association Conference (ASERA). Rotorua, New Zealand: ASERA.

Compton, V. J., & Harwood, C. D. (2003). Enhancing technological practice: An assessment framework for technology education in New Zealand. *International Journal of Technology and Design Education, 13*(1), 1–26.

Compton, V. J., & Harwood, C. D. (2005). Progression in technology education in New Zealand: Components of practice as a way forward. *International Journal of Design and Technology Education, 15,* (3), 253–287.

Compton, V. J., & Harwood, C. D. (2006). *Discussion document: Design ideas for future technology programme.* Retrieved from http://www.tki.org.nz/r/nzcurriculum/draft-curriculum/technology_e.php.

Compton, V. J., & Jones, A. T. (2003). Visionary practice or curriculum imposition?: Why it matters for implementation. In *Proceedings of technology education New Zealand* (pp. 49–58). Fourth Technology Education Biennial conference, Hamilton, New Zealand: TENZ.

Davies Burns, J. (2000). Learning about technology in society: Developing liberating technological literacy. In J. Ziman (Ed.), *Technological innovation as an evolutionary process.* Cambridge: Cambridge University Press.

Dinning, N. (2007). Enabling technological literacy in New Zealand: The GIF-Technology education initiative. In *Pupils' Attitudes Towards Technology (PATT) International Design & Technology education conference: Teaching and Learning technological literacy in the classroom* (pp. 273–281). Glasgow: University of Glasgow.

Harlow, A., Jones, A. T., & Cowie, B. (2002). The implementation of the technology curriculum in New Zealand: The results of the national school sampling study. In H. Middleton, M. Pavlova & D. Roebuck (Eds.), *Technology Education Conference Proceedings: Learning in technology education: Challenges for the 21st century* (Vol. 1, pp. 161–169).

Harwood, C. D. (2007). Implementing technological practice in New Zealand: A foundation for technological literacy. In *Pupils' Attitudes Towards Technology (PATT) International Design & Technology education conference: Teaching and Learning technological literacy in the classroom* (pp. 290–296). Glasgow: University of Glasgow.

Harwood, C. D., & Compton, V. J. (2007). *Moving from technical to technology education: Why it's so hard! Sixth Technology Education biennial conference New Zealand,* Biennual conference. Auckland, New Zealand: TENZ.

Jones, A., & Moreland, J. (1998). *Assessment in technology education: Developing an initial framework for research. Australasian Science Education Research Association.* Darwin: ASERA.

Keith, G. (2007). *Supporting technological literacy in New Zealand: Connecting policy with vision. Pupils' Attitudes Towards Technology (PATT) International Design & Technology education conference: Teaching and Learning technological literacy in the classroom* (pp. 282–289). Glasgow: Univeristy of Glasgow.

Ministry of Education. (1995). *Technology in the New Zealand curriculum.* Wellington: Learning Media.

Ministry of Education. (2002). *New Zealand ministry of education research project curriculum stocktake - national school sampling study of teachers' experiences in curriculum implementation. General curriculum, mathematics and technology.* Retrieved from www.minu.govt.nz/web/document.

Ministry of Education. (2005a). *Connected 2 meeting seating.* Wellington: Learning Media.

Ministry of Education. (2005b). *Connected 2 our pataka.* Wellington: Learning Media.

Ministry of Education. (2006a). *Connected 2 sniff, swing, swipe.* Wellington: Learning Media.

Ministry of Education. (2006b). *Draft technology curriculum materials for consultation.* Wellington: Learning Media.

Moreland, J., Jones, A., & Northover, A. (2001). Enhancing teachers' technological and assessment practices to enhance student learning in technology: A two year classroom study. *Research in Science Education, 31*(1), 155–176.

Pacey, A. (1983). *The culture of technology*. Oxford: Blackwell.

Petrina, S. (2000). The politics of technological literacy. *International Journal of Technology and Design Education 10*, 181–206.

Vicki Compton
Faculty of Education
The University of Auckland
New Zealand

WENDY J DOW

4. TECHNOLOGY IN THE SCOTTISH PRIMARY SCHOOL

A Twenty-Year Retrospective Study

THE BACKGROUND

In order to understand the development of technology education within the primary sector in Scotland, it is necessary to first consider it within the context of the secondary school. Traditionally technology education or 'technical' as it has generally been known in Scotland, was a vocational, predominantly craft-based subject considered most suitable for non-academic boys in secondary schools. It was prescriptive in its delivery and involved engineering based technical drawing and learning how to operate industrial machines. Girls were taught the skills involved in cooking and needlework in domestic science, which later became known as home economics. Boys therefore were in effect being trained to take up employment in the various trades while girls were educated in the art of homemaking. Although both subjects were later made available to both boys and girls, and also to those pupils who were considered more academic, they remained as two separate and distinct subjects within the Scottish secondary curriculum.

At this time, the primary curriculum was essentially holistic in nature. The Scottish Education Department's *Primary Education in Scotland (or The Primary Memorandum)* (SED, 1965), on which the primary curriculum was based, was considered revolutionary in its child-centred approach. Its rationale was underpinned by the belief that children should be active participants in learning, and was grounded on the premise that children have a natural curiosity about the world and, most importantly, a natural desire to learn (Paterson, 2003).

This child-centred approach provided teachers, in theory at least, with a high degree of autonomy in the teaching and learning process. This combination of teacher autonomy and a perception of what was known as 'technical education' as a vocational subject, ensured that it was firmly located in the secondary sector. Technical education comprised woodwork, metalwork and technical drawing, subjects which were considered to be beyond the capability of the primary schools to deliver (Dakers & Dow, 2004).

This remained the prevalent model until the beginning of the 1980s when a major survey into primary education was conducted by the Scottish Council for Research in Education (SED, 1980). The findings of this survey demonstrated that, rather than the broad, active, child-centred curriculum proposed in the 1965 policy, in practice

C. Benson and J. Lunt (eds.), International Handbook of Primary Technology Education: Reviewing the Past Twenty Years. 39–50.

a very narrow curriculum had evolved. There was found to be very little, and in some cases absolutely no technical or science education delivered in primary schools. Contrary to the spirit of the *Primary Memorandum*, moreover, very little discovery learning and very little curricular integration was in fact in evidence (Adams, 2003).

In 1985 the Committee on Technology set up by the then Scottish Consultative Committee on the Curriculum (SCCC) published a set of recommendations relating to the introduction of technological activities in primary schools with support provided in the form of appropriate resources. Despite this initiative, by 1993 the incorporation of technology into primary schools was found to be uneven and largely unsatisfactory, with an HMI Audit report (Scottish Office Education Department (SOED), 1993) finding that a minority of pupils were achieving a satisfactory understanding in the area.

The concept of technical or technology education in primary schools therefore did not have a strong focus in policy in Scotland until the early 1990s when, in response to the survey mentioned above, primary education underwent a systemic change with the introduction and implementation of the 5–14 national guidelines for the curriculum (Scottish Executive Education Department (SEED), 1991).

1990: 5–14 GUIDELINES

The 5–14 national guidelines were first introduced into Scotland in 1991. These were intended to provide breadth, balance, continuity, coherence and progression to pupils from the first stage of primary school (at age 5) to the end of the second year of secondary education (at age 14). In this way, both the deficiencies identified in the primary curriculum and the problems of transition from primary to secondary school were to be addressed. Unlike the National Curriculum in England and Wales, however these were intended as guidelines only, although in effect they were implemented by all schools in Scotland, at least in the core areas of the curriculum. The curriculum from age 5 to 14 was conceived as comprising 5 broad areas, namely language, mathematics, expressive arts, religious and moral education and environmental studies (which was added after the other areas in 1993).

At this stage, what now became predominantly known as technology education in the primary curriculum in Scotland was introduced as part of the curricular guidelines for environmental studies. This also encompassed the curricular areas of history, geography, science, Information and Communication Technology (ICT) and health education. The rationale behind this particular grouping was that these were all areas within which children could observe and learn from their complex physical and social environment and apply their understanding of these various areas to their own lives (McClelland, 1993). Two attainment outcomes were identified for technology. These were 'understanding and using technology in society', and 'understanding and using the design process'.

Minimum recommended time allocations were provided for each of the five broad areas of the 5–14 curriculum. Language had a minimum time allocation of 20% and mathematics of 15%. Environmental studies also had an allocation of 15% but of course this had to cover the six separate subject areas which it comprised.

Also, whilst teachers generally welcomed a structure to the curriculum and especially the detailed guidance which was offered in the guidelines, there were strong concerns about the ability of primary teachers to cope with the content and delivery of such a broad range of subject specific areas. This was particularly true in the case of the science and technology components (Pickard, 2003). While teachers were already familiar with, and were used to teaching, geography and history, they were clearly less familiar and therefore less confident with these newer areas. The small time allocation for the group of subjects and the flexibility of the guidelines, moreover, made it possible for these less familiar areas to be avoided or overlooked.

TECHNOLOGICAL CAPABILITY

Despite the provision for the first time of a specification for technology education within the primary curriculum, it soon became apparent that there was no generic view of the purpose or indeed nature of technology curriculum across schools. Even the title given to the subject area was open to debate. Secondary schools, for example, continued to use the title 'technical' while the 5–14 documentation referred to the subject area as 'technology' throughout.

A review of technology education in Scotland, conducted soon after the implementation of 5–14, moreover, identified continued inequities between the experience of boys and girls and pointed to

> the often insufficient attention given to the impact of technology on people's lives, the physical environment and the world of work (SCCC 1996, p. 1).

In an attempt to address these issues, in 1996 the Scottish Consultative Committee on the Curriculum, in association with a wide ranging panel of consultants, (although significantly with only one technology teacher representative in the main review), published its seminal position paper 'Technology Education in Scottish Schools' (SCCC, 1996). This paper outlined the four interconnected aspects of the underpinning concept which was considered important in informing all aspects of technological education in Scotland: that of Technological Capability. This capability is acquired, it was argued, through the realisation of technological perspective, technological confidence, technological sensitivity and technological creativity. This was to form the bedrock upon which all aspects of the modern technology curriculum was to stand. The rationale for technology education in this new model therefore involved:

> ... learning about the social and physical conditions that influence, or have influenced, the lives of individuals and communities and which shape, or have been shaped by, the actions, artefacts and institutions of successive generations. Acquiring, interpreting and using evidence and information about the world they [pupils] live in is part of a sequence of discovery and rediscovery for every generation. (LTS, 2000a, p. 3)

Policy during the 1990s therefore suggested that technology education should have less concern with the development of vocationally-oriented craft skills which had traditionally formed such an important part of the secondary curriculum and a greater concern with both the interpretation of the made world, and the impact of technologies on humans and the environment. In keeping with the intentions of the 1965 *Primary Memorandum*, moreover, the SCCC position paper was intended to help primary teachers realise the integrative nature of technology in developing creativity, collaboration and critical thinking skills across the whole of the primary curriculum.

In 1998, in response to a Ministerial request to review the 5–14 guidelines for Environmental Studies, the SCCC in its new guise as Learning and Teaching Scotland (LTS), launched an extensive consultation process for the environmental studies component of the 5–14 curriculum. Input to this consultation was sought from a wide range of interested parties including schools, higher education institutions, local authorities, professional associations and other interested parties. Thus all interested parties were given the opportunity to have a stake in shaping policy. Teachers who had found the existing guidelines difficult to follow, and therefore difficult to implement, had the opportunity to play a particularly important part in influencing the new guidelines that emerged.

In recognition of their growing status and importance within the curriculum, separate guidelines were developed at this stage for both ICT and health education. This meant that technology education was left as part of environmental studies alongside history, geography and science. In order to support teachers in the translation of policy into practice, the SCCC established, in addition, a Technology Education Development Programme to facilitate, through support and exemplification, school-based developments within technology education in both the primary and secondary sector.

Revised 5–14 guidelines for Environmental Studies were subsequently introduced in 2000 (LTS, 2000b). Two policy documents relevant to technology education were produced at this stage: one for environmental studies subtitled 'Society, Science and Technology', and one specifically for the technology component of the curriculum. Whilst in the former, only 11 of the 75 pages were devoted to the technology component of the curriculum, the latter provided a detailed, specific guide to planning, teaching, learning and assessing technology across the 5–14 age range. Within these documents attention was paid to the clarification of the aims of technology education and the language and terminology were simplified to take account of teacher criticisms and earlier evidence of weaknesses in implementation. These new guidelines were, moreover, rewritten in conjunction with a pack entitled 'Primary Technology in Scottish Schools: Education for Technological Capability'. New study tasks were provided as a result of a joint project between Learning and Teaching Scotland and the Nuffield Foundation, and these reflected the findings and recommendations of the 1996 SCCC report. This model for technology education delivery was, in addition, underpinned by research (e.g. Murphy, 1999; Barlex, 2000), and sought to develop technological capability where

> 'children are [seen to be] building a repertoire of design & technology problem-solving strategies whilst engaged in creative activity that makes sense to them and interests them' (SCCC, 1996, p. 13).

Despite these important revisions in policy, a subsequent report by HMI (2002) still identified major weaknesses in the delivery of technology in both primary and secondary schools with a majority of schools being awarded a rating of either 'fair' or 'unsatisfactory'. Significantly, a particular weakness identified was the limited understanding of the 1996 SCCC definition of technological capability and a resultant focus on the production of artefacts as opposed to the processes of creativity and critical thinking which had been identified as key contributions of the subject to the curriculum.

A CURRICULUM FOR EXCELLENCE

The latest Scottish initiative to have an impact on the curriculum began in 2002 with the 'National Debate on Education'. This was an extensive consultation exercise launched by the Scottish Executive to consider the purposes of education, to identify strengths, to address the need for change and improvement and thus ultimately to produce 'Ambitious, Excellent Schools' (SEED, 2004). A Curriculum Review Group was established in 2003 and 'A Curriculum for Excellence' published in November 2004. This framework for a single cohesive curriculum to cover the age range of 3–18 replaced the curriculum guidance for 5–14. A major aim of 'A Curriculum for Excellence' was to 'declutter and redesign' the curriculum, significantly starting with science which was regarded as an area of priority at this stage.

As with 5–14, learning within A Curriculum for Excellence framework is organised into eight broad areas. In this case, the areas included are health and wellbeing, languages, mathematics, sciences, social studies, expressive arts, religious and moral education, and technologies. Significantly both science and technology have now both been removed from the environmental studies umbrella, with the other components of environmental studies now becoming social studies. Whereas science has been given the status of an area of its own, however, the area of the technologies incorporates not only craft, design, technology and graphics but also computing. In addition, for the first time, home economics, which has traditionally been regarded as a separate subject within the Scottish curriculum, is included within the technologies, although it is also regarded as having a contribution to make to the area of health and wellbeing.

Whilst the focus behind the 5–14 policy was in developing a curriculum characterised by 'breadth, balance, coherence, continuity and progression', A Curriculum for Excellence has a greater emphasis on the development of the pupil. The main focus now is therefore to produce

> successful learners, confident individuals, responsible citizens and effective contributors to society, (SEED, 2004, p. 12)

with all eight curricular areas having important contributions to make to achieve these ends. Principles for curriculum design are now seen to be underpinned not only by the breadth, progression and coherence, of 5–14 but also importantly by

> challenge and enjoyment, relevance, personalisation and choice, and depth (SEED, 2004, p. 14).

In relation to technologies, three main lines of development have been identified. These are 'investigating and designing', 'producing', and 'evaluating'. Although there is still a clear emphasis on creativity and on the promotion of technological capability in activities such as 'learning about technologies and their effects on society' and 'evaluating the impact of products, systems and processes' there is also the reintroduction of a focus on the controversial vocational aspect which traditionally informed and shaped the Scottish technology curriculum prior to the 1990s in the statement:

> Experiences and outcomes will be expressed to promote innovative, creative learning approaches with a strong emphasis on practical activities and where appropriate vocationally-relevant learning.
>
> http://www.ltscotland.org.uk/curriculumforexcellence/buildingthecurriculum/

As the new policy is designed to cover the age range 3–18, however, it is probably safe to assume, that this particular element will not apply to the technologies delivered within the primary curriculum.

POLICY INTO PRACTICE?

Technology education in primary schools in Scotland has clearly had a chequered history in relation to policy during the past twenty years. Policy changes, however, are only part of the picture. Perhaps more important issues relate to a consideration of where and to what extent the many positive changes in policy have actually been translated into practice within the primary classroom.

There is some early evidence of the positive impact of the 5–14 guidelines in a small scale study of four Scottish primary schools funded by the SOED shortly after the introduction of the 5–14 guidelines (SCCC, 1996). Although the findings from this study were encouraging, there were, however, certain factors identified as important to success which had implications for the effective development of technology education across all Scottish primary schools. One of the most important of these was the perceived necessity of

> the presence of at least one energetic enthusiast for technology among the teaching staff, combined with supportive and effective management of the head teacher (Stewart, 1997).

This implies that the quality of delivery was in fact equally or even more dependent upon the vagaries of individual interest, than on the development of a coherent and cohesive policy for technology education.

This has indeed proven to be the case throughout Scotland during the past twenty years, with schools where there have happened to be interested, confident and enthusiastic members of staff demonstrating excellent practice in the area, whilst in other cases, very little provision for technology in the primary curriculum has been in evidence.

As Black & Atkin noted as early as 1996, the fact that the whole 5–14 programme comprised guidelines rather than prescription meant that

> …there could be no way of tightly controlling the detail of what went on in the classroom (p. 23).

A study of 85 primary Bachelor of Education (B.Ed.) student teachers' experience of the technology curriculum whilst on school placement across six local authorities in Scotland (Dow, 2003) provides illustration of the inconsistencies in provision. 31% of participants stated not only that they had had no experience of observing technology education taught but also that there was no evidence of technology education featuring at any point on the timetables in the schools in which they were placed. Where the teaching of technology was observed, the quality was equally inconsistent, with practice ranging from the complex and innovative to the simple and prescriptive with a high reliance on commercially produced kits and work cards from which pupils followed instructions. The quality of resources available was also mixed. Whereas some local authorities had purchased commercially produced packs, resources in other areas had been produced by local authorities or by individual teachers.

Integration

Even when technology was perceived as an important part of the curriculum, the extent to which it was integrated has been equally mixed. Dakers & Dow (1998), in an evaluation carried out for Glasgow City Council of the introduction of Nuffield design and technology materials into nine primary schools, found that despite the clear opportunities within the materials for cross curricular links, technology was treated as a discrete subject area within the curriculum. Literacy and numeracy activities arising from the projects were therefore taught within the time devoted to technology, rather than as part of the time devoted within the curriculum to literacy and numeracy activities.

In instances where technology was integrated into other areas of the curriculum, practice was again found to be inconsistent, with the technological outcome of activities not always clear.

> In a number of cases, cross-curricular links were apparent with the activity clearly designed to fit in with the social studies topic being covered. Thus, in a number of schools, the technology lessons included such activities as making parachutes, Anderson shelters (for protection from World War II bombs), Victorian toys, Roman chariots, Jacobite puppets and Medieval catapults, according to the period of history being studied. Whilst the use of cross curricular links is valuable, however, the technological outcome behind the products was not always apparent, since the product appeared to take precedence over the processes involved in its manufacture. (Dow, 2003, p. 32)

Whilst the introduction of the 5–14 Environmental Studies curriculum guidelines clearly heralded a significant change in the provision of technology education in Scotland from the earliest stage of primary to the end of the second year of secondary school, there were clearly issues which made the straightforward translation of policy into practice fraught with difficulty.

Teacher Confidence

One important issue in this respect was clearly teacher confidence. Research conducted in both Scotland and England during the 1990s indicated a low level of confidence in primary teachers in their ability to teach technology education. (Harlen & Holroyd, 1996; Stables, 1997). When asked to rate their confidence in teaching various areas of the curriculum, for example, 60% of primary teachers gave the lowest rating for technology, with only 6% rating themselves as fully confident in this area (Harlen & Holroyd, 1996).

Using a similar methodology to Harlen & Holroyd, Dakers & Dow (1998) identified practical issues such as obtaining and maintaining equipment, understanding and using the design process, and health and safety issues arising from the management of tools and activities, to be major factors in primary teachers' lack of confidence.

Part of the problem clearly arose from the lack of training received by primary teachers in the area. As Stables so rightly pointed out:

> Very few primary teachers have received formal training in the teaching of technology education. Even those countries that have decided to introduce compulsory technology education into their primary curriculum and who have set up training programmes to facilitate this, have a back log of unprepared technology educators teaching in primary schools. (Stables, 1997, p. 60)

Indeed, in the Dakers & Dow (1998) evaluation, none of the primary teachers involved had any qualification or training in technology.

The lack of confidence of primary teachers in teaching technology, moreover, can be exacerbated by the attitudes of their colleagues in the secondary sector.

Primary Secondary Liaison

> Significant weaknesses in primary-secondary liaison were identified with secondary teachers unable to appreciate the nature of work in primary and some primary teachers asking secondary teachers to tell them what to teach in upper primary leading to advice which was ill advised and which has had a restricting effect on the curriculum of the primary school. (Adams, 2003, p. 370)

The lack of knowledge on the part of primary teachers of the nature of the secondary technology curriculum as well as attitudes on the part of secondary teachers which were dismissive of attempts of their primary colleagues to teach what they clearly regarded as highly specialised areas are issues of particular concern which have been identified. Research carried out by Dakers & Dow (2001; 2004) found, for example that teachers in both sectors professed to be unfamiliar with the SCCC (1996) Position Paper which had been developed to form the bedrock of technology education in Scotland. Most were, indeed, completely unaware of its existence. There was, moreover, a clear uncertainty on the part of primary teachers regarding the precise content of the secondary technology curriculum, with the majority being unable to make any clear distinction between science and technology. Most had had no experience of technology education, either in their initial teacher education or in subsequent continued professional development. This reinforced findings by Eggleston, who suggests, moreover that this is not confined only to teachers, but that bewilderment and confusion about the precise nature and content of technological education is prevalent among parents, employers and the public at large. (Eggleston, 1994).

In addition although, traditionally, secondary schools in Scotland have had a maximum of five or six associated primaries, over the years a combination of the effects of school closures and the introduction of placing requests has resulted in this number rising significantly, with the result that it is no longer unusual for the number of primary schools who send at least some pupils to one secondary may be in the high teens or even twenties. This presents an obvious barrier to curricular continuity and progression and to effective liaison both within and across sectors.

Identity and Status of Technology

Another part of the problem in translating policy into practice relates to the perceived identity and status of technology education within the curriculum at both primary and secondary level. Although the 5–14 documentation consistently used the term 'technology,' for example the HMI report published in 1999 continued to use the earlier term of 'technical education'. This same problem may indeed be seen today in the different labels for the subject adopted by different local authorities across Scotland. Thus whilst a decision was taken by schools in some local authorities to refer to 'Design and Technology' others have retained the old label of 'Technical' with its connotations of a craft-based non-academic orientation.

Even when the same terminology is in use, moreover, there is no guarantee that a shared understanding of the concept of technology or technical education exists:

> ... there are different views, even within the established field of technology education, of what important aspects should be. Despite the undoubted support of the guidelines in Environmental Studies 5–14 and because of the position of technology as an innovation in the primary curriculum, there remains a question of interpretation to be resolved. (CCC, 1995, p. 3)

Perceptions of status may also be traced to the messages inherent in the original guidelines for environmental studies themselves. As part of these guidelines, a topic grid was provided outlining five topics for each year of the primary curriculum from Primary 1–7 (P1–7). Each topic identified a major and in some cases a minor focus for each activity from the three broad areas of environmental education: social subjects, science and technology. While science was identified as a major focus for fifteen of the thirty-five topics, technology was suggested as a major focus in only five of these (and in one instance the major focus was shared with science). Technology was indeed not identified as a major focus in any of the topics outlined for the P2, P4 or P5 stages. Thus it was possible for pupils in primary schools to encounter technology at only certain stages of their primary education.

Technology education was identified as the main focus topic only in 'Toys' in P1; 'Things we eat' in P3; 'Coping with climate' in P6; and in a 'Mini-enterprise' topic in P7. In addition there was a shared major focus for technology in a topic on the senses in P2. It is clear therefore that a very limited range of technological activity might actually be encountered throughout the primary years and that the quality and extent of this will be determined largely by the level of interest and confidence of the individual teacher involved in topics covering such aspects as materials, weather, houses and homes, light and sound, magnetism, water, pollution and conservation, electricity, exploring space, materials from our planet, and getting about. On the other hand, science was identified as the major focus, with technology being identified as only a minor focus in three of these areas, namely materials, electricity and getting about.

Science and Technology

Two main issues arise from this in relation to both science and technology and the relationship between them. One is the message portrayed about the relative status and importance of science compared to technology in the primary curriculum. The other concerns conceptions about the relationship between science and technology education which is encouraged.

This is a situation which has been compounded by the consistent emphasis given to the importance of science within the primary curriculum by the Scottish Executive, an emphasis which has resulted in secondary science specialists working alongside primary colleagues in primary classrooms to promote effective teaching in the area. This importance is further reflected in the selection of science as a priority area of 'A Curriculum for Excellence'. It is interesting to note, however, that although the areas traditionally associated with the earlier technology or technical curriculum are grouped in a separate area of 'the technologies', part of the rationale for the revised science 3–15 curriculum includes the following statement:

> Science is part of our heritage and part of our everyday lives at work at leisure or in the home. It is important that all young people experience a sound science education given our rapidly-evolving, *technologically driven* world. (My emphasis)
>
> http://www.scotland.gov.uk/Publications/2006/03/22090015/11

As Frey (1991) warns:

> As educators advocate, promote, and implement technology education in schools, they may find that the new curriculum is equated with science or competes with science programmes. In either case the distinctive character of technology is misunderstood. (Frey, 1991, p. 1)

Assessment

The status of technology education has also been affected by the emphasis which has until recently been placed upon assessment within the primary curriculum. Thus, whilst 5–14 for the first time introduced technology education into the primary curriculum in Scotland, it also served to conceal the intent of government to deliver a more centralised and prescriptive curriculum which would clear the way for the introduction of national testing. Although this emphasis on testing met with strong resistance nationally and was consequently never fully implemented (Paterson, 2003), there nevertheless continued to be, throughout the 1990s, a strong emphasis on attainment at all stages of the curriculum. Each subject area, including technology, was given a grading structure for reporting which was designed to ensure coherence, continuity and progression both within and between primary and secondary sectors. In reality, only in literacy and numeracy was this ever actually achieved. The emphasis (or lack of emphasis, in the case of technology) on assessment tended to drive the curriculum and was to a large extent, responsible for a polarisation of the primary curriculum into separate subject domains, each with its own assessment arrangements. The effect of this was to ensure that technical education, which has always been a controversial subject within primary schools, was perceived as a problematic subject area for primary teachers to deliver and assess.

The Way Ahead

There have clearly been problems over the past two decades that have militated against the successful implementation of technology into all primary schools across Scotland. Some of these problems may, however, be addressed by new policies and initiatives which have recently been introduced. There is evidence, for example, that the implementation in 2002 of the Assessment is for Learning (AifL) initiative with its emphasis on formative assessment in all areas of the curriculum may be having an important impact. By freeing teachers, at least to some extent, from the perceived constraints of assessment for measurement and accountability, it appears to be allowing a greater emphasis to be placed on problem solving, creativity and the development of the kind of technological capability which over the last decade has been at the heart of policy for the technology curriculum (see for example, McClaren, Stables & Bain, 2007).

A Curriculum for Excellence has similarly been developed with the intention of providing teachers with a greater degree of professionalism and autonomy in the delivery of the curriculum than the past two decades have provided. Whether these new policy measures will alleviate or exacerbate the problems of the last twenty years concerning the position, status and role of technology within the Scottish primary curriculum remains to be seen.

REFERENCES

Adams, F. R. (2003). 5–14: Origins, Development and Implementation. In T. G. K. Bryce & W. M. Humes (Eds.), *Scottish education (2nd edition, Post devolution)* (pp. 369–379). Edinburgh: Edinburgh University Press.

Barlex, D. (2000). Resources for Technology Education in Scottish Primary Schools. *Journal of Design and Technology Education*, *5*(1), 45–46.

Black, P., & Atkin, J. (1996). *Changing the subject: Innovations in mathematics, science and technology education*. London: Routledge.

Consultative Committee on the Curriculum. (1985). *The place of technology in the secondary curriculum.* Dundee: Dundee College of Education for the CCC.

Dakers, J., & Dow, W. (1998). *An evaluation of nuffield design and technology materials used in a two year pilot scheme in glasgow schools*. Unpublished Report to the City of Glasgow Education Department.

Dakers, J., & Dow, W. (2001). Attitudes of secondary teachers of technical subjects and Home Economics towards curricular continuity at the transition between Primary 7 (11 years) and Senior 1 (12 years). In C. Benson, M. Martin & W. Till (Eds.), *Third international primary design and technology conference 2001* (pp. 46–49). Birmingham, UK: CRIPT at University of Central England.

Dakers, J., & Dow, W. (2004). The Problem with Transition in Technology Education: A Scottish Perspective. *The Journal of Design and Technology Education, 9*(2), 116–124.

Dow, W. (2003). Student Teachers' Perceptions of Technology Teaching in Scottish Primary Schools. In C. Benson, M. Martin, & W. Till (Eds.), *Fourth international primary design and technology conference 2003* (pp. 31–34). Birmingham, UK: CRIPT at University of Central England.

Eggleston, J. (1994). What is Design and Technology Education? In F. Banks (Ed.), *Teaching technology*. London: Routledge.

Frey, R. (1991). Another Look at Technology and Science. *Journal of Technology Education, 3*, 1.

Harlen, W., & Holroyd, C. (1996). *Primary teachers' understanding of concepts in science and technology*. Interchange No. 34. Research and Intelligence Unit.

HMIE. (2002). *Standards and quality in primary and secondary schools: 1998–2001*. Author.

Learning and Teaching Scotland (LTS). (2000a). *Environmental studies: Society, science and technology 5–14 guidelines*. Edinburgh: Author.

LTS. (2000b). *Environmental studies: technology. Guide for teachers and managers*. Edinburgh: LTS.

McClaren, S., Stables, K., & Bain, J. (2007). Exploring creativity and progression in transition through 'Assessement is for Learning'. In J. Dakers, W. Dow & M. de Vries (Eds.), *Teaching and learning technological literacy in the classroom: PATT 18 International conference on design and technology* (pp. 143–151). Glasgow.

McClelland, S. (1993). *Scottish education 5–14: A Parents' guide*. Edinburgh: HMSO.

Murphy, P. (1999). *Evaluation of the nuffield approach to primary design and technology*. London: Nuffield Foundation & DATA.

Paterson, L. (2003). *Scottish education in the twentieth century*. Edinburgh: Edinburgh University Press.

Pickard, W. (2003). The History of Scottish Education, 1980 to the present day. In T. G. K. Bryce & W. M. Humes (Eds.), *Scottish education: (2nd edition, Post devolution)* (pp. 229–238). Edinburgh: Edinburgh University Press.

Scottish Consultative Committee on the Curriculum (SCCC). (1996). *Technology education in scottish schools: A statement of the position from the Scottish CCC.*

Scottish Education Department (SED). (1965). *Primary education in scotland*. Edinburgh: HMSO.

SED. (1980). *Learning and Teaching in primary 4 and primary 7*. Edinburgh: Scottish Education Department.

Scottish Executive Education Department (SEED). (1991). *5–14 National guidelines for the curriculum*. Edinburgh: Author.

SEED. (2004). *A curriculum for excellence: The curriculum review group purposes and principles of education 3–18*. Edinburgh: SEED.

SEED. (2006). *A curriculum for excellence: Progress and Proposals*. Retrieved from http://www.scotland.gov.uk/Publications/2006/03/22090015/11.

Scottish Office Education Department (SOED). (1993). *Standards and quality in scottish schools*. Edinburgh: Scottish Office.

Stables, K. (1997). Critical issues to consider when introducing technology education into the curriculum of young learners. *Journal of Technology Education, 8*(2).

Stewart, D. (1997). Primary Technology Education in Scottish schools: Past, present and possibilities for the future. In R. Ager, & C. Benson (Eds.), *International primary design and technology conference 1997* (pp. 50–53). Birmingham, UK: CRIPT at University of Central England.

Wendy Dow
Independent researcher

GILL HOPE, ZANARIAH MAHYUN YUSEF
AND RAMACHANDRAN VENGRASALAM

5. TECHNOLOGY IN MALAYSIAN PRIMARY SCHOOLS

INTRODUCTION

The *Rukunegara* (national ideology) declaration on which Malaysian civic principles and societal development are founded, includes a commitment to:

> ...ensuring a liberal approach to her rich and diverse cultural traditions,

whilst also...

> ...building a progressive society which shall be oriented to modern science and technology.
>
> This country must seriously enhance the production and supply of information, knowledge and wisdom and ensure their accessibility to all our people. (Prime Minister Dato' Seri Mahathir bin Mohammed, quoted in Ministry of Education, 2001, p. 4)

Since independence from Britain in 1957, a unified system of education for all has been an ongoing priority, in order to ensure Malaysia's economic future, political stability and civil cohesion. Geographically, Malaysia consists of West (peninsular) Malaysia and East Malaysia (on the island of Borneo). West Malaysia is more economically developed; its towns and cities have a strong manufacturing heritage, whereas within East Malaysia's large tracts of jungle are many villages which follow a traditional lifestyle far from roads and cities. The challenge, especially for Technology education, has been to provide a relevant curriculum that will move the nation forward towards its goal of gaining developed nation status by 2020.

The population is predominantly Muslim Malay, whose first language is Bahasa Melayu, with large Chinese and Indian minorities, with smaller populations of tribal peoples who live traditional lifestyles in the more remote areas. Bahasa Melayu is the official language of instruction in national primary schools. Most pupils from non-Malay homes attend vernacular primary schools and learn Bahasa Melayu as a second language ready to transfer to Bahasa Melayu medium secondary schools at age 13. All pupils learn English. Arabic is offered to all Muslims; Mandarin Chinese and Tamil to the major racial minorities.

C. Benson and J. Lunt (eds.), International Handbook of Primary Technology Education: Reviewing the Past Twenty Years. 51–60.

EDUCATION IN MALAYSIA

Education in Malaysia is firmly founded on the moral values that underpin Malaysian society and citizenship:

> A nation may grow and prosper as a result of economic and new technological achievements, but this prosperity is meaningless if it is not supported by a solid value-based foundation. Therefore, in education, it is imperative that we nurture and strengthen our value base while we seek economic advancement and technological supremacy. (Dato' Asiah bt. Abu Samah, Director General of Education Malaysia, 1991–1993, quoted in Ministry of Education, 2001, p. 22)

The solid value-based foundation is expressed in the Philosophy of Education statement:

> Education in Malaysia is an on-going effort towards further developing the potential of individuals in a holistic and integrated manner, so as to produce individuals who are intellectually, spiritually, emotionally and physically balanced and harmonious, based on firm belief in God. Our efforts are focused towards creating Malaysian citizens who are knowledgeable and competent, who possess high moral standards, and who are responsible and capable of achieving a high level of personal well-being and able to contribute to the harmony and prosperity of the family, the society and the nation at large. (Ministry of Education, 1982)

This is provided through the internalisation and practice of noble values which permeate every school subject at primary level, including Technology education.

Compulsory primary schooling begins at age 7, building on one year of pre-school provision, and is split between Primary Level 1 (Years 1–3; 7–9 years old) and 2 (Years 4–6; 10–13 years old) and every child is entitled to at least 11 years of schooling. Thus, pupils in Malaysian Primary Level 2 are aged 10–13 years and Malaysian Primary Level 2 should be considered as more equivalent to English Key Stage 3 than to English primary school levels.

Several objectives of the primary school curriculum are especially relevant to Technology teaching:

- to develop and improve intellectual capacities which include rational, critical and creative thinking;
- to be sensitive towards man and his environment;
- to master scientific and technical skills;
- to master the basics of entrepreneurship and productivity;
- to develop talent and creativity.

Technology is placed within the Living Skills subject area, both within Primary Level 2 (ages 10–13 years) and the Lower Secondary School Curriculum (ages 14–16 years).

In the Upper Secondary School level (ages 16–17 years), technical and vocational education is available. These prepare young people either for technical and scientific tertiary education or for careers at the technician and semi-skilled level (for instance, welding, catering, crop production, commerce).

NINTH MALAYSIA PLAN

The *Rukunegara* (national ideology) declaration on which the Malaysian civic principles and societal development are founded, includes a commitment to:

> ...ensuring a liberal approach to her rich and diverse cultural traditions

whilst also,

> ...building a progressive society which shall be oriented to modern science and technology. (New Economic Policy, 1990)

This permeates all aspects of education and is especially pertinent to Technology.

This has informed three key national policy frameworks, the New Economic Policy (NEP), 1971–1990, the National Development Policy (NDP), 1991–2000 and the National Vision Policy (NVP), 2001–2010. Vision 2020, launched in 1991, outlined the national aim of attaining developed nation status. This contributed to the Ninth Malaysia Plan, 2006–2010, which consisted of five major thrusts:

- to move the economy up the value chain;
- to raise the capacity for knowledge and innovation and nurture first class mentality;
- to address persistent socio-economic inequalities constructively and productively;
- to improve the quality and sustainability of life;
- to strengthen the institutional and implementation capacity.

Clearly, education, and in particular Technology education, has an important role to play in achieving these goals. The specific goals for improving educational provision lie within 'Thrust 2'. These are wide-ranging, from the introduction of pre-school education, the provision of electricity to schools in remote rural areas, to ambitious plans for world class higher education that will attract international students:

> In education, measures will be intensified to promote Malaysia as a regional centre of excellence for tertiary education. (Ninth Malaysia Plan – Malaysian Government, 2006, §1.10)

The importance of sound foundations within primary education for these national aspirations is clearly recognised. Technical education is considered particularly important, since the plans for economic development specify roles for higher technology, biotechnology (including agro- and aqua- technology) 'for wealth creation' (§1.11). The government's aim to create a knowledge economy in Malaysia strongly impacts on teaching. Traditional pedagogy is didactic and based on rote learning and repetition of factual information rather than on developing knowledge through understanding. Introducing Technology into the primary school curriculum is part of this pedagogical revolution. The focus of teaching input is shifting from identifying and naming parts of machinery or learning hand-skills such as embroidery stitches, with the aim that there will be a shift towards using a greater range of creative design skills.

Many urban primary schools have dedicated workshops for practical work but the activities remain more often teacher-led than utilising an exploratory problem-solving pedagogy. For instance, whole class sets of baseboards can be found with electrical components already mounted, to which pupils attach wiring to learn the difference between series and parallel circuits, without applying this knowledge in a practical way. Predominantly, pupils will learn about the occurrence of these circuits in daily life but as factual knowledge rather than to inform a solution to a hands-on problem. The new emphasis on technological development should mean that such activities provide the knowledge required for creative problem-solving rather than being an end in itself. This pedagogical transition is difficult: teachers feel under pressure to cover curricular content and new entrants to the profession frequently opt to teach as they were taught rather than in new and innovative ways. If placed in a school with a traditional ethos, a new teacher cannot swim against the tide.

Although the Living Skills curriculum has done much to spread awareness and knowledge of modern technology, it has been realised that the acquisition of facts about the technological systems of the modern world is not enough and that to be competitive against Malaysia's neighbours in South East Asia, a skilled and knowledgeable workforce is needed. In order to achieve this goal,

> ...Measures will be undertaken to enhance the quality of education and training to be at par with international best practices. (Ninth Malaysia Plan – Malaysian Government, 2006, §11.03).

At the same time, there is a strong promotion of traditional Malaysian crafts and skills, especially of batik. On the twice monthly batik days, government employees (including teachers) are expected to wear batik garments, thus helping the batik industry to flourish.

THE INTEGRATED LIVING SKILLS CURRICULUM

Pupils in Primary Level 2 (ages 10–13 years) and Lower Secondary (ages 14–16 years) learn Integrated Living Skills (Ministry of Education, 2002) of which Technology forms a part. This is a combination of the previous Manipulative Skills Curriculum with Business and Entrepreneurship (see Figure 1).

1989	Manipulative Skills was launched in 100 primary schools
1991	Manipulative Skills was implemented in 1000 primary schools
1991	Integrated Living Skills was introduced in Form 1
1992	Manipulative Living Skills was implemented in 3000 primary schools
1993	Integrated Living Skills was introduced in Year 4 in all primary schools. Schools that had implemented Manipulative Skills expanded its implementation to Years 5 and 6 of primary school.

Figure 1. Timetable of introduction of the living skills curriculum in primary schools.

The subject content of Manipulative Living Skills was conceived as hands-on learning of important practical and vocational skills. This focused on learning about technology and developing life skills, rather than encouraging creative designing. Integrated Living Skills incorporated this curriculum with Business and Entrepreneurship (see Figure 2). The content is strong on product and systems analysis and the teaching of specific practical skills and techniques. From an English viewpoint, there appears to be less emphasis on encouraging pupils to generate innovative designs than in the English Technology curriculum. However, the range of specific skills taught within Malaysian Living Skills is greater than in England, equipping pupils with a sound knowledge base for innovation later. Where English practice may be criticised for encouraging young children to design products that they cannot make, Malaysian pupils engage more frequently with real materials and learn real hands-on skills that will be useful in daily life. This can be seen in the content of *Kemahiran Hidup Bersepadu Teras Tingkatan 1* (Kementerian Pendidikan Malaysia, 1990), the textbook that supports lower secondary Technology in which strong subject knowledge is promoted. The correct way to wire a plug or a lamp, for instance, is shown through a series of detailed step-by-step diagrams. Labelled diagrams of electrical components, such as resistors, capacitors, relays and so on, are shown and pupils are expected to learn the names and applications of these components.

Most Malaysian primary schools have dedicated Living Skills workshops, equipped with wooden benches and stools, tool racks and a range of appropriate materials. Money has been given by the Malaysian government to establish the subject as part of pupils' primary school curricular entitlement developed from existing practice within Lower Secondary Schools.

Figure 2 provides an outline of the content of both the Technology and the Business Entrepreneurship strands of the Integrated Living Skills curriculum. The aim of the Living Skills Curriculum is to enable the pupils to have basic skills and knowledge of technology, entrepreneurship and self-management. There is a symbiotic relationship between the two strands, although the subject areas are frequently taught separately as discrete subjects. Recently, some schools have been experimenting with an integrated topic-based approach to teaching Living Skills, but this does not include other subjects in the curriculum. The provision of dedicated workshops and specialist teachers has made the subject into a stand-alone area of the curriculum.

Technology:	*Business entrepreneurship:*
Workshop management and safety Designing projects Electrical and electronics Maintenance work Textiles and sewing Ornamental plants Ornamental fish Vegetable cultivation	Planning and preparing goods for sale Promotion of goods Management of sales Record keeping Personal budgeting Risks in business Being a good consumer

Figure 2. Structure of the integrated living skills curriculum.

The 14 learning objectives for the Integrated Living Skills curriculum demonstrate this integration of Technology with Business Entrepreneurship:

- to use the basic form to generate creative ideas in producing projects;
- to explain the relationship between the form and function of the product;
- to dismantle and fix products to identify the function of each component;
- to produce a project or article using wood and non-metal recycled material;
- to choose appropriate hand tools;
- to identify and differentiate series and parallel circuits;
- to produce electrical based projects;
- to maintain hand tools after use;
- to produce ornamental plants;
- to identify products that could be sold;
- to plan for a sale;
- to identify good packaging for products;
- to be able to count the profit and determine sales price;
- to practise safety and responsible behaviour.

Guidance for teaching Integrated Living Skills stresses the importance of inculcating the values of independence, innovation and creativity, and of activity based teaching and learning. A wide range of teaching strategies are recommended, including using stories, simulations, and role play whenever this is suitable in order to motivate the pupils. Teachers are expected to develop pupils' thinking towards the future and the contribution that their own work will make to the future of their community and country. This is linked to the positive values of how to achieve excellence, be open minded, be careful, and think how much they have to spend on a specific project. There is an emphasis on the development of interpersonal skills to enable the pupils to appreciate the opinions of others, to communicate effectively and learn co-operatively. The use of Information and Communication Technology (ICT) as a teaching and learning tool is also important.

The Malaysian approach to the Technology curriculum is essentially vocational. This does not mean that the Malaysian curriculum does not value creativity and innovation but that these dispositions and skills are set within the needs of the individual to earn a living and the goal of Malaysia gaining developed nation status.

Agro-technology is an essential aspect of the Malaysian Technology curriculum. Agricultural products such as palm oil and rubber play a major part in the Malaysian economy. The breeding and sale of ornamental fish also contributes considerably to national earnings. The inclusion of agro-technology in Malaysian Technology demonstrates both the vocational nature of the curriculum and the valuing of the contribution of the workforce to national income generation. The Technology curriculum is designed to implement the commitment of the Malaysian government to improving their people's lifestyle and life chances.

CONTENT OF THE TECHNOLOGY CURRICULUM

Health and safety within the workshop is of paramount importance. Before the pupils undertake any practical activities, they learn about organisation of the workshop, care of tools and safe storage of hand tools, materials and equipment, and safety rules for workshop practice. The teacher must demonstrate the safe use of hand and machine tools, equipment and materials, respect for others, care of the workshop, including the floor. The pupils must learn to keep themselves safe, through thinking about safe procedures and responsible behaviour to avoid accidents and to ensure hygiene after working with plants and fish.

The process skills to be developed within all areas of the Technology curriculum are seen as the:

- application of creativity;
- consideration of the form and function of the product;
- ability to plan a project;
- development of problem solving capabilities;
- documenting the design process;
- use of wood, metal and non-metallic recycled materials.

Additionally, pupils are taught the maintenance of hand tools and parts of buildings, as well as to repair simple machines, clothing and furniture. Both boys and girls are taught to sew and to make simple textile products, such as pencil cases and embroidered handkerchiefs. They are also taught how to repair certain simple machines, to produce and repair sewn items (including mending clothes) and to care for ornamental plants and fish.

Learning about ornamental plants includes learning about the different ways in which plants are propagated, different methods of planting, the care of plants and the maintenance work required. Learning to care for ornamental fish includes the preparation and management of the aquarium as well as caring for the fish themselves. The breeding of ornamental fish is an important area of export business in Malaysia, hence its inclusion in the school curriculum. Vegetable cultivation (including soil preparation, planting, cultivation, managing and harvesting of crops) is considered important since much of Malaysia's land supports heavy crops of palm oil, fruits such as pineapples and bananas, and rubber trees. The Business Entrepreneurship curriculum overlaps here, as pupils are taught to model business activities, create sales plans, the promotion and management of sales, record keeping, and personal financial management. They also learn about the risks in business and about consumerism.

Teachers are provided with a syllabus that outlines the topics to be covered, learning outcomes, levels of achievement, activities to be carried out, moral values to be incorporated, resources to be used, and the thinking skills that pupils will learn. From this year plan the teacher will decide the structure of the work across the two semesters, which are then turned into weekly and daily plans. These are written in the teacher's record book and submitted to the Head Teacher. Each school receives Rm40 per annum per pupil for resources for Living Skills (equivalent to approximately £6 sterling). Pupils frequently provide appropriate recycled resources from home. For example, 5 litre plastic cooking oil bottles may be cut down to make tubs for planting banana cuttings.

The pupils undertake design projects, for which they will make a product based on creative ideas. These may be generated through drawing. The pupils need to consider the function of the product that they make. Knowledge of materials, structures and electrical control technology is utilised to build pupils' innovative and creative thinking. Hands-on activities within project work are taught to the whole class and pupils work together as groups sharing tools and resources. This encourages social skills, such as negotiation and awareness of the needs of others within the group and providing peer support within the learning environment.

Pupils are tested monthly in all subjects, including Living Skills. These are pencil and paper tests of their recall of factual information, rather than of practical competence or design capability. This has a direct and necessary effect on the focus of teaching and pedagogy. In order to perform well, pupils need good recall skills, which could detract from the encouragement of exploratory creativity. Subject knowledge and hand skills are frequently taught well, but the assessment regime may not always encourage individual creativity, either for teachers or pupils. For Year 6 pupils, the drive is to do well in Ujian Penilaian Sekolah Rendah (UPSR) examinations in Bahasa Melayu, English, Mathematics and Science. As in all school systems, subjects that are examined will take higher priority in the minds of teachers, pupils and parents. This inevitably means less focus on Living Skills in Year 6.

COLLABORATIVE INITIAL TEACHER EDUCATION COURSE DEVELOPMENT

Under the Malaysia Ninth Education Plan (2006–2010) trainees for primary schools will achieve B.A. degree level academic standard and teaching qualification in order to achieve a world class education system for all Malaysian children, for which the government rightly identifies the need to train world class teachers. The government's aim is that by 2010 50% of primary school teachers will be university graduates.

From 2005–2009 Canterbury Christ Church University (CCCU), England and Institut Pendidikan Guru Kampus Tun Hussein Onn, Batu Pahat, Malaysia (IPGKTHO) have been working collaboratively to develop a degree level programme for Initial Teacher Education with Technology as subject major. This is part of a larger initiative in which several non-Malaysian universities are involved. Institut Pendidikan Guru Kampus Tun Hussein Onn is a well respected provider of primary Living Skills courses at postgraduate diploma level for practising teachers.

The collaborative programme ran with a single cohort of 80 students, which has provided the platform for the development of a Malaysian National ITE degree level programme which began in January 2007. From the beginning of the programme, CCCU personnel have been committed to a fully collaborative model of course development alongside IPGKTHO colleagues. Fears of intellectual colonialism existed in the minds of colleagues on both sides of the partnership at the beginning of the working relationship. Within two years into the four year programme, these fears subsided as effective working relationships and genuine mutual respect for each others' professional capabilities developed.

IPGKTHO is an ideal environment in which to site a new degree-bearing course for primary Technology trainees. There are large spacious workshops, modern library facilities and well qualified and knowledgeable lecturers. Batu Pahat town and surrounding area provides a rich environment for Technology students. Its industrial base is as a quarrying town, with a strong textile industry, now supplemented by the manufacture of electrical and electronic goods. The town's architecture reflects the history and continued economic success of the town. Houses are a mix of modern and traditional vernacular styles; recent developments of shopping malls sit alongside more traditional shop lots. This has provided a rich environment for the trainees, who have been able to use the buildings and industries of the town for case studies and resources both for personal design activities and for developing curriculum resources.

At the beginning of the CCCU/IPGKTHO programme, there was scepticism about its collaborative nature. Tutors at IPGKTHO were concerned that CCCU staff did not come with the whole course ready written, as a week-by-week guide. As the project developed, the joint planning and discussions enabled the development of shared understandings across the cultural differences, about the embeddedness of knowledge and learning within cultural contexts. The way in which the built environment or manufacturing traditions can be used as learning resources for pupils has led on to deeper level discussions about the transmission of knowledge and understanding. The tutors have become concerned that the way in which this single cohort has been taught to teach may be dissipated in the short term through pressure from teachers in school who are facing the challenges of on-going change. Newly qualified Technology teachers may find themselves working with established teachers of Living Skills who are in the process of developing new approaches in pedagogy and may see the new colleague as a threat to their own professional competence.

To alleviate this, CCCU and IPGKTHO hosted an in-service training event for teachers who would be receiving the trainees. Like teachers everywhere, new initiatives cause anxiety and stress and the aim of the event was to inform and build relationships. The day consisted of a discussion about the programme and a creativity workshop in which the teachers were asked to design and make houses using just newspaper and sticky tape. Several teachers commented how this activity had reminded them that creativity could be developed using simple resources, readily available even in remote villages. This enhanced their confidence in their underlying belief that design skills could be developed without the need for expensive, new or advanced materials or techniques. The event appeared to alleviate fears that the CCCU/IPGKTHO students would arrive in schools expecting high-tech facilities and advanced knowledge of the latest techniques. That the programme promoted an approach to teaching Technology that encouraged pupil independence and creativity clearly won the approval of the teachers.

IN SUMMARY

The Malaysian educational system and pedagogical tradition is of respect for authority and of learning from those who know more, which may act against inculcating individuality and innovative responses leading to radical creativity. This is recognised by the government and part of the Ninth Malaysia Plan (Malaysian Government, 2006) is to encourage entrepreneurship, especially techno-entrepreneurship. The dual factors of the country's colonial past and the conservative Islamic tradition of many of its people, present challenges to these aspirations. The Malaysian government's embrace of *islam hadari*, a form of Islam which encourages engagement with modern, forward-looking modes of thought and action may in practice be difficult to reconcile with the deep-rooted tendency to look back to established tradition and authority. The tutors and their students at IPGKTHO are very conscious of their pivotal role in developing their nation's future. Students enrolled on the new Malayasian degree-bearing programme have already been sent for school experience placements to work alongside the newly qualified graduates from the CCCU/ IPGKTHO cohort. It has been a privilege to work together during this time of rapid change.

REFERENCES

Kementerian Pendidikan Malaysia. (1990). *Kemahiran hidup bersepadu teras tingkatan 1*. Kuala Lumpur, Malaysia: Tropical Press Sdn. Bhd.

Ministry of Education. (1982). *The philosophy of education in malaysia*. Kuala Lumpur, Malaysia: Ministry of Education.

Ministry of Education. (2001). *Education in malaysia: A journey to excellence*. Kuala Lumpur, Malaysia: Ministry of Education.

Ministry of Education. (2002). *Malaysian living skills curriculum*. Kuala Lumpur: Curriculum Development Centre, Ministry of Education.

Malaysian Government. (2006). *Ninth malaysia plan 2006–10*. Putrajaya: Economic Planning Unit.

Gill Hope
Canterbury Christchurch University
Kent
England

Zanariah Mahyun Yusef and Ramachandran Vengrasalam
Institut Pendidikan
Guru Kampus Tun Hussein Onn
Batu Pahat
Johore
Malaysia

STEVE KEIRL

6. CAUGHT IN THE CURRENTS - THE SHAPING OF PRIMARY TECHNOLOGY EDUCATION IN AUSTRALIA

INTRODUCTION

This chapter documents the principal curriculum developments in Technology Education for the Australian primary school sector over the past two decades. For reasons of space, scope and data availability, it is necessarily broad in its approach.

It is timely to report on a twenty-year period for a couple of reasons. First, significant educational reorganisation occurred in the late 1980s across all education sectors in Australia. This change was potentially positive for Technology Education. Second, at the time of writing, Australia stands at the dawn of its first national curriculum and the situation regarding Technology Education is neither clear nor assured.

This chapter is constrained by two principal factors. First, there are eight differing educational jurisdictions within the country, each with its own curriculum agenda. Second, there are no substantial research repositories available on the topic and, while rich data undoubtedly exist, they are spread across the jurisdictions. Each has its own history, circumstances and vision.

The opening section addresses the political and demographic organisation of education in Australia. The chapter then describes the significant national collaboration begun twenty years ago and the arrival of something called 'Technology Education' with its associated constructs. A section is devoted to the legacy that these constructs have given and examples of primary technology curriculum from around the country are presented. In closing, the chapter portrays the current situation – a threshold of change, uncertainty and new challenges.

DEMOGRAPHY, POLITICAL AND EDUCATIONAL ORGANISATION

Australia is a vast continent of approximately 7.7 million square kilometres yet with a population of only 22 million (Australian Bureau of Statistics (ABS), 2009) – 3 people per square kilometre (global average, 51; Netherlands, 401; UK, 255; US, 33 (United Nations, 2008)). The majority of the population lives in urbanised, coastal locations.

Lying between latitudes 10 and 44 degrees south, Australia has a broad range of climate conditions yet is the driest inhabited continent. With rich mineral, agricultural and maritime resources as well as a thriving arts culture, a full spectrum of occupations is represented. Currently, the economy is comparatively strong and unemployment is 5.8% (ABS, 2009).

C. Benson and J. Lunt (eds.), International Handbook of Primary Technology Education: Reviewing the Past Twenty Years. 61–76.

While Australia can be described as a multicultural society, the country's history is grounded in tens of thousands of years of Aboriginal existence intimately linked to the natural environment (Horton, 1994). European invasion began in 1788. The Aboriginal proportion of the population is 2.5% (ABS, 2009).

National government is through a federal system (federation of six self-governing colonies occurred in 1901). There are now six states and two territories, each with its own parliament. At any time the political persuasion of a state or territory government may match or differ from that of the national government.

While education is constitutionally the prerogative of each state or territory, the federal government has the power to give financial assistance to them as it sees fit. This can mean the shaping or influencing of curriculum through federal funding. Under various governments, initiatives over the years have included: vocationalism; civics and citizenship; values education; and the testing of literacy and numeracy.

There are approximately 3.45 million full-time school students of whom 65.9% attend government schools with the remainder at non-government (e.g. religious or independent) schools. There are 9,562 schools – 71.5% government and 28.5% non-government. The data show 67.4% of all non-special schools to be primary and 13% to be combined primary/secondary (typically in rural settings) (Ministerial Council on Education, Employment, Training and Youth Affairs (MCEETYA), 2008a).

In 2004, the state and territory education ministers' report profiling the teaching profession stated, 'Gender trends remain a matter of concern' (MCEETYA, 2004, p. 5). This trend continues to be upward with figures for the primary sector showing a profile that is 80.4% female. The staff-student ratio has dropped from 1:19.2 in 1983 (MCEETYA, 2004) to 1:16 - also the Organisation for Economic Cooperation and Development (OECD) average - in 2008 (MCEETYA, 2008a). A much-discussed statistic concerns the ageing of the profession with one third of all primary teachers having been 40–49 years old in 2004 (MCEETYA, 2004).

NATIONAL COLLABORATION ON CURRICULUM – 1986 TO 2009

To date, Australia has never had a *national* curriculum although the event is close (see below). The prospect of this, for some, is anathema and, while many support the principle, it means different things to different people. 'National collaboration on curriculum' began in 1986 to help maximise resource use and to reduce differences across jurisdictions (Australian Education Council) (AEC, 1994b). (For critiques and details of the period, the developments and the political arguments, see e.g. Marginson, 1997; Harris & Marsh, 2005; Marsh, 2005).

In the 1980s the peak body of the state and territory ministers of education was the Australian Education Council (AEC) later becoming the Ministerial Council for Employment, Education, Training and Youth Affairs (MCEETYA). By 1987 the AEC had identified five priority areas for collaboration – numeracy, literacy, science, languages other than English, and English as a second language. In 1988 it initiated development of a statement of national goals and purposes of education and these came to fruition in 1989 in the Hobart Declaration (AEC, 1989). A decade later the national goals were revised in the Adelaide Declaration (MCEETYA, 1999).

In 1989 the AEC launched mapping projects – analyses of curriculum documentation and literature within and beyond Australia – for Technology, Science and English literacy. The Technology mapping was completed in ten months and, by the end of 1990, approval was given for a *statement* on Technology to be developed. A statement sits within the nationally agreed goals and describes '…the knowledge and skills to which all students are entitled (and the) agreed areas of strength in curriculum development which might be shared and built upon' (AEC, 1994b, p. 150).

There were also to be developed a set of *profiles* which described student outcomes at a number of levels (across their years of schooling). By 1991,

> …the AEC launched the projects in their final form by deciding that statements and profiles would be developed for eight broad learning areas, forming a template of the knowledge and processes to be taught in Australian schools. Most States and Territories had already adopted their own sets of key learning areas, which generally clustered around the eight areas of learning adopted by the AEC. (AEC, 1994b, p. 151).

Behind this comfortable piece of authorship lay the federal agenda, state and territory agendas and multiple competing professional (subject association) agendas. While the Statements and Profiles shuttled between AEC meetings and were eventually agreed by all ministers, they were never endorsed in any binding way. Publication was to be '…at the prerogative of each State and Territory.' (AEC, 1994a, p. iv). As one respected curriculum researcher comments:

> The most ambitious attempt at national curriculum collaboration in Australia's history had foundered on the old rock of state-Commonwealth suspicion.' (Reid, 2005, p. 43).

A key aspect of the developments was the use of *outcomes* as a basis of reporting and assessment. Used well, these contribute to rich portrayals of students' learning and development. Outcomes-based education (OBE) has today become the focus of intensified attack from a neo-conservative lobby keen to promote accountability of teachers and schools as well as a return to 'basics' and regular, formalised testing of students.

THE ARRIVAL OF 'TECHNOLOGY'

Despite the chequered background described, there were significant developments that came with the Statement (AEC, 1994a) and Profile (AEC, 1994b) for Technology Education. A curious blend of circumstances brought many challenges but, importantly, a new beginning.

First, there had been enough discussion and broad agreement over the approach that the learning areas model of curriculum became the norm. To describe curriculum in eight broad areas was, for secondary education, a potential merging (or erosion) of a multiplicity of subjects. For primary there was, potentially at least, some policy rationalisation although the innovation was daunting for many teachers. The eight learning areas were: the Arts; English; Health and Physical Education; Languages other than English; Mathematics; Science; Studies of Society and Environment; and, Technology. Variants were either in place or emerged in some jurisdictions but, broadly, they followed this arrangement. Some States adopted the Statements and Profiles as they were.

Second, within this broad Australian curriculum development, the introduction of an area called 'Technology' was truly innovative. However, many took this to mean 'computers' or, more recently, 'information and communications technologies' (ICTs).

Third, there were signs that the time was right in a global sense. Emerging from the same period was a UNESCO report (Layton, 1994) - its first dedicated solely to Technology Education. There soon followed a study from the OECD (Black & Atkin, 1996) – one that affirms one of the orthodoxies of misunderstanding about Technology Education, namely, that it must be compounded with Maths and Science.

Nonetheless, each international study acknowledged the climate of the times. Black & Atkin (1996, p. 53) write of 'The New Subject: Technology' while Layton opened his review saying:

> It is rare today to be able to observe the emergence of a new subject in the school curriculum…
>
> In many education systems around the world, irrespective of whether the country is low income and developing or high income and industrialised, the case for technology as a component of general education is under examination and is impelling specific curriculum innovations. (Layton, 1994, p. 11)

Fourth, and no doubt as a contributor to the international climate, the growth of activity in classrooms in Australia was noted in the Technology Mapping Project (Owen & Abbott-Chapman, 1990). The authors noted that:

> …the explosion of knowledge and practice in this area has been quite phenomenal over the last few years (p. 1), and
>
> …we are seeing an increasing trend of the introduction of technology and technological problem solving at the primary school level. This trend will continue… (p. 3).

While much activity of a craft or making nature had in the past been happening in primary classrooms, there were practices emerging that resulted from primary teacher-innovation – so often successful because of teachers' capacities to respond to current social thinking and concerns (e.g. values issues in relation to technologies), to develop innovative, student-centred pedagogies (as are needed with design), and importantly, to recognise and use design- and technological-focused activities to integrate cross-curricular work.

As is often the case, it was the innovative work of primary school teachers which had contributed to the climate of change and adoption in Technology Education. Dedicated teachers were now finding an identity and a common language for their work and were able to shape the emerging area in ways that were far more than applied science, computing, or 'making things'. Such teachers now had both a potential home and an identity for their work.

THE PRIMARY SITUATION IN 1990

The Mapping Project identified several phenomena in relation to Technology curriculum and its primary dimensions:

- Technology *was* happening but in a scattered, cross-curricular way, that is it was present but often in the background;
- because of its breadth, Technology Education in the future might become *both* a subject and integrated across other areas;
- primary schools were not embracing social, economic, environmental and political aspects of Technology;
- teacher preparedness for teaching Technology was a concern. There were shortfalls in their pre- or in-service education and teachers may have lacked confidence, knowledge or may have not understood the rationale for the innovations; and
- there was a relative lack of support material for primary schools.

At its outset, the review signalled an aspect of the climate of the time that is borne through subsequent State and Territory policy documents:

> ...we see as an overriding concern the need to maintain a humane and humanistic perspective in design and implementation of technology education which will enrich life rather than impoverish it, and will acknowledge and incorporate the deep ethical and moral issues which technological developments and technology in the curriculum invariably raise. (Owen & Abbott-Chapman, 1990, p. 1)

The review also highlighted a key procedural dimension:

> In moving towards the integration of technology education within the curriculum there is a need to plan for and encourage the development of technological imagination and technological capability among students. This is best expressed through the processes of *designing, making* and *appraising*. (Owen & Abbott-Chapman, 1990, p. 3. Emphasis added)

If all the aspects of what has been presented above as Technology Education were to be successfully addressed in primary schools then the approach would have to be cross-curricular, comprehensive and very well resourced, not least in the ways teachers were supported. As the AEC itself had said, 'Making decisions about technology often involves a complex mixture of consensus, conflict and compromise.' (AEC, 1994b, p. 2). Such would be the case for Technology curriculum too.

THE 1994 NATIONAL TECHNOLOGY STATEMENT AND PROFILE

The *statements* were *frameworks* for curriculum development and were to define and describe the area – its distinctive qualities and a sequence for developing knowledge and skills. The sequence is expressed through four developmental *Bands* (A–D) with A and B generally covering primary years. Importantly, the framework approach meant that these statements were neither details of subjects nor associated syllabi – the intention was to respect local circumstances in schools and to facilitate teachers' professional curriculum innovation.

The *profiles* were intended to support teaching and learning and to provide a common reporting language for a learning area. Each area was organised through *strands* - groupings of its content, processes and concepts. The strands were expressed as more detailed *outcomes* at each of eight *levels* which span the compulsory school years (1–10 i.e. approximately ages 5–16). The arrangement of the Technology outcomes is illustrated in Table 1. (Only the Level 1 and Level 4 outcomes are presented. These broadly reflect the lower and upper primary years. The outcome statements of the same suffix (e.g. 1.1/4.1; 1.3/4.3; 1.10/4.10) are related and indicate progression in richness/complexity – culminating at Level 8 in Year 10 of schooling (age 16). Thus, in each numbering, the first digit indicates the level and the second the organiser's outcome.)

Technology Education was organised round four strands. In common with all other learning areas, these strands were dually described as 'process and/or conceptual understanding strands' and 'content strands'. Technology Education emerged with a process strand of 'Designing, Making and Appraising' (DMA) and three content strands of 'Information', 'Materials' and 'Systems'. The strands are described as being interdependent. However:

> All learning in technology involves the Designing, making and appraising strand. The relative emphasis on the Information, Materials and Systems strands varies according to the needs of the students and the nature of the programs and activities. (AEC, 1994a, p. 5)

Within the four strands were a total of ten *strand organisers* (sometimes, *substrands*) – four for DMA (Investigating, Devising, Producing, Evaluating - IDPE) and a pair (Nature, Techniques) used for each of the three content strands. For each organiser, at each level, there was an outcome and it is here that the core philosophy of OBE is articulated. As the documents state, the levels do not *equate* with the years of schooling nor any cohort of students. The individual student is the focus and is assessed against the range of outcomes. Thus, a cohort of students in Band B - upper primary, may be broadly performing at Level 4 and, within the cohort, student X may perform at Level 3 on a couple of outcomes and Level 5 on another. Meanwhile student Y may be strong across the level and reaching two Level 5 outcomes, and so on. Here, the student is central and each student's profile is composed to inform teachers, parents, the student, and the system of their performance.

For primary teachers unfamiliar with Technology Education trying to manage ten organisers and outcomes for reporting students' progress was a daunting prospect – especially when taken against the reporting needs for seven other learning areas. Thus, the device of strands – with its potential to support the novice to the area – was weakened by its accompanying organisational complexities.

Table 1. Statement and profile level 1 and level 4 technology outcomes (AEC, 1994b)

Technology Statement and Profile - Levels 1 & 4 Outcomes (1994)			
Strand	**Organiser**	**Level 1 Outcomes** - *The student:*	**Level 4 Outcomes** - *The student:*
Designing, Making and Appraising	*Investigating*	1.1 Investigates the form and identifies the uses of everyday products.	4.1 Determines the appropriateness of products and processes for communities and environments.
	Devising	1.2 Generates ideas for own designs using trial and error, simple models and drawings.	4.2 Creates and prepares design proposals that include: – options considered and reasons for the choices made – images used to visualise ideas and work out how they might be realised.
	Producing	1.3 Undertakes simple production processes with care and safety.	4.3 Organises and implements production processes to own specifications, recognising hazards and adopting safe work practices.
	Evaluating	1.4 Describes feelings about own design ideas, products and processes.	4.4 Assesses the effectiveness of own designs, products and processes in relation to design requirements including social and environmental criteria.
Information	*Nature*	1.5 Identifies the different ways information can be used and presented.	4.5 a) Identifies the form, structure, style and presentation used in particular information products and processes. b) Describes how processing and transmitting information have evolved and are continuing to change.
	Techniques	1.6 Uses simple techniques to access, record and present information.	4.6 Selects and uses recognised procedures, conventions and languages to process information and create information products.
Materials	*Nature*	1.7 Identifies common materials and some of their uses.	4.7 Identifies the characteristics of materials and relates them to the functional and aesthetic requirements of own designs.
	Techniques	1.8 Uses equipment to manipulate and process common materials.	4.8 Applies a range of techniques for safely working materials to the functional and aesthetic requirements of own designs.
Systems	*Nature*	1.9 Identifies some common systems and their uses.	4.9 Identifies the relationships between elements in systems (people and components) and some of the sequences through which the elements work.
	Techniques	1.10 Carries out a short sequence of steps to operate or assemble systems.	4.10 Selects and uses techniques to organise, assemble and disassemble systems to manage, control and assess them.

THE LEGACY OF THE TECHNOLOGY INNOVATIONS

I use the term 'legacy' because there has been a mixed inheritance. Despite the criticisms of the 1994 AEC initiatives, there have been clear advantages for the field – not least the gaining of an identity called Technology Education. One principal drawback has been the continuing failure of the field itself to establish nationally common terminology, networks and resources.

There has been only one national Primary Technology Education conference in Australia – in 1998. It was highly successful but too difficult for the organisers, with their limited resources, to repeat. There is no single national professional teachers' association for the Technology learning area. There were half a dozen subject-based secondary associations in various states of health united as a loose federation (TEFA, 2007, site last updated Jan 2000) which has been relatively impotent in influencing any national curriculum policy initiatives.

There have been sorties into the field of the kind that Layton identified. Notably, one body has set out its views on what primary education could be doing to better service the engineering professions (ASTEC, 1993; 1997) – a worthy claim but failing to acknowledge the myriad other professions and occupations which could also argue a case.

One major research project into the status of Technology Education in Australian schools was sponsored by the Federal government and lead by Edith Cowan University. While a final report was never published by the government, some general findings were documented (Williams & Keirl, 2001). The research noted that, for primary schools:

- there were disparate understandings of what was meant by Technology with interpretations involving science, art, craft, design, and computers;
- there were real equipment shortages and a lack of specialised facilities for Technology Education;
- there was inadequate pre-service and in-service teacher professional training and development for the field;
- Technology teaching activities and pedagogies addressed multiple learning styles with challenging and meaningful tasks and were broadly welcomed by primary students.

The paper called for better equipping of primary schools, the establishment of at least one Technology Education specialist in every school, ongoing professional development of all teachers, and the establishment of a core course of Technology Education in all university teacher education programmes (Williams & Keirl, 2001).

Another aspect of the legacy has been the embedding of some terms in the professional discourse of primary Technology Education across the country. The national conference showed the extent to which 'DMA' had become the jargon of the day across the country. Serving as both facilitator and inhibitor, the catch-cry of design-make-appraise was an ideal scaffold to the teacher-stranger to Technology Education. It communicated essence effectively. However, for many it was to become a linear, lockstep approach – three distinct steps always to be addressed in that order - a far cry from more meaningful cyclical and iterative models that reflect the realities of design and creative technological activity.

The sub-strands or organisers (IDPE) of the process strand of the Statement and Profile (Table 1) have, in cases, similarly served a linear approach when it is the integration of these that is the most fruitful. Here has been the challenge to the best of primary pedagogy – educating children about the dimensions of technological practice (e.g. *each* component of D-M-A and I-D-P-E) while also reaching for the interplay and integration of *all* of these into a purposeful whole.

It has also been the case that the curriculum language and organisational structures adopted by the various Australian jurisdictions continue to vary considerably. That is, there would still seem to be a search for meaning (Layton, 1994; Williams & Keirl, 2001) for primary Technology Education - for the field in general, for methodology, and indeed, for its naming.

The learning areas in their primary phases now include: Technology; Technology and Enterprise; Design and Technology; Science and Technology. Victoria has moved to a system of three *Strands*, one of which is '*Interdisciplinary Learning*' within which lies the *domain* of 'Design, Creativity and Technology' (VCAA, 2005). In Tasmania, 'Vocational and Applied Learning' and 'ICT' are identified (DoE, 2009).

Curriculum frameworks have varied across the country too. Within them, terminology varies but all have assumed the professional judgement of the teacher will be key. They describe the anticipated development of children across levels or standards. Teachers are to help students, within local contextual considerations, to achieve the outcomes.

A few examples illustrate the diversity of approaches and language. Interestingly (and importantly) verbs continue to serve this area of curriculum well. These examples consider only the primary components of the policies. Bearing in mind that Table 1 showed the nationally agreed position in 1994, it is possible to compare later iterations from two States. Table 2 shows the South Australian development with an attempt to break the DMA sequence and to establish a strand called 'Critiquing'. This policy includes an elaboration of its use of the term *technological literacy.* Table 3 shows the Western Australian descriptors for the *Technology Process* outcome. The language (IDPE) has its lineage back to the 1994 Statement and Profile and the document also uses the original Information, Materials and Systems terminology.

In the Victorian arrangement, a nuanced variation of IDPE occurs in standards expressed for students only from Year 3 (approximate age 8) onwards. Here, the dimensions at all levels are presented as *Investigating and designing; Producing;* and *Analysing and evaluating* (VCAA, 2005). Meanwhile Queensland's New Basics curriculum introduced in 2001 the curriculum innovation of *Rich Tasks* which are transdisciplinary or holistic in nature, are an ideal home for design and technological activity, but which give no explicit mention of it (EQ, 2001). New South Wales uses the traditional linkage of 'science and technology' for its K-6 curriculum with 'Learning Processes (of) Investigating; Designing and Making; and, Using Technology' (BoS, 2006).

A curriculum phenomenon of the late 1990s has been the introduction in some States of *essential learnings.* Originally these were identified as vital to all educational practice and were to interweave all learning areas and pedagogy. It is possible

Table 2. South Australian Curriculum, Standards and Accountability (SACSA) framework–D&T outcomes standards 1–3 (essential learnings and key competencies omitted. Adapted from DETE, 2001)

Strand	Standard 1 *Towards the end of Year 2 the child:*	Standard 2 *Towards the end of Year 4 the student:*	Standard 3 *Towards the end of Year 6 the student:*
Critiquing	1.1 Makes judgments about the significance of different characteristics of products, processes and systems made by themselves and others.	2.1 Identifies a range of ways in which the design of everyday products, processes and systems is related to those who use them.	3.1 Describes the significance to diverse groups of people of the various criteria used in the design of particular products, processes and systems.
Designing *Design strategies*	1.2 Demonstrates an initial variety of design practices and recognises design as a tool for change.	2.2 Develops a range of design skills and uses them to effect change.	3.2 Understands and uses the relationship between different design skills to become better designers.
Communicating designs	1.3 Shares a variety of ways of communicating their design ideas and thinking.	2.3 Uses a range of communication forms and technologies, as a means of self-reflection and to describe their design ideas, thinking and planning.	3.3 Selects appropriate communication forms and technologies to document and convey clearly design ideas, thinking and organisation.
Making *Making techniques*	1.4 Acts confidently through using materials and equipment to make products, processes and systems.	2.4 Demonstrates effective use of a broad range of materials and equipment, and reflects on their personal interaction with resources they use.	3.4 Demonstrates skills and confidence in creating products, processes and systems which respect personal and collective identities.
Resources for making	1.5 Explores current and alternative uses of materials and equipment in creating products, processes and systems.	2.5 Identifies the characteristics of a range of materials and equipment, and explains the relationship of those characteristics to designed and made products, processes and systems.	3.5 Investigates the characteristics of materials and equipment used in design and production in order to achieve sustainability.
Responsible management	1.6 Understands the importance of simple organisation and safety issues in terms of their consciousness of people and fairness.	2.6 Identifies the reasons for managing resources effectively and for working in personally and socially safe and responsible ways.	3.6 Identifies and articulates a range of responsible strategies for managing resources and working safely.

Table 3. Western Australia technology and enterprise technology process outcome level descriptors (DETWA, 2005)

Evaluating: Students evaluate intentions, plans and actions.		
Foundation	The student:	TP F Explores the form of familiar products and their uses in everyday life, uses production processes and expresses feelings about the results.
Level 1	The student:	TP 1 Uses an awareness of the form of familiar products and their uses, applying a trial-and-error approach when creating or modifying technologies, and expressing feelings about the result.
Level 2	The student:	TP 2 Uses an awareness of how existing products and processes affect people, applying a more methodical approach when creating or modifying technologies that meet human needs, communicating ideas and comparing the result with the original intention.
Level 3	The student:	TP 3 Uses an understanding of the relationship between aesthetics, social and environmental effects when generating and communicating designs and when creating and modifying technologies and evaluates results using functional and aesthetic criteria.
Investigating: Students investigate issues, values, needs and opportunities.		
Foundation	The student:	TP F.1 Explores the form of products and their everyday use.
Level 1	The student:	TP 1.1 Investigates the form and identifies the uses of everyday products.
Level 2	The student:	TP 2.1 Investigates and identifies the uses and effects of products, systems, processes, services and environments.
Level 3	The student:	TP 3.1 Examines and identifies key design features, including aesthetic features, and environmental effects of products, systems, processes, services and environments.
Devising: Students devise and generate ideas and prepare production proposals.		
Foundation	The student:	TP F.2 Indicates, suggests or describes their ideas verbally or by gestures.
Level 1	The student:	TP 1.2 Generates ideas for own designs, using trial-and-error, simple models and drawings.
Level 2	The student:	TP 2.2 Generates designs and recognises some practical constraints using text, drawings or models and introducing related technical terms.
Level 3	The student:	TP 3.2 Generates designs that take into account some social and environmental implications and communicates using a range of graphical representations, models and technical terms.
Producing: Students produce solutions and manage production processes.		
Foundation	The student:	TP F.3 Participates in production processes.
Level 1	The student:	TP 1.3 Undertakes simple production processes, using trial-and-error, with care and safety.
Level 2	The student:	TP 2.3 Plans production processes and makes products, systems, processes, services and environments using resources safely.
Level 3	The student:	TP 3.3 Plans and carries out the steps of production processes, making safe and efficient use of resources.
Evaluating: Students evaluate intentions, plans and actions.		
Foundation	The student:	TP F.4 Expresses feelings about own processes and products.
Level 1	The student:	TP 1.4 Expresses feelings about own design ideas, products and processes.
Level 2	The student:	TP 2.4 Compares own products, systems, processes, services and environments with original intentions.
Level 3	The student:	TP 3.4 Assesses how well the ideas, products, systems, processes, services and environments used meet design requirements, including consideration of functional and aesthetic criteria.

to speculate that the essential learnings might supersede learning areas (Keirl, 2002). Ironically, but rather differently, in Queensland and in Tasmania the learning areas have *become* the essential learnings (in name) (DoE, 2009; QSA, 2009). In South Australia they are Futures; Identity; Interdependence; Thinking; and, Communication (DETE, 2001).

While the curriculum picture is diverse across the country, there have been varying degrees of change expected of primary practitioners within each state or territory. Good primary teachers are well used to integrating curriculum and have used Technology Education successfully to achieve this. The fact that broadly-understood Technology Education is also well placed to deliver such dimensions as the essential learnings, values education, and literacy and numeracy strategies has meant that these teachers have been able to adapt and be creative in their pedagogical work as change took place. However, it is equally true to say that teachers without good preparation in Technology Education have found the innovations challenging. In this situation, neither the students nor the field have benefitted.

AN EMERGENT NATIONAL CURRICULUM – W(H)ITHER TECHNOLOGY EDUCATION IN THE TIDES OF CHANGE?

Following the Hobart and Adelaide Declarations, MCEETYA met again in 2008 and produced the 'Melbourne Declaration on Educational Goals for Young Australians' (MCEETYA, 2008b). In this, there are telling changes which aroused (with just ten months passing since the Declaration and this chapter) justifiable concerns for Technology Education in Australia. Of the eight learning areas that have been in place for fifteen years, seven remain. The one that differs is Technology which became 'Information and Communications Technology and design and technology' (ICT/d/t). There are, arguably, issues around the placing of the conjunctions and the lower case 'd' and 't' for 'design' and the latter 'technology' in this lengthy name (one which, in its entirety, would never be adopted in schools).

For Technology Education, other concerns emerge from the very brief (two paragraphs) rationale given for the learning areas which states:

> The learning areas are *not of equal importance* at all year levels. English and mathematics are of fundamental importance in all years of schooling and are the primary focus of learning in the early years. However, humanities and social sciences, for example, take on greater *scope* and increasing *specialisation* as students move through the years of schooling. Each learning area has a *specific discipline base* and each has application across the curriculum. (MCEETYA, 2008b, p. 14. Emphasis added)

The first concern here is the differential 'importance' rating of learning areas 'across the years'. Second is the 'example' given for one learning area valorising scope and specialisation. Third, the claim of 'a specific discipline base' cannot be demonstrated in the case of ICT/d/t. In other words traditional disciplines are more valued. The latest curriculum developments have proved the various concerns to be justified.

Several months *prior* (May, 2008) to the Declaration, the National Curriculum Board had already

> ...commenced work on a national curriculum in English, mathematics, the sciences and history. For *each learning area* the board... (NCB, 2009, p. 4. Emphasis added)

Thus science was at the forefront (as one might anticipate given the OECD agenda) and history had become a learning area in its own right. Further, all four of these are to be delivered from Kindergarten to Year 12.

The sole mention of ICT/d/t in the fifteen-page NCB statement in a section entitled 'Curriculum content: Knowledge, understanding and skills', prescribes a particularly restrictive and utilitarian role for the combined field:

> As a foundation for further learning and adult life the curriculum will include practical knowledge and skills development in areas such as Information and Communications Technology and design and technology, which are central to Australia's skilled economy and provide crucial pathways to post-school success. (NCB, 2009, p. 9)

The most recent developments have come from the Australian Curriculum, Assessment and Reporting Authority (ACARA). Their current (October, 2009) website states:

> ACARA is responsible for the development of Australia's national curriculum from Kindergarten to Year 12, starting with the learning areas of English, mathematics, science and history, for implementation from 2011.
>
> As a second phase of work, national curriculum will be developed in languages, geography and the arts.
>
> The development of continua for literacy and numeracy skills and ICT will be a foundation of the curriculum. (ACARA, 2009)

In this pronouncement, (Design and) technology is now invisible. At the time of writing, the Technology education community awaits news of its fate.

From what has been presented it can be seen that Technology Education in Australia, having been given an opportunity to establish identity, integrity and momentum, now finds itself challenged by those very criteria. For many years it swam well in the mainstream but it now faces the rapids of change.

The field has already undergone some erosion having disappeared in name in one State and become increasingly a part of interdisciplinary approaches elsewhere. That said, the interdisciplinary approach is one that primary Technology Education is good at and this is where it can, where good pedagogical practices exist, maintain its integrity. It may be that the momentum is still rolling though there is no substantial evidence available to support or refute this. As Keirl & MacGregor (Chapter 7) report, at least in one state, there can be some optimism that reasonable numbers of well-prepared primary teachers are entering the profession.

However, currents far stronger than those that teachers can control are clearly running. The national curriculum agenda cannot be separated from the fact that

Australia is a member of the OECD which '...brings together the governments of countries committed to democracy and the market economy...' (OECD, 2007). As Galbraith (2004) has shown, 'market economy' is simply a benign name for capitalism and, increasingly, education is being shaped as a tool to enhance the market economy. OECD policy is now a major shaper of educational policy in countries such as Australia. As Apple says,

> (E)ducation is seen as simply one more product...Rather than democracy being a *political* concept, it is transformed into a wholly *economic* concept. (Apple, 2001, p. 39)

Primary Technology Education in Australia may now find itself once more being reduced to crafts and skilling, as the curriculum is wagged by a vocational tail. In recent years, as has happened in other countries, a climate of 'failing schools' has been promulgated by the right wing of politics, and pressures for change abound in the name of standards and meeting jobs shortages. Further, the MCEETYA view on national curriculum has affirmed the default 'disciplines'. Unsurprisingly, English, Maths and Science (EMS) remain ensconced and History is the new member of the first division quadrivium.

Furthermore, one might expect a role for Technology Education in the area of *problem solving* which the OECD has promulgated. The Programme for International Student Assessment (PISA) – the OECD's assessment league – assesses EMS but also includes problem solving (OECD, 2003; 2004) which takes place '...where the content areas or curricular areas that might be applicable are not within a single subject area of mathematics, science or reading (sic)' (OECD, 2004, p. 26). Yet deeper investigation shows that the 'problems' cited in the 2004 report claim to be 'real world' yet are all paper-based, and stress logic, analysis and reason. Design and critique are barely present, as are the in-vogue catchcries of the capitalist economies: creativity, innovation and enterprise. It is as though these attributes are potentially drowning.

From what has occurred nationally and internationally it can be said that primary Technology Education in Australia is facing some considerable challenges. It would seem that Australia's emergent national curriculum is far from imaginative and, further, has eschewed a field that celebrates imagination as a practice. The very field that can integrate, in meaningful and stimulating ways, curriculum areas through design and problem solving would seem to be destined to a loss of identity and integrity. It is as though Technology Education is being pulled down by the rip current of orthodoxy.

However, something that is good for children, which children enjoy, and which can celebrate much successful practice will not disappear because of curriculum policy change. Such change will not discourage good primary educators who know the realities of what works so well, and who will continue to teach well, and in ways that embrace the existential and societal realities of the technological world. There are strong swimmers who read the currents and know that, in time, Technology Education practices will resume their place in the mainstream of curriculum practice.

REFERENCES

Apple, M. W. (2001). *Educating the "Right" way: Markets, standards, God and inequality*. New York: Routledge Falmer.

Australian Bureau of Statistics (ABS). (2009). *Year Book Australia 2008*. Retrieved from http://www.abs.gov.au/ausstats/abs@.nsf/mf/1301.0.

Australian Curriculum, Assessment and Reporting Authority (ACARA). (2009). *Australian curriculum*. Retrieved from http://www.acara.edu.au/curriculum.html.

**Australian Education Council (AEC). (1989). Sixtieth Australian Education Council, 14–16 April 1989. *Hobart declaration on schooling: Common and agreed goals for schooling in Australia*, No 7. Retrieved from http://www.mceecdya.edu.au/mceecdya/hobart_declaration,11577.html.

Australian Education Council (AEC). (1994a). *A statement on technology for Australian schools*. Carlton: Curriculum Corporation.

Australian Education Council (AEC). (1994b). *Technology - A curriculum profile for Australian schools*. Carlton: Curriculum Corporation.

Australian Science and Technology Council (ASTEC). (1993). *Bridging the gap*. Canberra: Commonwealth of Australia.

Australian Science and Technology Council (ASTEC). (1997). *Foundations for Australia's future: Science and technology in primary schools*. Canberra: AGPS.

Black, P., & Atkin, J. M. (1996). *Changing the subject: Innovations in Science, Mathematics and Technology education*. London: Routledge.

Board of Studies (BoS) (New South Wales). (2006). *Science and technology K-6: Outcomes and indicators*. Retrieved from http://k6.boardofstudies.nsw.edu.au/files/science-and-technology/k6_scitech_outcomes.pdf.

Department of Education (DoE) (Tasmania). (2009). *A focus on curriculum areas*. Retrieved from http://www.education.tas.gov.au/curriculum/focus.

Department of Education and Training, Western Australia (DETWA). (2005). *Outcomes and standards framework – Technology and enterprise*. East Perth: Western Australia.

Department of Education, Training and Employment (DETE). (2001). *South Australian curriculum standards and accountability framework (SACSA)*. Retrieved from http://www.sacsa.sa.edu.au.

Education Queensland (EQ). (2001). *Rich tasks*. Retrieved from http://education.qld.gov.au/corporate/newbasics/html/richtasks/richtasks.html.

Galbraith, J. K. (2004). *The economics of innocent fraud*. London: Penguin.

Harris, C., & Marsh, C. (Eds.), (2005). *Curriculum developments in Australia: Promising initiatives, impasses and dead-ends*. Adelaide: Openbook.

Horton, D. (1994). *The encyclopaedia of aboriginal Australia*. Canberra: Australian Institute of Aboriginal and Torres Strait Islander Studies.

Keirl, S. (2002). Against the provincialism of customary existence: Issues arising from the interplay of 'essential learnings', design and technology and general education. In H. Middleton, M. Pavlova, & D. Roebuck (Eds.), *Learning in technology education: Challenges for the 21st century, Proceedings of the 2nd biennial international conference on technology education research*. Queensland: Centre for Technology Education Research, Griffith University.

Layton, D. (Ed.), (1994). *Innovations in science and technology education* (Vol. V). Paris: UNESCO.

Marsh, C. (Ed.), (2005). *Curriculum controversies: Point and counterpoint 1980–2005*. Deakin West, Australian Capital Territory: Australian Curriculum Studies Association.

Marginson, S. (1997). *Educating Australia: Government, economy and citizen since 1960*. Cambridge: Cambridge University Press.

Ministerial Council on Education, Employment, Training and Youth Affairs (MCEETYA). (1999). *National goals for schooling in the twenty-first century*. Retrieved from http://www.mceecdya.edu.au/mceecdya/adelaide_declaration_1999_text,28298.html.

Ministerial Council on Education, Employment, Training and Youth Affairs (MCEETYA). (2004). *Demand and supply of primary and secondary school teachers in Australia*. Retrieved from http://www.mceetya.edu.au/verve/_resources/-DAS_teachers-PartsA-d.pdf.

Ministerial Council on Education, Employment, Training and Youth Affairs (MCEETYA). (2008a). *National report on schooling in Australia 2008*. Retrieved from http://cms.curriculum.edu.au/anr2008/index.htm.

Ministerial Council on Education, Employment, Training and Youth Affairs (MCEETYA). (2008b). *National declaration on educational goals for young Australians, December 2008*. Retrieved from http://www.mceecdya.edu.au/verve/_resources/National_Declaration_on_the_Educational_Goals_for_Young_Australians.pdf.

National Curriculum Board. (2009). *The shape of the Australian curriculum*. Commonwealth of Australia, Barton: ACT.

Organisation for Economic Co-operation and Development (OECD). (2003). *The PISA 2003 assessment framework: Mathematics, reading, science and problem solving knowledge and skills*. Paris: Author.

Organisation for Economic Co-operation and Development (OECD). (2004). *Problem solving for tomorrow's world – First measures of cross-curricular competencies from PISA 2003*. Paris: Author.

Organisation for Economic Cooperation and Development (OECD). (2007). *About OECD: Our mission*. Retrieved from http://www.oecd.org/pages/0,3417,en_36734052_36734103_1_1_1_1_1,00.html.

Owen, C., & Abbott-Chapman, J. (1990). *Technology in the curriculum: Trends and issues emerging in the 1990s*. Hobart: Education Participation Studies Unit, University of Tasmania.

Queensland Studies Authority (QSA). (2009). *Essential learnings*. Spring Hill, Queensland: QSA. Retrieved from http://www.qsa.qld.edu.au/learning/7261.html.

Reid, A. (2005). The politics of national curriculum collaboration: How can Australia move beyond the railway gauge metaphor? In C. Harris, & C. Marsh (Eds.), *Curriculum developments in Australia: Promising initiatives, impasses and dead-ends*. Adelaide: Openbook.

Technology Education Federation of Australia (TEFA). (2007). Retrieved from http://www.pa.ash.org.au/tefa/ (last updated Jan 2000).

United Nations (UN). (2008). *World population prospects database*. Retrieved from http://esa.un.org/unpp/.

Victorian Curriculum and Assessment Authority (VCAA). (2005). *Victorian essential learning standards: Interdisciplinary learning strand: Design, creativity and technology*. East Melbourne: VCAA.

Williams, P. J., & Keirl, S. (2001). The status of teaching and learning of technology in primary and secondary schools in Australia. In E. W. L. Norman & P. H. Roberts (Eds.), *International conference on design and technology educational research and curriculum development 2001*. Loughborough: Loughborough University.

Steve Keirl
Centre for Research in Education
University of South Australia
Australia

STEVE KEIRL AND DENISE MACGREGOR

7. THE GROWTH OF PRIMARY DESIGN AND TECHNOLOGY TEACHER EDUCATION IN SOUTH AUSTRALIA

More Head, Less Hands, Always with Heart

INTRODUCTION

As a major Australian Primary Design and Technology Education provider, the University of South Australia's School of Education has undergone a decade of considerable change - of the kind affecting all universities (e.g. globalisation, markets, new policy directions). In the same period, Design and Technology curriculum design has also moved significantly.

In this chapter we describe some of the principal changes and innovations that have occurred in Primary Design and Technology education at the University. Some context – historical, curricular and political – is presented and we address matters of demography, programme (degree) and course (subject) design, pedagogy, innovation, influences, challenges and opportunities. Having described the evolution of the Design and Technology developments, the chapter concludes with summary reflections and it speculates on what the next fifteen years might bring.

CONTEXT

University Provision

There are three universities in the state of South Australia. While each offers some education programmes the Design and Technology components vary. The University of Adelaide has no primary teacher education programme. Flinders University offers a broad range of teacher education programmes. All its primary programmes include a single-semester course based on Expressive Arts addressing English, visual arts, Design and Technology, drama, media and music.

The University of South Australia is the largest of the three universities in the state with over 33,000 students, over 1,000 academic staff and five campuses. The School of Education offers undergraduate and postgraduate education programmes, ranging across early childhood, primary, secondary, and adult education. Three years ago the University closed one of its largest campuses (Underdale) and relocated staff and students. This paper reports on the work begun in Design and Technology Education at Underdale which now continues at the Mawson Lakes campus.

C. Benson and J. Lunt (eds.), International Handbook of Primary Technology Education: Reviewing the Past Twenty Years. 77–88.

In response to national (AEC, 1994a,b) and international curriculum developments, the School of Education (Underdale) introduced Australia's first dedicated suite of Technology Education courses for pre-service primary teachers. These courses were delivered across a four-year degree period from 1995 onwards. Reflections on this significant innovation were reported at the international Centre for Research in Primary Technology (CRIPT) conference in 1999 (MacGregor, 1999).

The Changing World and Education

The past two decades have witnessed significant change in the political, social and economic spheres in which universities operate. Globalisation has taken its course. The 'knowledge economy' has driven the 'knowledge society' and, in turn, new forces seeking to shape education in schools (Hargreaves, 2003). Universities have continued to compete for students in new markets while also seeking to be efficient and to maintain standards.

In line with some countries (though not with others) Australia is witnessing growing engagement of governments at state and national levels in education (to the extent that 'standards', civics, literacy, and numeracy have been foregrounded on a daily basis in the media and in current federal electioneering). 'Curriculum wars' are engaged with a battle raging between Outcomes Based Education (OBE) and a 'back-to-basics' return to 'traditional disciplines' and the testing of knowledge of them (Killen, 2006).

Curriculum development across Australia has taken both professional and political paths. Professionally, teachers and academics have together contributed to rich and purposeful debate around curriculum design (see, for example, Harris & Marsh, 2005). Politically, education now finds itself at the centre of party politics – more so than it has ever been. However, where the professional and the political have met for almost two decades, and with some degree of harmony, has been in the area of curriculum. There are signs that this harmony is under threat of erosion.

Curriculum Influences Nationally

In 1989 Australia's State and Territory Education ministers agreed to national goals for education across the country. The innovations (AEC, 1994a&b) included the proposal that curriculum be structured around eight 'Learning Areas' (one was Technology). As MacGregor (1999) reported:

> A major contributing factor in the conception and development of Primary Technology Education courses at the University...was the introduction of Technology Education into Primary Curriculum. (p. 86)

Here was a long-sought recognition which, in turn, gave identity to the field in the bigger curriculum picture. Being new has its accompanying difficulties but these are compounded when set against such elements as the 'big three' of English, maths and science. As Williams & Keirl (2001) reported from a national research study into Technology Education across Australia:

> …in the case of primary education, technology had not generally been part of school programmes, and primary teachers have little experience to draw on to develop programmes. (p. 154), and,
>
> Technology education was traditionally an 'elective' area in secondary schools and is a 'new' area in primary schools. Because of this it is often perceived as a less important learning area and this perception has been slow to change. (p. 155)

Curriculum in South Australia

South Australia adopted the 1994 innovations and the Technology Statement and Profile (AEC, 1994a&b) informed both curriculum in schools and the University's teacher education programmes (MacGregor, 1999). Design was given greater emphasis in a process of *design-make-appraise* (DMA) and the underpinning theory of the curriculum was outcomes- rather than content-focused (see, for example, Griffin, 1998). The new attention to outcomes and the profile of the individual student contrasted the prior emphasis which was teacher-centred and system- and grade-focused.

In 2001, a new curriculum policy - the South Australian Curriculum, Standards and Accountability (SACSA) framework (DETE, 2001) was introduced. Being a framework, this policy again respected the professional judgement of teachers in assessing the outcomes of students' learning. Building on the Statement and Profile, SACSA developed a new emphasis on *critiquing* along with *designing* and *making*. Interwoven with these three 'strands' were five cross-curricular Essential Learnings – Communication; Futures; Identity; Interdependence; and Thinking (Keirl, 2001b). The new policy also introduced a name change for the Learning Area: Technology Education it became Design and Technology Education.

University Restructuring

Institutional reorganisation happened at various levels in this period. The former Faculty of Education became one school in a much larger (8000+ students) Division of Education, Arts and Social Sciences. In 1997 the University introduced its Graduate Qualities – generic dispositions deemed to be of value to all students regardless of their study focus. (These 'qualities' are now expected to be thoroughly integrated with course delivery.) A review of the whole of the University's education programmes and practices (Reid & O'Donoghue, 2001) considered markets, offerings and locations, and opportunities came to restructure teacher education.

What had previously been an undergraduate Junior Primary/Primary programme (hereafter BEdJPP) became a Primary/Middle years programme; the Bachelor of Education Primary/Middle (hereafter BEdPM). This change influenced the content and number of Design and Technology courses offered, the gender balance in the programme, staffing, and resourcing. Course offerings in Design and Technology have become more diverse in addressing middle years' needs, for example, offering food and textile technologies, futures technology and information and computer technology (ICT).

The new BEdPM programme was underpinned by core principles such as: professional competence; wellbeing; social justice; futures thinking; sustainability; education for community living (place-based learning); and, sound pedagogical reasoning that is enquiry-based. (The BEdJPP programme had no such explicit principles). These principles were not difficult to embed in the Design and Technology courses which had been initially shaped by issues of environmental, cultural and human concerns. In fact discussions regarding these principles make the underlying philosophies of Design and Technology more visible to a wider audience.

COURSE DESIGN AND DEVELOPMENT

The influence of all the factors reported above has led to extensive revision of the Design and Technology courses offered. Both degrees are/were of four years duration with a total unit (points) value of 144 units. The courses under discussion are of 4.5 units value. Within the degrees, two course arrangements exist, each with its own role. *Core* courses are compulsory for all students and each subject area within the degree has (had) only one such course. Further *General Studies* courses are specialised suites of courses and may relate to a particular Learning Area as happens with Design and Technology.

BEdPM students complete a combined Professional Pathway and 'specialist' teaching practicum experience. Design and Technology students teach Design and Technology as a specialisation in a primary or middle school setting during their final practicum for half their teaching load. In such a specialist role students teach Design and Technology to five classes assuming responsibility for planning, delivering and assessing the work of over 120 students. Many also assume responsibility for establishing a specialist teaching area and purchasing resources. The high regard in which these students are held has often meant they are sought after to establish new programmes in schools, to lead in-service professional development during the teaching practicum, or to take up offers of employment in the school once the placement is complete.

Core Courses

Design and Technology Education is currently offered as a compulsory 4.5 unit curriculum core course. (Previously this was linked to the teaching of Arts Education, was delivered in Year One and was afforded only 2.25 units.). It occurs in the second year of the degree and is 'linked' to a practicum placement in schools thus providing greater pedagogical relevance and context. This school experience occurs in a primary school.

For the majority (75%) of students the core course is their only exposure to Design and Technology in their degree. For now the core course continues to be primary school focused in content. Students who choose to teach in primary/middle school settings are encouraged to undertake further study in the Design and Technology General Study.

As a direct reflection of the current political climate, there has been an increase in the number of core courses in Numeracy and Literacy. Central to current course re-development is the need to increase students' capabilities to explicitly teach numeracy and literacy relevant to the year levels they are teaching as well as subject specific literacy and numeracy demands. Mandated language enrichment courses and additional language electives have impacted on the number of Design and Technology courses offered. Increased emphasis has also been placed on explicitly identifying where literacy and numeracy concepts are taught in Design and Technology courses. While these changes could provide opportunity to highlight the cross-disciplinary and contextual approach that teaching in Design and Technology can afford; this has not been the case. The value of the place of Design and Technology in the curriculum continues to be overlooked and reinforced emphasis is given to the importance of Maths, Science and English.

General Studies

In the BEdPM programme students can study from two to six Design and Technology courses. The majority enrol in six which gives them a valuable teaching specialisation. It is the General Study courses that deepen understanding and strengthen pedagogy in Design and Technology. General Study courses aim to introduce students to recent innovation and ideas related to teaching and learning in Design and Technology.

Over 100 students are currently enrolled in Design and Technology General Studies. Originally six General Study courses were offered. Several years ago this grew to eleven courses, but with increased student numbers and less teaching staff, five of the General Study courses are now combined with courses offered in a new undergraduate Secondary Design and Technology programme. This combination has proved to be a very positive move as it has facilitated the sharing of Design and

Table 1. Comparative general study courses 1995 and 2009

1995 *Technology Education* via Design-Make-Appraise	2009 *Design and Technology Education* via Critiquing, Designing and Making
Imagineering-Creative Construction Awareness of the concepts of Technology, Technology Education and Technological Practice. Process of Technology developed through problem solving tasks. This included Story Book stimulus, an assessment task through which students developed interactive models that integrated a technological system to support the telling of a children's story. The models and stories where shared with children in schools. Issues of safety, resources and skill development.	Design and Technology 1 This course provides a broad and holistic introduction to D&T. It is the foundation course for all Design and Technology students in primary, middle school, secondary, undergraduate and postgraduate programs. The course explores Design and Technology curriculum; change and futures issues; skills and values; assessment problems; design processes and outcomes; occupational health, safety and welfare issues through design based tasks.

Table 1. (Continued)

Materials for Design, Make and Appraise Reflecting the DMA pedagogical framework via activities around a breadth of materials and their properties to enable successful production of design solutions. Integrated use of materials were encouraged through assignments such as The Biscuit Factory, where student worked with a range of materials such as metal- to produce the biscuit cutter, food technology – to produce biscuits, information and communication technology to produce packaging and media studies to market the final product.	Materials Technology A greater diversity of materials and critiquing, designing and making techniques. Greater emphasis on integration of resistant materials, systems and control technologies. The Biscuit Factory remains an assessment task for the current course as a result of positive student feedback. Emphasis is placed on the sustainable use of materials; students have the opportunity to work with a range of organisations that reflect a philosophy of sustainable practice to produce educational resources.
Technology and Us Explorations of how technology interacts with culture, society and the environment. Many of the technological interactions explored were at a local level. Students produced Design and Technology units of work in collaboration with schools, visited wetlands and presented and discussed possible solutions for technologically based issues raised in local newspapers.	Technology and Society Broadly similar with greater emphasis on developing authentic learning experiences through Place-Based projects. In this course technology continues to be explored through three perspectives: social, cultural, and environmental. Students are involved in community based projects that include; working collaboratively in schools with students with special learning needs to produce sensory teaching aids, they exploring traditional and contemporary indigenous technologies, and attend field trips to investigate the use of technology to support sustainable practice, e.g. wet land development, recycling industries.
Information Highways Computers and related technologies for primary and secondary: multi-media applications; relationship between information and learning; computer as a learning tool. This course was very skills based and focused on developing students' abilities to work with a number of computer programs used in schools. Little emphasis was given to critiquing the use or educational effectiveness of the introduced programs.	Computers in Education This course provides an examination of the use of computers as a tool to facilitate learning. In this context, students explore the educational applications of various software such as word processing, desktop publishing, database management, spreadsheet, adventure games, telecommunication and the Internet, multimedia, and software used in special education settings. Attention is also given to the integration of theory and practice, in particular, the application of research findings to computer usage in education.

Table 1. (Continued)

Technology By Design/Through Invention Practical project-based design work through liaison with schools, community, industry. Students were assigned to an organisation which had identified a specific technological problem that needed to be addressed. Students worked in teams to solve problems and presented their solutions to the group for analysis.	Technology-Innovation and Invention This course is no longer offered. The content of the course is embedded throughout other Design and Technology General study courses.
Technology: Negotiated Study A negotiated activity matching student interest. Research project or community/industry design project work. This course provided a showcase for final year students to develop a design-based project that reflected an area of personal interest. For example designing and producing solar vehicles and boats to enter the national solar challenge or producing Design and Technology units of work for students in remote and rural areas of Australia.	Professional Pathway/Professional Application and Reflection 4 Application of knowledge, skills and values developed throughout D&T General Studies in school practice. Deepening insights of quality D&T for general education of all students. Students as advocates and change agents for D&T. Students complete a classroom based critical inquiry that is linked to an area of concern in the teaching of Design and Technology education, e.g. gender inclusive practice, student engagement in design based activity, classroom resource management etc. Students present their research at a Conference day, together with Design and Technology honours students.

Technology pedagogy from a grade 3–12 perspective, and as a consequence has strengthened students' understanding of the sequence and progression of student learning. The collaborative approach has also enabled a shared sense of advocacy for the place of the Design and Technology Learning Area to develop. Students from the primary, middle and secondary programmes work together in areas such as food and textile technology, computer aided design and workshop knowledge and safety. Table 1 illustrates the six original courses and their current counterparts.

These Design and Technology General Study courses are new to the BEdPM programme:

Foundations in Design and Technology Workshop Knowledge (School of Education). This course has a middle school focus as students develop practical Design and Technology techniques for years 7–10. Emphasis is based on Occupational Health, Safety and Welfare (OHS&W) as students are introduced to basic machinery, tools and skills. Students are encouraged to work with recycled materials in an integrated way. Students work with open design briefs to encourage innovative outcomes.

Computer Graphics for Engineers (School of Mechanical Engineering). Engineering and education students collaborate in problem solving experiences. Techniques for visualisation and drawing including sketching; orthographic projection; auxiliary projection; pictorial drawing; dimensioning practices are developed. Simple geometric constructions are introduced as well as computer aided design and drafting using 2D and 3D modelling software.

Food and Society (School of Health Sciences). This course introduces students to issues associated with food production. Students are encouraged to explore ways in which to achieve an ecologically sustainable food supply. Food legislation, choice, social and cultural influences, health are also examined. This course is a combination of lectures and practicals. Through the practicals students develop skills in food handling, cooking and budgeting.

Human Nutrition (School of Health Sciences). Students explore the relationships between food, nutrition and health. They plan a healthy diet; explore human nutritional needs and dietary sources. They are introduced to Energy concepts (energy balance; weight control; obesity; and methods of weight loss diets) and research the link between nutrition and lifestyle including nutritional issues in lifestyle diseases: diabetes, heart disease and cancer; designer foods – functional foods, healthy fats, genetically engineered foods; and food safety.

Idea Generation Methods for Designers (School of Art and Design). In this course Design is explored as core pedagogy for Design and Technology. Theories of idea generation, creative thinking techniques, and relationships of ideas and image development are investigated. Students develop design journals in which they document their learning throughout the course. Innovation and creative thought is encouraged and celebrated.

Many of the General Study courses, particularly those offered through the School of Education, provide students with not only content knowledge; they provide opportunities to develop pedagogical knowledge. Education lecturers in the General Study aim to model, and introduce students to, pedagogies that are transformative and facilitate the critique of current approaches to teaching the Learning Area, particularly in middle years of schooling. This includes problem posing rather than problem solving methodologies, together with cross- and inter-disciplinary approaches to teaching. Students are encouraged to see themselves as 'educators first and Design and Technology teachers second' this enables students to adopt a more generalist, rather than specialist, view of the educator in primary, middle school and secondary settings.

STAFFING AND STUDENTS

The increase in courses offered has had major implications for staffing and resource funding. As courses have become more specialised in content, greater numbers of sessional staff (often practising teachers) and lecturers from other schools (see above) within the University have been employed. In her 1999 review, MacGregor noted that school-based sessional staff

> ...brought with them a wealth of knowledge of current technological practice. (p. 87)

Collaboration was also highlighted:

> The opportunity to develop and maintain long-term rich professional relationships with practising teachers has impacted greatly on the content and delivery of the courses that were offered to Technology Education pre-service teachers. (MacGregor, 1999, p. 87)

Sessional staff continue to bring with them rich knowledge, and collaboration is still central to the effective planning and delivery of Design and Technology courses. However, it must be acknowledged that the content of courses taught elsewhere in the University are not generally delivered using a Design and Technology pedagogy (by which we mean one modelling critiquing, designing and making in a holistic way and which is transformative rather than transmissive in style).

Significantly, over the decade, Design and Technology lecturers have deepened their knowledge and research – through exploring new pedagogies, pursuing advanced studies, publishing, and attending Design and Technology, and other, education conferences.

Students and their Prior Knowledge/Experience

Several of the General Studies courses are also offered to students in other pre-service education programmes. The 'mixing' of students from a range of programmes enables a shared and more holistic understanding of Design and Technology to develop, breaking down some of the traditional approaches that depended on the 'making' focus of Technology Education of the past.

The change in programme focus from primary to primary/middle schooling has enriched the student experience base and profile. There has been a significant increase in the number of mature-age students and males enrolling in the programme. This has meant the type and depth of prior knowledge that students bring is richer. Broader life experiences also add to this level of richness. The level of computer literacy continues to rise. More students have extensive computer knowledge and have used a range of software programs to design and present ideas. Many students are well beyond basic word-processing and are proficient in web-page authoring, 3-D drawing, robotics and clay animation. These technologies are growing in popularity in both primary and secondary schools and are readily adapted to design-based pedagogy.

Greater numbers of students entering the programme have trade qualifications and related work experience. Originally most entrants were year 12 school leavers and were generally passive learners who listened intently and questioned little. Innovative critiquing and designing pedagogies both validate students' prior learning and life experience as well as lead to much enquiry-based learning.

In the mid-1990s Technology as a Learning Area was just emerging in the primary curriculum. Until this time, Technical studies, with a focus on skill development in the use of a range of materials, had been taught in Secondary schools only. Now, the majority of teacher education entrants arrive with a more informed understanding of Design and Technology, and they generally have a greater understanding of the concept of design and the importance of critiquing.

For the promotion of Design and Technology today, the greatest advocates are the graduates of the General Study courses. The level at which students value their learning resulted in the highest Course Evaluation Instrument (CEI) of all the courses offered in the BEdPM. When these students graduate they take with them a broad and holistic view of Design and Technology, and help to ensure that the Learning Area remains relevant, vibrant and, most importantly, valued as a domain of learning.

Over the last fifteen years there has been an increase in the number of students who choose to take Honours. Each year fifteen BEdPM students in their final year of study are offered Honours. Half of these students have completed courses in the Design and Technology General Study. Each year two or three students complete their honours research in the area of Design and Technology. In more recent years several students have enrolled in the Masters programme, specialising in Design and Technology education. This research contributes to the growing body of knowledge in, and the status of, the Learning Area. It also informs both lecturers' and students' knowledge, understanding and pedagogy.

LOOKING BACK…

In our teaching we use a Venn model depicting the integration of *head*, *hands* and *heart* to illustrate their co-dependence in quality Design and Technology practice. We reflect on such a schema here to show how, a decade ago, Design and Technology courses were much more 'practical' – concerned with *making*. This is not to decry making, rather, it is to signify the affirmation of *designing* as central practice and *critiquing* as vital for the necessary questioning of technologies (Keirl, 2001a). This is the direction that Design and Technology has taken in South Australian education.

Another three-set Venn diagram would show the interplay of the university Design and Technology team, the SACSA developments, and the pre-service course developments across the period. Events have not unfolded either randomly or in isolation from each other. They have evolved as a synergy. Each area of growth, whether human, policy or planning, has fed another. Professional knowledge growth, curriculum evolution from DMA to CDM through Essential Learnings, and university course development, have all fed one another in dynamic ways.

The profile of the students has changed and they arrive not only as part of a much richer cohort, but also with knowledges and with perceptions of technology different from those held by their counterparts ten years ago.

Not only has course development continued to value the input of the professional knowledge of practising teachers but it has opened new cross-programme dialogues, the fruits of which are only beginning to appear. The innovations embrace many positives: the closer interplay of Design and Technology with reflective practice in school placements and through community based experiences; the CEI successes and the growth in Honours and Masters activity through Design and Technology; the long-overdue appointment of another permanent lecturer; and the first iteration of a new foundation course for all Design and Technology students.

LOOKING FORWARD...

If the past fifteen years could be described as *responsive development* – responsiveness to political climes, to professional reflection, to curriculum development, and to social change – then so might the next fifteen. But that is not to say *reactive* – after the fact. If head, hands and heart apply to future Design and Technology practice, how can these be qualified? We would suggest that the *head* warrants not only analysis of the political but also professional contributions to policy within and beyond the university. The *hands* are about practical action in course design and delivery. Meanwhile the *heart* must be about confidence in Design and Technology's curriculum place and about vision for what it has to offer in the future.

Design and Technology has achieved a position of some strength and viability but there is no room for complacency. We would argue that some of this strength is because of its capacity to defend a place in general education. Thus, as each of the Graduate Qualities, Essential Learnings and the BEdPM core principles have appeared, Design and Technology, because of its very nature, has had no difficulty in articulating them. Such adaptability is necessary for a field which is without the privilege of English, maths or science. In times of curriculum wars between Outcomes Based Education (OBE) and a 'back-to-basics' fundamentalism, astute determination will be needed for the survival of Design and Technology.

The future need not be seen as a lottery or beyond our control if Design and Technology can be continuously redesigning itself in response to astute reading of political and curriculum trends. Perhaps...it may move from *responding to* to *informing* such trends. The Design and Technology journey reported here shows that such redesign is both practical action as well as worthwhile (curriculum) engagement of Design and Technology principles.

REFERENCES

Australian Education Council (AEC). (1994a). *A statement on technology for Australian schools*. Carlton: Curriculum Corporation.

Australian Education Council (AEC). (1994b). *Technology – A curriculum profile for Australian schools*. Carlton: Curriculum Corporation.

Department of Education, Training and Employment (DETE). (2001). *South Australian curriculum standards and accountability framework (SACSA)*. Retrieved from http://www.sacsa.sa.edu.au

Griffin, P. (1998). Outcomes and profiles: Changes in teachers' assessment practices. *Curriculum Perspectives. 18*(1).

Hargreaves, A. (2003). *Teaching in the knowledge society: Education in the age of insecurity*. Maidenhead: Open University Press.

Harris, C., & Marsh, C. (Eds.). (2005). *Curriculum developments in Australia: Promising initiatives, impasses and dead-ends*. Adelaide: Openbook.

Keirl, S. (2001a). Critical beginnings for Design and Technology education - Why and how might critiquing be a key component of children's learning in the early and primary years? In C. Benson, M. Martin & W. Till (Eds.), *Third international primary design and technology conference 2001*. Birmingham: CRIPT, University of Central England.

Keirl, S. (2001b). Design and Technology and the five 'Essential Learnings' of a new curriculum framework. In E. W. L. Norman & P. H. Roberts (Eds.), *Proceedings of the international conference on Design and Technology educational research and curriculum development 2001*. Loughborough: Loughborough University.

Killen, R. (2006). *Effective teaching strategies: Lessons from research and practice*. South Melbourne: Thomson Social Science Press.

MacGregor, D. (1999). 'Initial teacher education for primary technology education in South Australia: Innovations, reflections and futures'. In C. Benson & W. Till (Eds.), *Second international primary design and technology conference 1999*. Birmingham: CRIPT, University of Central England.

Reid, A., & O'Donoghue, M. (2001). *Shaping the future: Educating professional educators*. Report of the Review of Education at the University of South Australia.

Williams, P. J., & Keirl, S. (2001). The status of teaching and learning of technology in primary and secondary schools in Australia. In E. W. L. Norman & P. H. Roberts (Eds.), *Proceedings of the international conference on design and technology educational research and curriculum development 2001*. Loughborough: Loughborough University.

Steve Keirl and Denise MacGregor
Centre for Research in Education
University of South Australia
Australia

GARY O'SULLIVAN

8. PRIMARY TECHNOLOGY EDUCATION IN NEW ZEALAND

A Scenic Twenty Year Journey

INTRODUCTION

This chapter will introduce you to developments in primary technology education that have occurred in New Zealand over the last twenty or so years. Obviously this chapter is part of a series of international perspectives; therefore it will be just that - an introduction. I encourage you to contact the author and read the reference materials identified for further clarification. The focus will be on Ministry of Education policy directives and curriculum development because this is what drives education in the New Zealand primary classroom.

THE JOURNEY BEGINS

New Zealand has a long historical connection with technology education although for most of this time this association has been through a narrow technical education rather than what we think of as technology education today. Technical education in New Zealand was established in 1890 and played a major part in the educational history of New Zealand. It was not until the relatively recent curriculum reforms of the 1990s that we saw the emergence of a much broader technology education. These reforms were brought about because of the significant changes that were occurring in the political, economic and educational landscape of New Zealand during this period. The replacement of the old Department of Education and the creation of the Ministry of Education were to be the beginning of significant changes that are still impacting on the delivery of technology education today.

In 1990 the National Party released its election manifesto that included a new policy for education called the 'Achievement Initiative'. Part of this initiative was to develop learning objectives for children in English, Mathematics, Science and Technology. The National Government was elected in 1990 and the following year work began on the educational reforms. Papers were written and meetings held to discuss amongst other things the position of technology education. Dr. Smith, the then Minister of Education, decided that Technology education should be a separate area of the curriculum. Later a task force set up jointly by the Minister of Research Science and Technology and the Minister of Education supported the independent nature of Technology education, in their report published in February 1992 (Ferguson, 2009).

C. Benson and J. Lunt (eds.), International Handbook of Primary Technology Education: Reviewing the Past Twenty Years. 89–95.

In 1993 as a result of these reforms seven Essential Learning Areas were established to be a part of the New Zealand Curriculum Framework (NZCF). One of the named areas was Technology Education. This was a significant policy move and meant a significant change for many of the primary classrooms in New Zealand. A draft technology education curriculum was developed and trialled in schools in 1994. By 1995, submissions on the draft had been received and the final statement Technology in the New Zealand Curriculum (TINZC) was published in 1995. Implementation concerns were central to a decision taken in May 1997 by the Ministerial Consultative Group on workloads; subsequently the revised date for implementation became February 1999. 1997 also saw the establishment of Technology Education New Zealand (TENZ) a professional association for technology teachers. Since 1997 TENZ has run a very popular and successful biennial conference.

In February 1999 the New Zealand Ministry of Education *gazetted* the statement, i.e. made it mandatory in state schools. This meant that Primary schools years 1–6, Intermediate schools years 7–8 and Secondary schools years 9–10 were required to deliver programmes designed to implement this curriculum. The Ministry of Education through its publication Technology in the New Zealand Curriculum described technology as:

> ….a creative, purposeful activity aimed at meeting needs and opportunities through the development of products, systems, or environments. Knowledge, skills, and resources are combined to help solve practical problems. Technological practice takes place within, and is influenced by, social contexts. (Ministry of Education, 1995, p. 6)

The delivery of this creative, purposeful activity in technology education was to be through three interweaving strands.

Technological Knowledge and Understanding

- understanding the use and operation of technologies;
- understanding technological principles and systems;
- understanding the nature of technological practice;
- understanding strategies for the communication, promotion, and evaluation of technological ideas and outcomes.

Technological Capability

- identifying needs and opportunities;
- with reference to identified needs and opportunities:
  generating, selecting, developing, and adapting appropriate solutions;
  managing time, and human and physical resources, to produce technological outcomes, and products, systems, and environments;
  presenting and promoting ideas, strategies, and outcomes;
  evaluating designs, strategies, and outcomes.

Technology and Society

- the ways the beliefs, values, and ethics of individuals and groups:
  promote or constrain technological development;
  influence attitudes towards technological development;
- understanding the impacts of technology on society and the environment:
  in the past, present, and possible future;
  in local, national, and international settings.

In addition there were identified achievement objectives which were expressed at each of the eight progressive levels in line with The New Zealand Curriculum Framework. These level statements were designed to show progression of the technology curriculum from junior primary (J1=Year 1) to senior secondary (F7=Year 13) although the curriculum statement was only compulsory to Year 10. Years 11–13 were to follow assessment guidelines associated with the National Certificate of Educational Achievement (NCEA). NCEA is the new senior secondary school national qualification which was implemented in New Zealand schools in 2002.

Seven technological areas were identified to help students achieve the objectives of the curriculum. These areas were considered to be key areas of study particularly to New Zealand. In some ways this was a double-edged sword: positive in that it helped teachers, particularly primary teachers, identify areas of study, however, in reality many teachers saw these as separate subjects and upon reflection this was definitely seen as a miscommunication of the curriculum. The areas identified were:

- Biotechnology
- Electronics and Control technology
- Food technology
- Information and Communication technology
- Materials technology
- Production and Process technology
- Structures and Mechanisms

In addition there were nine contexts established in the statement; technological activities it was suggested are carried out in a variety of broad, overlapping contexts. Again these contexts helped to scaffold teachers' understanding but led to some misunderstandings. Schools struggled to meet requirements for coverage of these areas, due to perceptions of separation.

- Personal
- Home School
- Recreational
- Community
- Environmental
- Energy
- Business
- Industrial

Technological areas, contexts, strands, and achievement objectives combined together provided the framework for technology education. Technology education became a compulsory curriculum for all students from years 1–10. Although there had been some shift from the early days of technical education this was a giant and brave leap forward. The central thrust of the change had been the movement towards developing in students the notion of technological literacy. This literacy was much broader than the technical expertise which the previous historical offerings had encouraged and expected. Education in New Zealand was going through major reform during this period, not just in curriculum but also in assessment. A new secondary qualification was introduced. The National Certificate of Educational Achievement was launched in late 1998 under the umbrella of a project called Achievement 2001.

THE ROUTE CHANGES

Around the same time in 2001 a curriculum stock take was initiated by the Ministry of Education. This was to review all aspects of the compulsory curriculum including technology education. The stock take had a relatively wide remit in that reviews were sought from international experts as well as evaluations of teachers' experiences of curriculum delivery. As a result of the stock take a decision was taken to redefine the existing NZCF and the formation of a new framework called the New Zealand Curriculum and Marautanga Project (NZC&MP) was proposed.

As part of the stock take review of technology education a National School Sampling Study (NSSS) was carried out and this was to provide teachers with the chance to share their experiences of curriculum implementation. The sample was about 10% of the 2900 schools in New Zealand.

The data from this sample indicated that there was a reasonable amount of satisfaction with the structure and organisation of the new curriculum with only a third highlighting changes they would like to make. The upper secondary school teachers in years 11–13 were the most disgruntled of the sample collected. The secondary teachers were concerned with the amount of paperwork. This could be associated with the new assessment procedures for NCEA. Primary teachers were more concerned with, and asked for more guidance on, planning and assessment.

The findings from approximately 70% of respondents indicated that most primary school teachers were aiming for curriculum coverage and had moderate levels of confidence. Around 60% of primary teachers expressed concern over obtaining resources and appropriate equipment. Years 7 and 8 teachers who taught mainly in specialised intermediate schools were concerned about assessment. Further information of the findings from this research can be accessed in the work of Jones, Harlow & Cowie (2004).

Considering the relatively short timeframe that the compulsory technology curriculum had been in place, these findings might well be viewed as satisfactory. However, there was some evidence from both New Zealand and international research studies that students and teachers lacked a coherent conceptual understanding of the nature of technology. This, combined with a perceived lack of technological knowledge, was to be the thrust behind a revision of the curriculum statement for technology as part of the New Zealand Curriculum and Marautanga Project.

To address these concerns, it was suggested in 2004, that there needed to be a greater focus on the philosophical understanding of technology and further development of technological knowledge (Compton & Jones, 2004). As a result, technology education was to be restructured around three new strands:

- Technological Practice
- Nature of Technology
- Technological Knowledge

Just like the 1995 statement these strands were still seen as intertwining to realise the aim of developing and increasing student technological literacy. However, the new curriculum was re-conceptualised and had more of an emphasis on critical literacy. Less emphasis was placed on the separate technological areas and contexts.

The New Zealand Curriculum Draft for Consultation was distributed in June 2006 and the Ministry of Education asked for feedback. The section devoted to Technology Education was incomplete and a supplement had to be released in October 2006 (see Ministry of Education 2006). This had a negative impact on Technology Education and led to some debate as to the validity of feedback received. That said, there was significant and widespread consultation sought during this period ranging from schools through to universities, business and industry.

The current New Zealand Curriculum (NZC) was published late in 2007. The process of technology in the New Zealand Curriculum 2007 is described as:

> Technology is intervention by design: the use of practical and intellectual resources to develop products and systems (technological outcomes) that expand human possibilities by addressing needs and realising opportunities. Adaptation and innovation are at the heart of technological practice. Quality outcomes result from thinking and practices that are informed, critical and creative. (Ministry of Education, 2007, p. 32)

Students will develop their technological literacy, it is suggested, by developing their learning through work in the three identified strands. Teaching programmes can still integrate all three strands, although individual units of work may focus on one or two strands at a time. This is a significant shift in guidance from the previous 1995 curriculum. The change has been made to increase flexibility and manageability for teachers. Additionally it is hoped it will facilitate students' learning experiences by allowing them to develop greater progression in their context specific knowledge, skills and practice.

The Technological Practice strand which has been expressed as a combination of the earlier three strands identified in the 1995 curriculum should provide opportunity for students to examine the practice of others as well as undertake and critique their own. Further explanation to match the more critical literacy sought is provided through the supplementary material which included additional justification. It is now identified that the students' practice should include identifying, investigating issues and existing outcomes. It also includes consideration of ethics, legal requirements, protocols, codes of practice, and the needs of, and potential impacts, on stakeholders and the environment. Through technological practice, students may design, develop and communicate a range of outcomes, including concepts, plans, briefs, technological models and fully implemented technological outcomes. In the

2007 curriculum there are three components identified within the practice strand. These are *Planning for Practice, Brief Development* and *Outcome Development and Evaluation.*

The Technological Knowledge strand provides opportunity for students to develop understandings of '*how things work*' and develop technological knowledge specific to technological endeavours. So the rationale here is that students should have access through technology education to 'key' concepts and knowledge which is generic to any technological context. The components of this strand are *Technological Modelling, Technological Products* and *Technological Systems.* Compton & France (2007) put forward an interesting separation and description of technological modelling which includes both functional and prototype modelling. They suggest that functional modelling is the exploration of the feasibility of design ideas and concepts; prototype modelling is the exploration of fitness of purpose of the outcome.

Within the Nature of Technology strand there are two components *Characteristics of Technology* and *Characteristics of Technological Outcomes.* This strand provides opportunity for students to develop a philosophical understanding of technology, including how it is different from other domains of human activity. For this to occur, students must develop a shared understanding of the purpose of technology and also develop an appreciation that outcomes of technological practice are often a 'best fit' response. This strand supports the development of an understanding of technology that is critical in nature, and allows for informed debate of historical and contemporary issues and future case scenarios. Clearly this places technological activity in a 'critical moment' both from a timeframe and a societal perspective. This moment appreciation should give the students an understanding to be able to critically analyse what came before and what might occur in the future.

From 2010 teachers in New Zealand schools will be required to incorporate all three strands into their technology education programmes. Assessing and reporting on student achievement will occur using all eight achievement objectives identified in the 2007 curriculum.

THE JOURNEY THUS FAR

This chapter has attempted to highlight briefly some of the major changes occurring in New Zealand with regard to primary technology education over the previous twenty years. The main driver for these changes has been curriculum reform led by the Ministry of Education. What impact these major changes will have in the primary classroom is difficult to ascertain. What is certain is that the recent initiatives such as the numeracy and literacy projects have compacted an already over-crowded curriculum to such a degree that some schools may struggle to justify technology's inclusion. However due to the developments in this area over the last twenty years there are a number of extremely passionate teachers, teacher educators, advisors and researchers working in technology education in New Zealand. Good practice should continue to shine through.

The impact of the most recent curriculum changes will not be known for some time as the phased introduction of the new curriculum occurs. What is clear is that technology education in New Zealand has gone through a dramatic journey since its inception and the last twenty years has been a period of significant scenic change.

REFERENCES

Compton, V. J., & Jones, A. (2004). *The nature of technology*. Briefing paper prepared for the New Zealand Ministry of Education Curriculum Project. http://www.nzcurriculum.tki.org.nz/content/download/854/6044/file/nature-techn.doc

Compton, V., & France, B. (2007). *Discussion document: Background information on the new strand*. http://www.tki.org.nz/r/nzcurriculum/docs/technology-new-strands.doc

Ferguson, D. (2009). *Development of technology education in New Zealand schools 1985–2008*. Wellington: Ministry of Education.

Jones, A., Harlow, A., & Cowie, B. (2004). New Zealand teachers' experiences in implementing the technology curriculum. *International Journal of Technology and Design Education, 14*, 101–119.

Ministry of Education. (1995). *Technology in the New Zealand curriculum*. Wellington: Learning Media.

Ministry of Education. (2006). *Draft curriculum: Technology materials for consultation*. Wellington: Learning Media.

Ministry of Education. (2007). *The New Zealand curriculum*. Wellington: Learning Media.

Gary O'Sullivan
School of Curriculum and Pedagogy
Massey University
New Zealand

AKI RASINEN, PASI IKONEN AND TIMO RISSANEN

9. TECHNOLOGY EDUCATION IN FINNISH COMPREHENSIVE SCHOOLS

INTRODUCTION

Finnish technology education dates back to 1866 when craft education was accepted to be one of the compulsory subjects in the school curriculum. Uno Cygnaeus, founder of Finnish general education, considered 'technological' contents an important part of craft education. Cygnaeus emphasised dexterity, design and aesthetics but also consideration, innovation and creativity (Kantola, 1997). He also underlined the importance of realising the connections between natural sciences and craft education (Kantola et al., 1999).

There have been many pedagogical and administrative changes in general education since Cygnaeus' times, but one of the most remarkable changes took place in the beginning of the 1970s when the parallel school system (folk school and gymnasium) was abolished and comprehensive schools were introduced in the country. A significant reform was introduced in teacher education in 1979. Since then all comprehensive school teachers (grades 1 to 9, ages 7 to 15), both generalist class teachers and subject teachers, have been trained up to Master's degree level.

In this chapter, we will discuss the changes in Finnish technology education in comprehensive schools since 1970 from the point of view of changes in the curriculum, but also considering gender equality, pedagogy, teacher education, society, and the concept of learning.

TECHNOLOGY EDUCATION AND CRAFT EDUCATION

Handicraft teaching and technology teaching have seldom been compared in the research literature. Comparisons are mainly made between technology, science and mathematics. The reason for this is obviously that, for instance, in the United Kingdom (UK) and the United States (US), handicraft education has developed into technology education. According to Alamäki (1999) technology education has evolved from craft education in many countries. He also argues that, due to technology education still being in the process of evolution, many approaches from craft to applied science are being used in technology. Järvinen (2004) claims that technology education cannot be monopolised by either craft or science education because it involves mathematics, science, arts, handicrafts, and genuine innovative problem solving.

Kantola (1997) and Parikka (1998) define technology as an umbrella concept for handicraft education. A different view is taken by Peltonen (1988), Anttila (1993)

C. Benson and J. Lunt (eds.), International Handbook of Primary Technology Education: Reviewing the Past Twenty Years. 97–105.

and Suojanen (1993), who regard handicraft education as an umbrella concept for technology education. Alamäki explains that 'käsityö' (craft or handicraft) is the official name and overall term for a subject group that consists of the school subjects 'tekninen työ' (technical work) and 'tekstiilityö' (textile work).

> Käsityö in the Finnish educational context has no direct English equivalent but implies a combination of crafts, design and technology education. (Alamäki, 1999, p. 173)

He also notes that:

> The contents and processes of the Finnish 'tekninen työ' correspond to the international view of technology education. (p. 173)

Alamäki goes on to say that in many Finnish publications (e.g. Alamäki, 1999; Kankare, 1997) the English equivalent of the term 'tekninen työ' is technology education. By merely changing the title of the subject there is no change in learning. What matters are the contents and methods of teaching. Therefore, the objectives and contents of craft education have to be discussed and altered towards technology education.

Experts in craft education and technology education, whether Finnish or foreign, agree particularly on one aspect. Both groups see that an essential part of learning is the creative planning and production process (Anttila, 1993; Hill & Lutherdt, 1999; Eggleston, 1994; Lindfors, 1992; Peltonen, 1988; Suojanen, 1993). Kojonkoski-Rännäli (1995), mainly following Bunge (1985), distinguishes between the handicraft production activity and the technological production activity. According to her, hands-on methods are used in handicraft, whereas in technology, methods of modern technology are used. Dyrenfurth (1991) in turn claims that skilful operations are an essential part of technology.

In this article, thinking and use of the brain is considered to lead all work done by hand. Technology is seen as 'logos' of 'techné', where technology is not restricted to modern technology, but is seen from a wide perspective - from traditional to modern.

THE 1970 FRAMEWORK CURRICULUM AND THE 1970 CURRICULUM

Technology as a Concept is Not to be Found

In 1970, the Ministry of Education published two memoranda to guide teachers in transferring from the old parallel school system to the comprehensive school system. The 1970 Framework curriculum gave schools the rationale and philosophy, aims and objectives, information needed to implement and develop the curriculum, different methods, information about learning materials, information about differentiation, evaluation, extra mural activities, counselling, organising the work and co-operation between school and home (Peruskoulun opetussuunnitelmakomitean mietintö I, 1970).

The 1970 Curriculum stated the attainment targets and contents for different school subjects. In craft education it listed grade by grade the techniques (i.e. measuring, marking, sawing etc.), materials (i.e. planks, metal rod, plastics etc.), and objectives (mainly different techniques) with some ideas for different projects. It also gave information on different working, learning and teaching methods, evaluation and integration. Craft education was divided into two sub-areas: technical craft and textile craft. The document emphasised that the division should no longer be according to one's sex, but both girls and boys should study textile craft and technical craft. All pupils were supposed to study the same programme from grade one to three (ages 7 to 9), then choose one of the two subject areas for grades four to seven. During the spring term (January – May) grade six pupils (age 12) were supposed to change the subject area (Peruskoulun opetussuunnitelmakomitean mietintö II, 1970). Boys mainly went for technical craft classes and girls for textile craft classes. However, there was more variety in the choices for girls than those for boys.

Technology as a concept is not to be found in the 1970 Curriculum. In turn, the concept of technique is to be found under the title 'technical craft'. One of the general objectives in technical craft studies was to become acquainted with technical domains. The pupils' own design process was regarded as important and the contents of, for instance, machinery and electronics can be seen to be of a technological nature. The 1970 Curriculum and Framework Curriculum is a very radical, educationally professional, ambitious and future oriented document. Even today, after three national framework curricula, the text of the 1970 curriculum is very much up to date.

THE FRAMEWORK CURRICULUM FOR COMPREHENSIVE SCHOOLS 1985

The Concept of Technology was Introduced

Since the 1970 Curriculum document there has not been a national curriculum in Finland. The documents since then have been framework curricula, and the municipalities and schools have planned their own curricula following the national framework. The reasons for this are decentralisation of educational management, reform in teacher education, and the need to plan the curriculum to fit local circumstances. In the 1980s the inspection system was also abolished. Inspectors' posts at national and regional level were changed to instructors' and supervisors' posts. Their role was not to check if the teachers had done their job, but to assist and help teachers in planning, developing, and organising in-service education for teachers. Schools and municipalities were guided to develop their own curriculum following the national framework curriculum. Teachers were highly educated to Masters degree level and they were considered to be able to develop their own curricula.

In 1985, after 15 years experience of the comprehensive school-system, a Framework Curriculum for Comprehensive Schools (Peruskoulun opetussuunnitelman perusteet, 1985) was published by the National Board of Education. The document introduced six general objectives, one of which is gender equality. Enhancing equality at school means offering the same opportunities for both boys and girls (ibid, 1985). There are references to discussions in parliament about promoting gender equality. According to the law, schools should promote equality between the sexes.

The National Board of Education leaves it to municipalities to decide how to organise craft education. From grade one to grade three all pupils should study both textile work and technical work. However, the contents at grade one and two were mainly oriented towards textile work. From grade four to six part of the studies were common to all pupils but part was either technical or textile work. At grade seven technical work and textile work were supposed to be common subjects to all pupils. However, if the municipalities want they can, on top of the common studies, differentiate teaching into technical or textile work.

For the first time also the concept of technology was introduced - but not defined. The concept is to be found only under 'Craft, technical work and textile work'. Technology is the starting point of technical abilities, planning, and implementing. During technical work lessons pupils should also learn to manage technology.

In the curriculum the section on craft, technical work and textile work introduces first the general objectives and gives information on how teaching should be organised. After this, the objectives of technical work and textile work are introduced together with contents grade by grade. The contents are mainly different techniques (i.e. cutting, sawing, soldering etc.). The influence of this type of curriculum planning is still to be seen in some school curricula today. The document also gives information on how to differentiate the curriculum in different municipalities, how to evaluate, and what the opportunities are for integration. Although the general objectives are to develop problem solving and planning skills, the specific objectives are a mere list of different techniques (ibid, pp. 208–213). The approach in the curriculum can be characterised as behaviouristic. It has been written from the teachers' point of view rather than from the pupils' viewpoint. Such expressions as 'pupils will be taught to turn wood' and similar are used (ibid, pp. 208–213).

In practice, most of the schools continued to differentiate pupils after grade three in either textile or technical work groups. The groups were in most cases formed according to sex. Pupils were probably offered a chance for a short change of three to six weeks to study the other subject area of craft.

THE FRAMEWORK CURRICULUM FOR COMPREHENSIVE SCHOOLS 1994

Technology is Mentioned in the General Objectives

For the first time in the history of the curriculum development of Finnish general education schools, technology is clearly mentioned in the general objectives of the curriculum. For the comprehensive school the national guidelines state that the technical development of society makes it necessary for all citizens to have a new kind of readiness to use technical applications and to be able to exert an influence on the direction of technical development. Furthermore, it states that students without regard to sex must have the chance to acquaint themselves with technology and to learn to understand and use technology. What is particularly important is to take a critical look at the effects technology has on the interaction between humanity and nature, to be able to make use of the possibilities it offers and to understand their consequences. However, the document does not give any practical guidance on how to study technology.

Under chemistry, the concept technology is mentioned once:

> Pupils should be able to acquire such a terminology that they are able to discuss questions concerning nature, environment, and technology (ibid, p. 86).

Under craft, the technological objective is that pupils will acquire knowledge of traditional and modern technological materials, knowledge of tools and techniques that can be applied in daily life, further studies, jobs, and hobbies. Despite the stated objective at the end of 1990s woodwork was mainly taught during technical work lessons in the Finnish primary schools. Electricity and electronics tasks, plastics work, and service and repair were taught to a certain extent. Lack of financial resources and ideas were regarded as the most significant obstacles to the development of technology education (Alamäki, 1999). In informal discussions between teachers and teacher educators, technical work education in schools has been said to mainly consist of copying and reproducing processes, such as the copying of wooden and metal items, not modern, design-oriented processes. According to Kankare (1997) woodwork was mainly emphasised by the Finnish technical craft teachers, although most teachers did not want to divide the contents according to materials, but considered the subject area in a holistic manner. Also Sanders (2001) has found in the US that most technology education teachers still adhere to traditional general technology education and woodwork courses.

The 1994 Framework Curriculum documents states that:

> ... craft, technical work and textile work form an entity at primary and junior secondary level which is meant for all pupils regardless of gender (Peruskoulun opetussuunnitelman perusteet 1994, p. 104)

However, the document allowed the schools to emphasise one of the two craft domains. This meant in practice that most schools continued dividing pupils into either textile work or technical work after grade three (age 9).

The 1994 framework curriculum is the first document since 1970 where cross-curricular subject areas are introduced. The 1970 and 1985 curricula mention holistic teaching and integration but there are no clear cross-curricular titles.

FRAMEWORK CURRICULUM FOR COMPREHENSIVE EDUCATION 2004

The Human Being and Technology – a New Cross-Curricular Theme

For the first time in the history of Finnish general education curriculum planning the 2004 framework curriculum (Perusopetuksen opetussuunnitelman perusteet, 2004) introduces a cross-curricular theme: *the human being and technology*. In addition there are six more themes. There are no classes allocated to the themes. The idea of the cross-curricular themes is that all school subjects should consider the themes in their objectives and contents. Under the title 'The human being and technology' the meaning of technology in our everyday lives and the dependency of human beings on modern technology should be studied. This theme offers basic know-how of technology, the development of technology and the effects of technology, guides pupils to make reasonable choices and guides them to consider the ethical, moral and equality questions related to technology. Teaching should also improve the ability to understand how different devices, equipment and machines work and how to use them.

The aims are as follows:
A pupil will learn to:
- understand technology, the development of technology and its impacts on different fields of life, different sectors in society, and on the environment;
- use technology in a responsible and critical manner;
- use information technology equipment, programs and networks for different purposes;
- state one's opinion concerning technological choices, and to consider the effects of today's decisions about technology on the future.

The core contents are:
- technology in everyday life, in society and in local trade and industry;
- the development of technology and factors affecting the development in different cultures and different fields of life during different eras;
- the development, modelling, and assessing of technological ideas and the life-span of a product;
- the use of information and communication technology and information networks;
- the ethical, moral, well-being, and equality concerns related to technology;
- future society and technology.

(Perusopetuksen opetussuunnitelman perusteet, 2004, pp. 40–41)

Technological Studies can be Found to a Considerable Extent Under Craft (Particularly Technical Work)

In the framework curriculum, references to technological studies can be found only under science (particularly physics) and to a considerable extent under craft (particularly technical work). The subjects grouped together in other main subjects have not considered the cross-curricular theme 'the human being and technology' in their text. However, the instructions from the National Board of Education are that schools have to clearly indicate in their curricula how these cross-curricular themes are included in different school subjects and they have to be seen in the activities of schools. The framework curriculum does not give instructions how this should be done; this is left for schools to decide. By studying 50 Finnish municipal curricula (this covers an average of 400 schools) one notes that often 'The human being and technology' cross-curricular theme is mainly understood to be information and communication technology. This indicates that the theme has not been understood in its broad sense, but in a very narrow way.

Technology education objectives under craft education are as follows:
Pupils will:
- familiarise themselves with everyday technology;
- familiarise themselves with Finland's technology and to an appropriate extent also other nations', design, craft, and technology culture for building their own identity and their own design activities;
- familiarise themselves with the know-how connected to traditional and modern technology which can be applied in daily life, further studies, in future jobs, and hobbies;

– learn to state their stand on the development of technology and the meaning of it for the well-being of human beings, society and nature.

(Perusopetuksen opetussuunnitelman perusteet, 2004, pp. 241–242)

If one compares the objectives to the contents of technical work and textile work, it is obvious that by studying only one sub-area all technological objectives cannot be achieved. However, most municipalities (of the 50 municipal curricula studied) have decided (against the regulations of the framework curriculum) to differentiate pupils after grade four (age 10) into technical work or textile work.

The document suggests integration between different school subjects. It is based on a constructivist learning concept where the learner is active and goal-oriented. The objectives are stated from the learner's point of view, not as teachers' activities.

DISCUSSION

According to the latest national framework curriculum, technology has to be studied by all pupils at all levels. As long as technology is a cross-curricular theme, different subjects should consider how it should be studied. There should be continuous consultation between different subject areas and strong co-operation and, where it is advisable, integration should be applied. Technology education is mainly to be seen under the objectives and contents of craft education. Therefore, this subject area should take the main responsibility for making sure that all pupils will study technology and co-ordinate the activities at school level. Different studies (Alamäki, 1999; Kankare, 1997; Rasinen, 2000) show that to develop the subject area, learning materials and in-service education have to be improved. In fact an extensive in-service education programme with new learning materials should always be put in place when something new is introduced into the curriculum. Some everyday experiences show that the written curriculum does not always take place in the schools. Therefore, it is important to study how the curriculum is implemented in practice. This type of research has not been conducted in Finland for the past years. The analysis of the latest framework curriculum indicates that innovation should be studied during craft lessons (and particularly technical work) (Rasinen, Virtanen & Miyakawa, 2009). If the framework curriculum is not followed the objective to learn innovativeness will not be achieved. In future, to guarantee more efficient learning, the subject area of 'technology' should be introduced. However, even today by following the approved framework curriculum, schools can offer technology education to all pupils.

REFERENCES

Alamäki, A. (1999). *How to educate students for a technological future: Technology education in early childhood and primary education*. Publications of the University of Turku, Annales Universitas Turkuensis. Series B: 233.

Anttila, P. (1993). *Käsityön ja muotoilun perusteet (Principles of handicrafts and forming)*. Porvoo: WSOY.

Bunge, M. (1985). *Treatise on basic philosophy* (Vol. 7). *Epistemology & Methodology III: Philosophy of science and technology. Part II. Life science, social science and technology.* Boston: D. Reidel.

Dyrenfurth, M. J. (1991). Technological literacy: Characteristics and competencies revealed and detailed. In *Technology and school. Report of the PATT conference in Poland 1990.* Zielona Góra: Pedagogical University Press.

Eggleston, J. (1994). What is design and technology education? In F. Banks (Ed.), *Teaching technology.* London: The Open University.

Hill, B., & Lutherdt, M. (1999). Structuring innovation-oriented approaches to teaching technology. In T. Kananoja, J. Kantola & M. Issakainen (Eds.), *Development of technology education - Conference B98* (pp. 181–199). University of Jyväskylä. Department of Teacher Education. The Principles and Practice of Teaching 33.

Järvinen, E. M. (2004). Näkökulmia teknologian opetukseen. (Some views in teaching of technology.) *Dimensio 3/2004,* 44 – 45.

Kankare, P. (1997). Teknologian lukutaidon toteutuskonteksti peruskoulun teknisessa tyossä. (The context of implementing technological literacy instruction in comprehensive schools' technical work.) *Turun yliopiston julkaisusarja.* Sarja C: 139.

Kantola, J. (1997). Cygnaeuksen jäljillä käsityökasvatuksesta teknologiseen kasvatukseen. (In the footsteps of Cygnaeus: From handicraft teaching to technological education.) Jyväskylän yliopisto. *Jyväskylä studies in education, psychology and social research 133.*

Kantola, J. Nikkanen, P., Kari, J, & Kananoja, T. (1999). *Through education into the world of work: Uno Cygnaeus, the father of technology education.* University of Jyväskylä: Institute for Educational Research.

Kojonkoski-Rännäli, S. (1995). Ajatus käsissämme. Käsityön merkityssisällön analyysi. (Thoughts in our hands. Analysis of the concept of handicrafts.) *Turun yliopiston julkaisusarja.* Sarja C: 109.

Lindfors, L. (1992). Formgivning i slöjd. Ämnesteoretisk och slöjdpedagogisk orienteringsgrund med exempel från textilslöjdundervisning. (Forming in handicrafts. Orientation basis for subject orientation and handicraft pedagogy.) *Raporter från Pedagogiska fakulten vid Åbo Akademi* nr 1 1992.

Parikka, M. (1998). Teknologiakompetenssi; Teknologiakasvatuksen uudistamishaasteita peruskoulussa ja lukiossa. (Technological competence: Challenges of reforming technology education in the Finnish comprehensive and upper secondary school.) Jyväskylän yliopisto. *Jyväskylä studies in education, psychology and social research 141.*

Peltonen, J. (1988). Käsityökasvatuksen perusteet. Koulukäsityön ja sen opetuksen teoria ja empiirinen tutkimus peruskoulun yläasteen teknisen työn oppisisällöistä ja opetuksesta. (Principles of handicraft education. Theory of school crafts and how it is taught, and an empirical study of the contents and teaching of technical work in the upper classes of comprehensive schools.) *Turun yliopisto. Kasvatustieteiden tiedekunta. Julkaisusarja* A: 1332.

Peruskoulun opetussuunnitelmakomitean mietintö I. (1970). *Opetussuunnitelman perusteet. (Memorandum. Foundations of the comprehensive school curriculum.) Komiteamietintö 1970:* A 4OPS 1970 I ja II. Helsinki: Valtion painatuskeskus.

Peruskoulun opetussuunnitelmakomitean mietintö II. (1970). *Oppiaineiden opetussuunnitelmat. 1970. (Memorandum. Curricula of Different Subjects.) Komiteamietintö 1970*: A 4OPS 1970 I ja II. Helsinki: Valtion painatuskeskus.

Peruskoulun opetussuunnitelman perusteet. (1985). (Foundations of the comprehensive school curriculum.) Kouluhalitus. Helsinki: Valtion painatuskeskus.

Peruskoulun opetussuunnitelman perusteet. (1994). (Foundations of the comprehensive school curriculum.) Opetushallitus. Helsinki: Painatuskeskus.

Perusopetuksen opetussuunnitelman perusteet. (2004). (The Framework curriculum of comprehensive schools.) Opetushallitus. Vammala: Vammalan kirjapaino.

Rasinen, A. (2000). Developing technology education. In search of curriculum elements for Finnish general education schools. Jyväskylän yliopisto. *Jyväskylä studies in education, psychology and social research, 171.*

Rasinen, A., Virtanen, S., & Miyakawa, H. (2009). Analysis of technology education in the five EU countries and challenges of technology education - the Finnish perspective. In H. Miyakawa (Ed.), *Cross border: International cooperation in industrial technology education*. Aichi, Japan: Aichi University of Education.
Suojanen, U. (1993). *Principles of Handicraft Education*. Provoo: WYSO.

Aki Rasinen, Pasi Ikonen and Timo Rissanen
University of Jyväskylä
Finland

ANDREW STEVENS AND KATE TER MORSHUIZEN

10. THE IMPLEMENTATION OF PRIMARY TECHNOLOGY EDUCATION IN SOUTH AFRICA

INTRODUCTION

Since South Africa achieved democracy in 1994, there has been fundamental change in many spheres, not least in education. In order to present a coherent picture of primary technology education, it is necessary to provide a brief account of its history.

Prior to 1994, education in South Africa was organised on racial lines with separate schools, universities, teacher colleges and administration systems for each of the four racial groups as defined by the apartheid state, namely black, white, coloured, and Asian. Although the education was supposed to be 'separate but equal', this was far from the reality. This was clearly evident in terms of the provisioning for the resource heavy areas of the curriculum such as Science, Home Economics, Woodwork and other subjects with a practical component. It was also evident in that the more vocational subjects such as agriculture were more readily available to blacks than the academic subjects such as the Sciences and Mathematics. The insidious persistence of this apartheid legacy is revealed in the 'bottom of the class' performance by South African learners in internationally benchmarked studies such as Trends in International Mathematics and Science Study (TIMSS) and others.

The 'Soweto Uprisings' of 1976 were sparked by educational issues. Ongoing unrest in black schools led to a comprehensive investigation into education, published by the Human Sciences Research Council (HSRC, 1981) as 'Provision of Education in the Republic of South Africa' (also known as the de Lange Report). One of the central recommendations of the commission was an attempt to shift the focus of formal education based on the traditional academic Arts and Sciences curriculum towards a more skills-based vocational curriculum, particularly for the majority of black learners. Although the recommendations of the de Lange Report were largely ignored by the apartheid state at the time, many of the ideas, albeit couched in somewhat different language and terminology, have resurfaced in the new democratic South Africa.

During the eighties and early nineties, the language of the market assumed increasing prominence in South African education policy circles and market-driven analyses and policies began to gain ascendancy over the traditional race-based ideology of apartheid. The influential Walters Report of 1990 (South Africa, Department of Education and Culture, 1990) - a product of the old regime - was the first to recommend changing the former technical subjects to bring them more in line with the 'Design and Technology' approach of the English. The Educational Renewal

C. Benson and J. Lunt (eds.), International Handbook of Primary Technology Education: Reviewing the Past Twenty Years. 107–116.

Strategy (ERS) (South Africa, Department of National Education, 1991) was likewise substantially influenced by the demands of the rapidly globalising knowledge economy in which flexibility, adaptability, and the ability to respond to changing market forces are key skills. Thus the ERS recommended the introduction of a number of new subjects into the general formative curriculum. Amongst these were Technology, Economics and Arts Education, the rationale being that these three subjects would provide education relevant to the needs of learners and society as well as contributing to the person power needs of the country. The ERS was careful not to propose too clear a differentiation between academic and vocational pathways in the compulsory (Grades 1–9) phases of education for 6–14 year olds (Kraak, 2002). However, in the proposed post-compulsory phase (Grades 10–12), for ages 15–18 year olds, vocational education was to assume far greater significance. After 1994, the new government committed itself to eradicating the differentiation in status and privilege which a differentiated academic vs. technical/vocational system promoted. The new curriculum, named 'Curriculum 2005' (C2005) for the year in which final implementation was to be accomplished, was the first school curriculum to be introduced for all South Africans irrespective of race. For the first time, a learning area called 'Technology', in a form corresponding largely to the English Design and Technology model, was to be part of every learner's education to Grade 9.

THE TECHNOLOGY 2005 PROJECT

In order to spearhead the introduction of the new technology learning area, a national project, known as 'Technology 2005' (T2005) was formed late in 1994, and a National Task Team (NTT) comprising four members was appointed until 1999 to operationalise the plans of the committee. The Technology 2005 project was formed because, unlike other learning areas in the National Curriculum, Technology had no earlier form or history of development in South African schools or Adult Basic Education programmes. Although piloting of the Technology curriculum had not begun by the advent of the new Curriculum 2005 in 1996, the project had by that time, reached a position where it was able to make a significant contribution to the development of the national curriculum frameworks and other policy related to the new Technology learning area.

The Technology 2005 project focused on working with the national and provincial education departments to support the implementation of primary Technology through specific initiatives, which included:

- developing, piloting and refining teaching and learning materials capable of supporting teachers involved in the implementation of Technology;
- facilitating the development of pre-service and in-service teacher education programmes in colleges of education and supporting lecturers in the implementation of these programmes;
- developing Technology Unit Standards for Adult Basic Education and Training (ABET) and in the General Education and Training Band (GET Band); and
- conducting a detailed evaluation of the project's development and the implementation of Technology in pilot primary schools so that strategies for wider implementation of Technology in the GET Band could be properly developed.

The primary schools' Technology pilot in 1998 resulted in teaching and learning material being piloted in 60 schools in the provinces of KwaZulu-Natal, Western Cape and Gauteng. Materials development was a collaborative exercise in which the National Task team coordinated a series of inter-provincial workshops. These had to double up as training workshops as most of the Provincial Task Team staff had no initial training or experience in Technology and much of the policy documentation was still evolving through 1997. Besides the assessment associated with the materials' development process, the National Task Team felt that it was important and necessary to investigate the broader implications of assessment for GET Technology. The Independent Examinations Board (IEB) was approached and a collaborative initiative to trial a national Grade 9 examination was launched in 1998 involving 1500 candidates in 25 schools in 5 provinces. Twenty three of the schools were Technology 2005 pilot schools and two others were private IEB schools. The IEB managed and administered the examination whilst a member of the National Task Team acted as an examiner and moderator.

Another component of the strategy developed by the National Task Team was to encourage the introduction of Pre-service Education and Training (PRESET) courses of training at colleges of education. This was done by developing and distributing a curriculum model for PRESET (Technology 2005, undated) and a two-year support and re-training programme for lecturers based on the curriculum model. The purpose of the strategy was to establish pre-service primary and junior secondary Technology education programmes in Colleges of Education as an important component of the infrastructure. These colleges became suitable sites to help meet the enormous need for on-going in-service training programmes for teachers (INSET). Seventy lecturers from forty-two colleges, representative of all nine provinces, were trained. Most colleges began offering courses by 1999. However, the effective implementation of this programme of pre-service and in-service training by college lecturers was severely interrupted by the 'rationalising' and merging of colleges which began in 1999. The result of this process was that most of the colleges of education were closed down: those that survived were incorporated into universities which themselves underwent extensive restructuring and merging operations. These processes have dramatically transformed South Africa's teacher education landscape.

The T2005 project was evaluated by the Foundation for Research Development (FRD) which published its extensive findings in March 1999 (Mouton et al., 1999). Since the pilot had only been implemented in three of the nine provinces, the report stressed that it was too early to judge the success of the introduction of the new learning area. However, it did find that the cascade model, which had been envisaged as a means of extending training into a wider and wider network of schools, had not worked as intended. Nevertheless, the report did note many positive aspects of the project, perhaps the most telling of which was the enthusiasm with which Technology was embraced by teachers and learners alike, as the following quote suggests:

> Teachers' enthusiasm for and dedication to technology is one of the most consistent and impressive findings from this evaluation. *The positive attitude of teachers was fed, in part, by the enthusiasm of their learners* (our emphasis). Most teachers indicated that they would like to continue technology. More than

> that, many seemed pleased to be able to break out of the old modes of teaching and reconceptualise their notions of what it means to be a teacher/facilitator.
>
> Technology was an introduction to OBE[1]-style teaching for most teachers, who found this approach to be a positive experience, and one that often gained the attention and recognition of their peers. Most teachers thus commented that they had benefited both professionally and personally from their participation in the project. (Mouton et al., 1999, pp. 157–8).

At the request of Higher Education Committee (HEDCOM), the members of the project team were asked to extend their work by a further year. This provided important time for the project members to finish off various aspects of their work and assist in the establishing of structures at provincial levels which would hopefully continue the task of promoting and developing GET Technology in the schools. Subsequently, three of the four Technology 2005 National Task team members formed a company known as 'Technology for All'. The company is contracted to the University of KwaZulu-Natal to continue offering courses for primary teachers in Technology.

IMPLEMENTATION OF CURRICULUM 2005

Curriculum 2005 attempted to transform South Africa's educational system from a teacher-centred, 'chalk and talk' approach to a learner-centred, activity-based approach to learning. Along with the eight new learning areas, a new philosophy of outcomes-based education was introduced. This brought with it a plethora of new jargon which the majority of teachers and teacher educators struggled to master. In the midst of this confusion, Technology was introduced throughout the primary and junior secondary schools so it is not surprising that the introduction of Technology was somewhat uneven throughout South Africa and remains problematic to this day. While the provincial departments provided basic two- or three- day workshops, this was clearly insufficient, particularly with a learning area such as Technology which was completely new to the vast majority of teachers. The numbers of teachers requiring training and the limited resources available for this purpose made it impossible to supply training on a scale sufficient to keep pace with the required time frames. In most provinces the lead teachers who had received training from the T2005 project were a tiny minority of the total in each grade. The cascade model which had been envisaged as a means of extending training into a wider and wider network of schools was largely a failure (Mouton et al., 1999) and the traditional teacher in-service courses offered by the tertiary institutions could only provide for a relatively small number of students.

After three years of the new curriculum, the new minister of education, Professor Kader Asmal, commissioned a review of the implementation of C2005 (South Africa, Department of Education, 2000) – through the Chisholm Committee. Although this Committee made many positive recommendations for simplifying and streamlining the curriculum, it also suggested that the two newest learning areas, namely Technology and Economic and Management Sciences, be scrapped, partly because the need for trained teachers seemed impossible to achieve. Fortunately, this recommendation was not accepted by the ministerial committee mainly due to three factors:

- the enthusiastic following that Technology had built up in the short period following its introduction, both among teachers and learners (see the quote from Mouton above) and among teacher educators at universities, colleges and non-government organisations (NGOs). All of these groups raised their objections in well considered submissions to the Minister;
- the strong support from a number of international Technology experts, including prominent academics from the United Kingdom and the United States whose submissions lent considerable weight to the arguments of local educationalists; and
- the nature of the Technology learning area itself, which of all the learning areas appears most able to meet the 'Critical and Developmental Outcomes' (which include problem solving, creative thinking, working in groups, time management, using Science and Technology effectively etc.) and which are the foundation stones of the new South African curriculum.

The result was that Technology survived, but it and all the other learning areas were completely revised, resulting in 'Revised National Curriculum Statements' being published in all eight learning areas for the nine grades of the compulsory GET Band (South Africa, Department of Education, 2002). While still written within an outcomes-based paradigm, these statements also specified some content and made important issues, such as progression, more explicit. The implementation time frames were also more relaxed than the previous curriculum changes and followed the following schedule:

2003	2004	2005	2006
Foundation Phase Grades 1–3 6–9 years	Intermediate Phase Grades 4–6 9–12 years	Senior Phase Grade 7 12–13 years	Senior Phase Grade 8–9 13–15 years

Once the new primary/GET school curriculum was launched, attention turned to the Further Education and Training (FET) band (Grades 10–12). In relation to Technology, the new FET curricula have abandoned the generalist and integrated approach of the GET Band, and the new technology offerings in the final three years of schooling have a decidedly specialised and vocational slant. Here learners will be required to choose studies in 'Electrical Technology', 'Mechanical Technology' or 'Civil Technology', 'Engineering Graphics and Design' and 'Computer Applications Technology'. These subjects appear largely to be reformulations of the technical subjects which were offered in vocationally-oriented technical schools and colleges of the past, albeit with some sociological and process elements grafted on. For learners who prefer a softer, more designerly approach to technology, there is the option of the new subject 'Design' which has emerged from the 'Arts and Culture' field and which most closely resembles 'Technology' from the GET Band. However, it remains to be seen whether these various offerings can provide the kind of learning which a general design and technology–type subject offers. There are some who feel that it is necessary for a general technology subject to be developed if the visionary goals of Technology education as a part of general, academic education are to be fully realised (Stevens, 2004).

TECHNOLOGY TEACHER EDUCATION IN SOUTH AFRICA

From Surplus to Shortage

A national audit in 1995 revealed that there were some 150 publicly funded institutions providing teacher education to over 200,000 students including 70,000 initial trainees at 93 contact colleges alone (Hofmeyr & Hall, 1995). A major restructuring has taken place and the number of contact colleges has been reduced to around 25 and the distance colleges to two. All these colleges have been incorporated into universities and the minimum qualification for a teacher entering the profession is now a bachelor degree. Prior to this a three-year college qualification was seen as sufficient and this remains the most common qualification of teachers in South Africa. As far as the subjects of Mathematics, Science and Technology are concerned, most teachers are under-qualified, and some are even unqualified. The drastic step of closing most of the colleges and incorporating the remainder into universities has been further complicated by the restructuring and merging of higher education institutions themselves - a process which has finally resulted in twenty three universities and universities of technology, most of which have had to restructure their teacher education faculties.

As a result of these transformations and the effects of HIV/AIDS amongst teachers, analysts predicted that South Africa would experience a shortage of supply of teachers from 2005 onwards (Crouch & Perry, 2003). The Western Cape Education Department estimated that total enrolment in teacher education declined from 70,000 in 1994 to only 13,000 in 2003 (Vinjevold, 2002). A shortfall of 15,000 new teachers by the end of the decade was predicted with scarce skills identified including Technology. In spite of these predictions, some provinces continued to make it difficult for new entrants to find employment by publishing erratic bulletins and making it difficult for schools to employ teachers of their choice. To compound the problem, many of South Africa's newly qualified young teachers are seeking employment overseas. At the time of writing (2009), South Africa was indeed experiencing a shortage of teachers in many subjects, including Technology, forcing the Department of Education to introduce bursary incentives to attract new entrants to the profession.

The Role of Tertiary Institutions

A superficial survey of the teacher education landscape undertaken recently revealed the following features:

- A majority of tertiary institutions are offering teacher education in primary Technology but there are significant exceptions where some of the prestige universities are tending to focus on research rather than initial teacher education since this is both more prestigious and more economic.

- Some institutions offer only in-service courses and some only pre-service programmes, but many offer both in an attempt to meet the critical shortage of qualified Technology teachers in primary schools.
- The most common initial teacher qualification is the Bachelor of Education (B.Ed.) in which students will take Technology as one of their 'teaching method' subjects. Also available in some institutions is the Post-Graduate Certificate of Education (PGCE) which is a one year capping qualification in which students will get little more than a 'Cook's Tour' of Technology, since few of these students will have taken any courses in the undergraduate training which remotely resemble the primary school subject 'Technology'. As far as in-service courses are concerned, the most common is the Advanced Certificate in Education (ACE) which is a two year, part-time course for qualified teachers who wish to 'reskill' in primary Technology. Within these courses there is considerable variation between institutions in relation to duration of course, scope and depth of content, development of practical skills, time allocated to ancillary subjects etc.
- The B.Ed. Honours in Technology education is offered at a minority of universities and is an opportunity for a more rigorous, research-orientated treatment of technology education. Most of these courses are still in their infancy. There have been very few Masters and Doctoral students in technology education in the period since its inception.
- Teaching staff on Technology Education courses represent a mix of technical / vocational and academically trained personnel. Some of them are science graduates. This provides an opportunity for a cross-fertilisation. A number of institutions employ part-time lecturers to deliver the courses. Some employ NGO partners in a range of creative arrangements to provide education to teachers nearer their places of work.
- Many of the programmes suffer from a severe shortage of staff with some institutions operating a 'one man band'. Considering the diverse nature of primary school Technology, this is not ideal in terms of expertise. There are very few institutions with large Technology departments employing more than three Technology education staff members.
- Although the school curriculum clearly underpins much of the content of the various courses, there is nevertheless a wide variation in emphasis between institutions in relation to the development of knowledge and skills. Few of the institutions have made extensive investments in technology-specific laboratories or workshops, for example. There is also a wide variety in length and duration of courses with some cutting contact time with students to a minimum.
- The role of Information Technology (IT) and the infusion of computers across the curriculum remains a goal but with so many schools being under resourced, there are other priorities before incorporating IT as a compulsory component of technology in schools.

The Contribution of Non Government Organisations

Technology education in South Africa owes much to the work of NGOs which performed much of the pioneering work to support teachers and developed many of the early curricula and learning support materials. The first of these was PROTEC (Programme for Technological Careers) which began by offering additional Mathematics, Science and Technology (MST) programmes to disadvantaged learners and has expanded into teacher education. It has a number of branches throughout South Africa and is an accredited provider of in-service education courses to teachers.

The ORT-STEP Institute was a large, national NGO which developed one of the first curricula for primary Technology for South Africa. It was very active in teacher education in the 1990s until problems with funding forced it to close its branches in all provinces except the Western Cape. Fortunately, many of the branches were absorbed by the universities with which the Institute had formed good working relationships and so the pioneering work in Technology education continues. In the Western Cape, the sole surviving branch (now named ORT-TECH) has flourished and is involved in a number of Technology education initiatives in partnership with the local universities and the provincial education department. A recent innovation by the organisation is the ORT Sustainable Educator and Empowerment Development (ORTSEED) project. This results from their finding that ordinary in-service training did not succeed in changing the classroom behaviour of the majority of teachers who needed more sustained school-based support. They attempt to provide this support by facilitating the development of Technology Centres in certain schools (called *anchor schools*) where teachers will be supplied with resources and additional expertise. It is hoped that these will form a model which will be replicated in other schools in time.

In addition to these non-profit organisations, there are a few private companies which provide services in the field of Technology education, the most prominent of which is '*Technology for All*', a group which was formed by three of the four National Task Team members of Technology 2005. This company is currently contracted to the University of KwaZulu-Natal and is responsible for most of the in-service courses in Technology (primarily the ACE) offered by this university. However, their work is not limited to the geographical area of the province in which they are located and they frequently run courses in other provinces as well. They are also actively involved in the writing and production of learning support materials for schools.

The Contribution of Teacher Organisations

The Technology Association of South Africa (TA) was founded by teachers in the Western Cape in 1995 to promote and stimulate the teaching of the new learning area. It grew rapidly and spread to other regions and provinces and is now a national association which holds an annual conference in different centres throughout South Africa. The association has provincial branches, the largest and most active of which are found in the provinces of the Western Cape, Gauteng and Kwa Zulu-Natal. These branches hold workshops for large numbers of teachers on a regular basis and have proved vital in ensuring that Technology takes root in all primary schools. The annual conference, which is organised by teachers for teachers, consists of a wide range of workshops on various topics and trade demonstrations and displays of primary learning support materials. A recent addition has been a parallel conference for teacher educators at which research papers are presented and issues of concern are discussed. This shows exciting potential for supporting and developing the research base of the fledgling discipline.

The South African Association of Science & Technology Educators (SAASTE) was established to support and promote the interests of Science and Technology education in South Africa. It has also held many conferences and workshops and has published the first Technology education journal, produced by teachers and provincial education department officials in the Western Cape, the province where it is most active. Here a skills development project, funded by the Shuttleworth Foundation, focused on a group of key teachers and developed their understanding of the new curriculum to the extent where they produced their own learning support materials. These teachers were then able to work at a higher level in the provincial system of extending Technology education to other groups of teachers and schools.

CONCLUSION

Funding remains a crucial stumbling block in South African education. The inequities of the past are proving difficult to erase in spite of increased expenditure on education, both in real terms and as a percentage of gross domestic product. In a country where water, sanitation, electricity and telephone lines are still not available in significant numbers of schools, Technology education facilities and resources unfortunately remain a low priority. Although South Africa has been the beneficiary of large sums of overseas aid in the post-apartheid era, much of this has gone into developing new systemic infrastructure, and spending on teacher education remains a difficult area to fund. Multinational and local companies have provided large amounts for teacher education in Technology, but significant numbers of practising teachers remain untouched by these efforts, particularly those in remote rural areas. None of the NGOs mentioned above could have survived without the provision of funding from the private sector and much of the innovative work in Technology education is supported and funded by private companies.

Important as economic resources are to the future development of Technology in schools in South Africa, it is the vision of the subject which sustains it. This vision sees the new learning area as contributing not only to the economic growth and development of the country, but also to the educational needs of its learners. For it is Technology's unique capacity to nurture the development of relevant knowledge, practical skills and essential values in young learners, that will ensure its place in the South African curriculum.

NOTES

[1] 'OBE refers to Outcomes Based Education'.

REFERENCES

Crouch, L., & Perry, H. (2003). In *Human resources development review 2003: Chapter 21 educators*. Pretoria: HSRC. Retrieved from http://hrdwarehouse.hsrc.ac.za/hrd/chapter.jsp?chid=138

Hofmeyr, J., & Hall, G. (1995). *The national teacher education audit. Synthesis report*. Johannesburg: CEPD.

Human Sciences Research Council (HSRC). (1981). *Provision of education in the RSA. Report of the main committee of the investigation into education.* Pretoria: Author.
Kraak, A. (2002). Discursive shifts and structural continuities in South African vocational education and training: 1981–1999. In P. Kallaway (Ed.), *The history of education under apartheid, 1948–1994.* New York: Peter Lang.
Mouton, J., Tapp, J., Luthuli, D., & Rogan, J. (1999). *Technology 2005: A National implementation evaluation study.* Stellenbosch: CENIS.
South Africa, Department of Education and Culture (DEC). (1990). *The evaluation and promotion of career education in south africa: Main report of the committee chaired by Dr S.W. Walters.* Pretoria: Government Printer.
South Africa, Department of National Education (DNE). (1991). *A curriculum model for education in South Africa.* Pretoria: Committee of Heads of Education Departments.
South Africa, Department of Education (DoE). (2000). *A South African curriculum for the twenty-first century: Report of the review committee on curriculum 2005.* Pretoria: DoE.
South Africa, Department of Education (DoE). (2002). *Revised national curriculum statement grades R-9 (Schools).* Pretoria: DoE.
Stevens, A. (2004). Getting Technology into the FET. In A. Buffler & R. C. Laugksch (Eds.), *Proceedings of the annual meeting of the Southern African Association for Research in Mathematics, Science and Technology Education (SAARMSTE)* (Vol. 12, pp. 983–987). Cape Town: University of Cape Town.
Technology 2005. (1996). *Draft national framework for curriculum development.* Pretoria: HEDCOM.
Technology 2005. (undated). *Curriculum model for a PRESET teacher education course in technology.* Unpublished circular compiled by H. Johnstone, National task team co-ordinator.
Vinjevold, P., quoted by *Business Day, 29 September 2002.* Retrieved from http://www.bday.co.za/bday/content/direct/1%2C3523%2C1095576-6096-0%2C00.html

Andrew Stevens
Department of Education
Rhodes University
South Africa

Kate Ter Morshuizen
Formerly Technology for All
Kwa Zulu Natal
South Africa

PART B:
ISSUES WITHIN DESIGN AND TECHNOLOGY EDUCATION

DAVID BARLEX

11. NUFFIELD PRIMARY DESIGN & TECHNOLOGY – A BRIEF HISTORY

INTRODUCTION

This chapter charts the history of the Nuffield Primary Design & Technology Project since its inception in 1996. It will describe four phases of activity:

- Phase 1 development, in which the approach to teaching and learning was devised and piloted with independent external evaluation in a variety of schools, and the first website was developed.
- Phase 2 publication and dissemination, in which the results of the evaluation were used to inform the content and design of the published materials and a revised website became a major vehicle for dissemination.
- Phase 3 after care, in which the project concentrated on working with various groups concerned with supporting the growth of primary Design & Technology in schools in England and contributed to the technology curricula in other countries, including those in the rest of the United Kingdom.
- Phase 4 recent developments involving the increasing importance of Information and Communication Technology (ICT) and putting this in the context of the Rose Review of the primary school curriculum.

PHASE 1 DEVELOPMENT

In the mid 1990s it was widely acknowledged that many primary school teachers found Design & Technology alien with regard to the established concerns and culture of the primary school. Design & Technology subject reports from the Office for Standards in Education (Ofsted) indicated that although some good practice could be found it was rare and that most primary school teachers were struggling in their attempts to teach Design & Technology (Office for Standards in Education 1995, 1996).

> Lack of non-teaching time for D&T co-ordinators, cramped accommodation, large group sizes and insufficient resources hinder attainment and progress and limit coverage of the Programme of Study. Whilst there continues to be small improvements in specialist facilities and resources for D&T, provision in nearly one-third of schools falls short of what is required for the National Curriculum. (*Ofsted, 1995, p. 8*)

> Standards in Design & Technology continue to improve in both Key Stages 1 and 2 although they remain lower than in most other subjects. (*Ofsted, 1996, p. 1*)

C. Benson and J. Lunt (eds.), International Handbook of Primary Technology Education: Reviewing the Past Twenty Years. 119–135.

> Many teachers are not sure what to assess or how to assess their pupils' attainment and progress in Design & Technology. As a result, in nearly half of schools, teachers are unable to plan activities which build successfully on earlier knowledge, or provide a programme which supports the progressive development of skills. ... Few teachers have any initial training in Design & Technology and arrangements for in-service professional development are generally inadequate. This results, overall, in low levels of teacher expertise. (*Ofsted, 1996, p. 2*)

In the autumn of 1996 the Nuffield Foundation Trustees agreed to allocate £150,000 to a first exploratory phase of a project designed to consider Design & Technology in the primary curriculum. This first phase, lasting one year, was required to explore the prevailing situation. The objectives of the project were to develop approaches to teaching and learning Design & Technology that were appropriate for the primary school and, in the second phase of the project, to produce associated resources that would enable primary teachers to be effective in the classroom. It was agreed that this effectiveness would involve meeting the needs of subject leaders who have the responsibility for training their colleagues and convincing head teachers of the value of Design & Technology. The third phase would be concerned with dissemination and training. David Barlex (Figure 1) was appointed Project Director. Although he had extensive experience of Design & Technology through his work on the highly successful Nuffield Secondary Design & Technology Project, it was particularly important to avoid secondary education assumptions in developing a primary school project. To ensure that secondary education thinking did not unduly influence the project Jane Mitra (Figure 2) was appointed as Co-director. Jane had extensive primary school experience and was a nationally acknowledged expert in the use of ICT in Design & Technology. Nina Towndrow (Figure 3) was appointed Project Administrator and was responsible for diary management, meetings organisation, financial management, production of trial materials and maintaining the project website. In the light of the progress made in the first phase the Trustees agreed funding for a further three years in the autumn of 1997.

Figures 1 and 2. David Barlex and Jane Mitra – the project directors for the Nuffield Primary Design & Technology project.

Figure 3. Nina Towndrow, the project administrator.

As the number of effective primary Design & Technology teachers was seen as very small, an attempt was made to build on the extensive primary school network established by the Nuffield Primary Science project (Black & Harlen, 1993). The Nuffield Curriculum Centre invited primary teachers, advisers and teacher educators who had been involved in this science project to a short seminar to discuss the place of Design & Technology in the primary curriculum. Several important features emerged. Even amongst primary teachers who had made good progress in developing their understanding of science and their confidence to teach science there was a considerable reluctance to engage with Design & Technology despite the apparently obvious connections between the technical understanding acquired in science and how this might be used in Design & Technology. This reluctance was understandable in the light of the amount of an initial teacher education programme that was likely to be dedicated to Design & Technology. In many cases this was a single day. There was a similar dearth of opportunities for in-service training. In the experience of those attending the seminar it appeared that most primary schools had put Design & Technology 'on the back burner'. One participant, Stuart Naylor summed it up when he said, "Design & Technology is an invisible subject in the primary curriculum; no one would notice if it disappeared" (Naylor, 1996). This was a more than depressing note on which to end the seminar and gave the project considerable food for thought. However even more bad news was not far away.

In May 1996 the government set up a task force to develop a strategy for substantially raising standards of literacy in primary schools in England over the following five to ten years. In February 1997 it published its preliminary findings in the report *A Reading Revolution - how we can help every child to read well.* This initial report was consulted on widely and led to a final report, *The Implementation of the National Literacy Strategy*, published in September 1997 (Department for Education and Employment, 1997). A key element of the strategy was the expectation that from September 1998 all primary schools in England would teach a daily literacy hour (Department for Education and Employment, 1998). The government then adopted a similar approach with regard to developing numeracy and from

September 1999 primary schools provided a structured daily mathematics lesson of 45 minutes to one hour for all pupils (Department for Education and Employment, 1999). So within two years of a seminar which indicated at best a low profile for Design & Technology in the curriculum the government had introduced two measures that put large time constraints on the primary curriculum and made it ever more difficult for schools to give so-called practical/creative subjects reasonable amounts of time. Where Design & Technology was taught it was often in lessons of short duration, approximately 1 hour in length, once a week for half a term, alternating with art on a termly basis. Some schools tried to meet their statutory obligation to provide Design & Technology lessons by means of a practical activity week at the end of the summer term.

Whilst the government was introducing the National Literacy Strategy the project commissioned a wide range of primary Design & Technology experts to develop units of work and used the following questions as starting points:

- What sort of products will children enjoy designing and making?
- Which people will children enjoy designing and making products for?
- What sort of products will children be able to design and make with the resources available in schools?
- What sort of products will teachers feel comfortable teaching children to design and make?

These units were piloted with independent external evaluation provided by Patricia Murphy (Figure 4) of The Open University. The findings were drawn from three classrooms in two schools involving KS1 and KS2 children. Data was collected through video and audio recording of classroom activities and interviews with teachers, head teachers, subject leaders and children. There were two important findings that indicated that the approach being taken by the project was appropriate. The evaluation reported:

> Children as young as five are well able to undertake design decisions if the context is appropriate to their experiences, if the decisions are not too extensive in number and type and if account is taken of their relatively limited manipulative skills. (Murphy, 1999, p. 13)

This was encouraging as the project had anecdotal evidence indicating that some teachers believed primary school children could not design. The evaluation (ibid) also reported:

> The initial conception of the materials and their structure is appropriate and effective. With further refinement the materials will represent both a major curriculum development but more importantly a major in-service support for use by teachers with colleagues in school. (Murphy, 1999, p. 13)

This too was encouraging as one of the major hurdles to enabling Design & Technology in primary schools was seen as the lack of in-service programmes. The materials were revised in the light of this evaluation. A unit of work was now organised in terms of Small Tasks that taught knowledge, understanding and skills likely to be needed in the successful completion of a designing and making Big Task.

Figure 4. Patricia Murphy of the open university undertook the independent evaluation of the pilot materials.

A further evaluation (Murphy, 1999) took place in the summer of 1998 using the revised materials and these findings were included in the final unpublished report. The findings were drawn from four case studies in four schools involving Key Stage 1 (KS1) (5–7 years) and Key Stage 2 (KS2) (7–11 years) children. Data was collected through video and audio recording of classroom activities and interviews with teachers and target children. The findings of this evaluation were very encouraging and included:

- all the activities were enjoyed by teachers and children;
- the layout was considered to be 'user friendly' and easy to use;
- teachers were pleased with the learning outcomes achieved through the activities;
- the structuring of the activities into Small Tasks to support a Big Task was highly recommended by experienced and inexperienced teachers alike;
- the Nuffield structured approach to D&T is making a significant contribution to primary pedagogy that is recognised and welcomed by experienced teachers;
- the materials are able to support less experienced teachers and raise challenges for experienced ones.

The evaluation also made suggestions for improvement. These were:

- ensure that exploration of user needs is integrated into the unit;
- ensure that children have the opportunity to reflect upon and justify their design decisions and can record their design decisions;
- ensure that teachers are aware of the design opportunities inherent in the unit.

These suggestions were used to inform the appearance and content of the final publication.

Whilst the second evaluation was taking place the project launched its website. This was envisaged as a way to provide support for those teachers who wanted to teach Design & Technology. It was used to make the trial units of work being developed by the project easily available to a wide range of schools and also to present examples of work carried out by schools using the trial units. The site quickly became heavily used, particularly with regard to downloading units of work. During a period September 1999 – December 2000 a total of almost 8,000 units were downloaded. The number of user sessions grew steadily during this time also - from 300 per month initially to just over 3,200 per month. Details of the use of the site were reported at the PATT New Media in Technology Education Conference (Barlex, 2001).

During the period of the second evaluation and website development there was concern that the literacy and numeracy strategies were having such an impact on the primary school curriculum that Design & Technology was to a large extent ignored by many schools (Barlex, 1999). An initial response from the project was to include in each unit a section showing how the unit could be related strongly to literacy and numeracy. The Trustees of the Nuffield Foundation were naturally concerned that the project, despite the positive evaluation and website use, might be ineffective because the prevailing climate in schools did not predispose head teachers to engage with the subject. It would have been understandable if the Trustees had decided to 'cut their losses' and close the project. However they did not do this. Instead they supported a joint initiative with the Scottish Consultative Council on the Curriculum (Scottish CCC, which became, in 2000, Learning and Teaching Scotland, LTS), to use the Nuffield approach to develop curriculum materials for the newly revised national guidelines for the Technology component of the Scottish primary curriculum.

There was already a strong congruence between the requirements of the Scottish Technology curriculum and the Nuffield approach although the terminology differed. The policy paper, *Technology Education in Scottish Schools* (Scottish Consultative Committee on the Curriculum (CCC), 1996) described a pedagogical framework that included 'Creative Practical Tasks' and 'Proficiency Tasks' which were, in Nuffield parlance, Big Tasks and Small Tasks. Denis Stewart (Figure 5) was Director of the Scottish Consultative Council. In addition the Scottish framework proposed the use of 'Case Study Tasks' to enable children to consider the relationship between technology and society. This collaborative endeavour between Nuffield and the Scottish CCC produced a suite of creative practical tasks complete with their own proficiency tasks as a set of 22 separate booklets and a set of 10 generic proficiency tasks as an additional single booklet. It also produced a set of 8 case study tasks in a single separate booklet and *The Essential Handbook for Teachers* that explained exactly how to use these resources to construct and teach a Technology scheme of work that meets the requirements of the revised Technology curriculum guidelines for Scottish primary schools (Barlex & Edwards, 1999). The materials were launched at a conference in November 2000 and subsequently made available to schools through LTS at the price of £25.00 for the complete package. Within one year of publication the materials had been purchased by about 1000 schools. As there are just over 2,200 primary schools in Scotland this was impressive market penetration in a short time. This uptake was further evidence that the Nuffield approach was sound and that materials aimed specifically at teachers as opposed to pupils were an important means of boosting teacher confidence and competence.

At the same time as collaborating with Scottish CCC/LTS the project took the opportunity to work closely with two head teachers who had indicated their interest in Design & Technology but were concerned with the lack of appropriate teaching

Figure 5. Denis Stewart, who, as a director of the Scottish Consultative Council on the curriculum, instigated collaboration with the Nuffield Foundation, and later as a Deputy Chief Executive of learning and teaching Scotland, played an active role in using the Nuffield approach to develop technology teaching materials in Scotland.

materials and support. The project worked closely with teachers in both schools and developed two interesting approaches to developing effective practice amongst teachers with little if any previous experience of Design & Technology.

Brian Mulroy, the head teacher of St Monica's Primary School in Bootle, Liverpool, has used Nuffield Design & Technology pilot units as the basis for a buddy system to help his teachers with Design & Technology. He opted for a flexible one session a week model in which two teachers (the Design & Technology subject leader and one other) were paired in order to support each other. As these teachers worked together and became confident he planned that this 'buddy' system could be extended to involve two more teachers and then four more so that within a relatively short space of time there would be seven teachers plus the subject leader, who could work with other teachers in the school in providing good Design & Technology lessons for all pupils. To quote Brian:

> This roll out and ripple approach enabled the school to start with confident staff and gradually impact on the whole school in a planned way. This will result in a whole school approach that is understood and implemented by everyone including teaching assistants.

This combination of a buddy system and the Nuffield units has become a major element of the continuing professional development provided by the school, which links directly with the School's Improvement Plan. In 2000 the buddy system incorporated four teachers and was so successful that the following year it was extended to include 8 teachers.

Margaret Lyn, head teacher of Our Lady of Compassion Roman Catholic Primary School in Formby, Liverpool was concerned that the impact of teaching literacy and numeracy on the curriculum had led to a steep decline in the amount of time available for Design & Technology. She was also aware that the time available was highly fragmented. This had led to a situation in which the standard of Design & Technology was well below an acceptable level.

Margaret decided that a radical approach was needed; one in which the children and their teachers had enough time to become intensely involved in designing and making and to have this experience often enough for the children to make progress. In October 2000 she chose to suspend the timetable for 3 consecutive days each term and dedicate those days to Design & Technology. The teachers in Margaret's school used pilot versions of the Nuffield Design & Technology units. The two Design & Technology subject leaders were able to use the units as a basis for discussion with teachers in deciding how best to adapt each unit to the needs of the children and the expertise of the teacher. Class teachers were able to use the units with classroom assistants and parent helpers. Margaret commented on the progress using this immersion approach as follows.

The first 3-day event was nerve wracking, tiring but successful. It was particularly rewarding to see the children become so involved. We learned three important lessons: avoid being over ambitious, target parent help where it is most needed, and take care to ensure all resources are in place. The second 3-day event went very much as planned with the children eagerly anticipating more designing and making. Staff took the third 3-day event at the end of the summer term, after an Ofsted inspection, in their stride and it was particularly pleasing to see that the children had made significant progress over the year; drawings of design intentions were matching made outcomes, manual dexterity had improved, and constructive evaluation of design decisions was becoming the norm.

The Design & Technology subject leaders from both schools had been instrumental in the success of these interventions and their professional growth was such that they were able to present their work at the Centre for Research in Primary Technology (CRIPT) Conference in 2001 (Barlex et al., 2001). However, a very clear lesson for the project in the work of the Liverpool schools was the importance of the head teacher in enabling Design & Technology to flourish in the curriculum.

PHASE 2 PUBLICATION AND DISSEMINATION

Given the proven success of the approach and associated materials, it was time to consider how to publish the project. Previously all Nuffield projects had used commercial publishers to produce the printed materials used by teachers and pupils. The project approached Longmans who had successfully published the Nuffield secondary Design & Technology materials (Barlex, 1995) but they were not interested. The project then approached Collins and had extensive talks. Collins were concerned that the average amount of money spent by a primary school on Design & Technology was so low that they could not see how they would recoup any significant investment. Their best offer was to ask the project to re-conceptualise itself as an Information and Communication Technology (ICT) project which incorporated Design & Technology. They reasoned that schools had money to spend on ICT and that this was the best way to infiltrate Design & Technology back into the curriculum. The project was not impressed with this proposal and decided to investigate the possibility of handling the design through consultation with a graphic designer and development of the final products in electronic format in-house. This still left the project with the problem of who would actually publish and sell the materials. To solve this, the project entered into an agreement with the Design &

Technology Association. The project would provide electronic copy of units of work using the feedback from the independent evaluation of the pilots plus the findings of the collaboration with head teachers, and the Design & Technology Association would finance the production of these units and sell them on a cost-recovering basis. This would enable the Design & Technology Association to reprint the materials if they were successful. This was the first time that a Nuffield project had taken this approach. This is an interesting example of a partnership between an educational charity (which does not need to make a profit) and a professional association (which although also a charity could, from a small investment, actually make a profit). The partnership enabled the provision of curriculum materials and professional support at a time when commercial publishers were unable to do so.

The project worked closely with the graphic designer David Mackerall. The project shared with him the findings concerning not only the success of the pilot materials and approach but also that the majority of primary teachers lacked confidence, had limited subject knowledge and were understandably reluctant to become involved in teaching Design & Technology. The project discussed with him at length the recommendations of the evaluations. The aim of his design was to provide an attractive, accessible, highly visual product that subject leaders could use to support their colleagues and that teachers would find easy to use with little in-service training. This was a significant challenge but he responded superbly. He conceived the product as consisting of a CD ROM (containing 24 units of work in full colour and grey scale which could be printed plus an interactive introduction to the project and guide to the website); a short teachers' guide; and a sample unit all enclosed in a slim, elegant card folder. This could be produced and sold by the Design & Technology Association for under £10.00 including postage and package. He developed a structure for the units of work that could apply to all units. Hence teachers need only use one unit to become familiar with the key features in all. The units were structured as follows:

- Cover
- Learning context
  Design context
  Learning purposes
- Tasks for learning
  The Small Tasks
  The Big Task
- Design decisions
- Teaching the unit
  Small tasks
  Big task
  Evaluation
  Unit review
- Resources and links
  Vocabulary
- Resources summary
  Links to other subjects

– Copiable sheets for pupils
  Always including a specification and evaluation
– Acknowledgements

The title for the pack sums up the designer's insight. He named it *Primary Solutions* in Design & Technology – a brilliant idea! Dave Mackerell (Figure 6) designed a template for unit presentation and Nina Towndrow was able to use this to produce the units of work for the CD ROM. This was the first time that a project at the Nuffield Curriculum Centre had produced high quality material for publication without the involvement of a commercial publisher.

Figure 6. Dave Mackerell - the graphic designer who helped shape the image of the final publications.

The *Primary Solutions* pack (Figure 7) was published in 2001 and launched at the Design & Technology Association conference in the summer of 2001. By 2007 approximately 3,000 copies of the pack have been sold through the Design & Technology Association. The Project took the decision to continue making the units of work available from the website. The units of work downloaded from the site grew steadily, peaking in 2005 and then diminishing as shown in Table 1. Overall there have been almost 640,000 downloads in this time.

Figure 7. The primary solutions pack.

Table 1. Units of work downloaded from the Nuffield Primary Design & Technology website www.primarydandt.org annually 2001–2007

2001 Sept – Dec	2002	2003	2004	2005	2006	2007 Months 1–4
41,690 units	97,972 units	106,363 units	133,223 Units	137,674 units	92,023 units	27,576 units

PHASE 3 AFTERCARE

The Project undertook dissemination of the Nuffield approach to teaching Design & Technology, the Project website and the *Primary Solutions* pack in England through the standard channels. The Design & Technology Association advertised the pack in its publication listings. The approach was strongly promoted on the Nuffield stand at the annual Design & Technology Show held at the National Exhibition Centre in Birmingham, England. The work of teachers using the approach was celebrated in the showcase on the website and informed a variety of conference papers (Barlex, 2003; Barlex, 2004; Barlex, 2007). The Project worked closely with the Qualifications and Curriculum Authority (QCA) using the Nuffield approach to support the adaptation of QCA Units of Work (QCA, 1998) with the classroom experience being documented and posted as reports on the Nuffield Primary Design & Technology website (see http://www.primarydandt.org/resources/reports,1162,NA.html). The Project Director regularly provided in-service training sessions and made presentations to pre-service (undergraduate) teachers.

The experience of the Project had indicated the importance of head teachers in securing a place in the curriculum for Design & Technology. In response to this, the Project forged strong links with the National Primary Headteachers' Association, presented at their annual conference in 2003 and set up a head teacher working group to articulate the benefits of Design & Technology for the primary curriculum from the head teacher perspective. As a result the group made a presentation to Maureen Lewis, Regional Director for Literacy for the Primary National Strategy. In response to this presentation Maureen invited a Nuffield pilot school to contribute their practice of teaching Design & Technology to the Primary Strategy Excellence and Enjoyment publications (Department for Education and Skills, 2004a & 2004b).

It is acknowledged that Design & Technology education is under researched (Harris & Wilson, 2003). In response the project organised two seminars for those involved in primary initial teacher education that had a special interest in Design & Technology. The reports of the seminars are available at http://www.primarydandt.org/resources/reports,1162,NA.html?pageNo=3. As a result of these seminars regional research groups were set up with the aim of using data about primary practice in Design & Technology collected by pre-service teachers in training. The most active group in the South East of England was able to present their findings at the Design & Technology Association International Research Conference (Rutland et al., 2006).

Plans were made for a further seminar in January 2008 to enable the regional groups to share their findings and develop publications.

An important aspect of the Project's work during development and in aftercare has been collaborating with Design & Technology educators in other countries. The Project worked closely with Esa Matti Jarvinnen (Figure 10) of the Oulu University in Finland. By means of in-service and pre-service activities he introduced teachers to the Nuffield approach through the *Primary Solutions* unit 'How will your roly poly roll?' He also adapted an early trial unit about the design of noise-makers resulting in one child designing a simple rattle that could be used to flush hares in the wild from hiding so that adult hunters could then shoot the hares. The children skinned, cooked and ate the hares. As Esa Matti put it, "Technology had given these children their dinner!" The Project also worked closely with Thomas Ginner (Figure 9) of the Centre for School Technology Education at Linkoping University, Sweden. Thomas, together with a group of teachers and teacher trainers from Sweden, visited Our Lady of Compassion School in Formby, Liverpool, during a Design & Technology three-day immersion event. They were so impressed that they have translated a number of units into Swedish and are using them for both in-service and pre-service activities. The closest international ties have been made with Malcolm Welch (Figure 8) of Queen's University, Kingston, Ontario, Canada. Malcolm was the Project Director for an elementary Science and Technology programme, which was charged with developing curriculum materials by means of in-service activity with teachers. This is an extremely demanding and difficult way to approach curriculum development. Malcolm used the Nuffield Small Tasks – Big Task approach to both the Science and Technology units. He renamed Small Tasks 'Support Tasks'. In Technology lessons the Big Task was to design and make a product of some sort. In Science lessons the Big Task was to answer a Big Question. Malcolm and his team of teachers were able to produce 14 units of works across grades 1–6 for the Ontario elementary Science and Technology curriculum and develop a powerful means of in-service training (Welch et al., 2000a; Welch et al., 2000b; Welch, Barlex, & Mueller, 2001; Mueller & Welch, 2006).

Figure 8. International collaborator, Malcolm Welch.

Figure 9. International collaborator, Thomas Ginner.

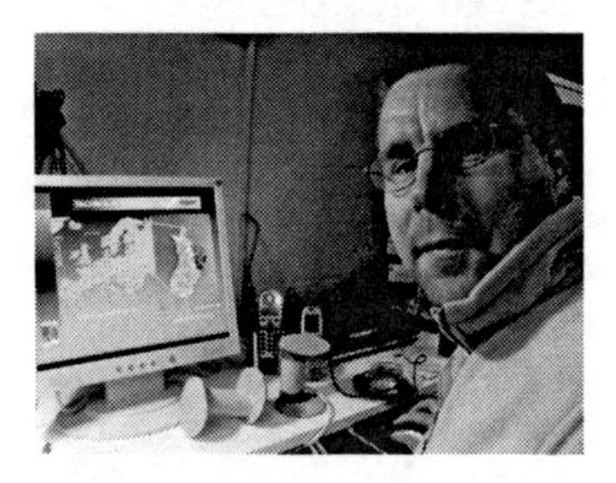

Figure 10. International collaborator, Esa Matti Jarvinen.

PHASE 4 RECENT DEVELOPMENTS

There is little doubt that the government in England places huge store on the potential of ICT to enhance pupils' learning in both primary and secondary school. In the forward to *Harnessing Technology for Next Generation Learning Children, Schools and Families Implementation Plan 2009–2012*, Jim Knight, Minister of State for Schools and Learners writes:

> My Department has a strong commitment to continued investment in technology to support improvement. Over £600 million is being distributed between 2008 and 2011 to schools through the Harnessing Technology grant, enabling every school to make strategic investment in the technology it needs for the future. (Department for Children, Schools and Families, 2009a, p. 1)

At the grassroots level British Educational Communications and Technology Agency (Becta) cites the Learner entitlement to ICT as follows:

> Pupils should be given opportunities to apply and develop their ICT capability through the use of ICT tools to support their learning in all subjects. (Becta Schools Update Volume 2, September 2009, p. 6)

The entitlement itself is clear about the importance of teaching in this regard.

> One of the challenges is for improved teaching and promoting a technology-related learner entitlement. Retrieved from http://schools.becta.org.uk/index.php?section=cu&catcode=ss_cu_ent_02

Hence it is clear that both teachers and pupils are expected to use ICT in subjects across the curriculum. Design & Technology has played an important role in using ICT to enhance pupil learning. In 1999 the Department for Education and Skills provided funds to support the launch of the CADCAM (computer assisted design and computer assisted manufacture) in Schools programme managed by the Design & Technology Association. This programme was in essence an attempt to modernise the Design & Technology curriculum in secondary schools by providing professional development for teachers in the use of an industry standard 3D solid modelling software (Prodesktop). Since that time the majority of secondary schools in England have introduced this software into their Design & Technology curriculum and this has been supported by the development of a website (www.cadcamcurriculum.org/) which provides a framework of progression for learning in CADCAM. Through the work of the Techlink project, funded by the Medlock Charity, primary schools in the south west of England were being introduced to CADCAM as part of their Design & Technology curriculum with the aim of ensuring a smooth transition from primary to secondary school with regard to this use of ICT (www.techlinkinschools.com). In 2007 the Nuffield Primary Design & Technology project made contact with Hayesfield School, which had been working with the Techlink project. A major consideration for the Nuffield project was the extent to which the introduction of CADCAM into the primary curriculum would empower pupils to make genuine design decisions as opposed to constraining, limiting or even eliminating such decision making due to the nature of the CADCCAM software. To explore the potential of the Techlink approach to CADCAM the Nuffield Project collaborated with Hayesfield School in their liaison work with partner primary schools. The collaboration involved the adaptation of proven Nuffield primary schools Design & Technology units of work (in which pupils were known to be able to make design decisions) to include a CADCAM approach to designing and making. An innovative approach involving peer-peer tutoring by pupils and an immersion experience for teaching CADCAM with partnership primary school pupils emerged (Barlex & Miles-Pearson, 2008 and 2009) illustrating that although there was the danger of pupils' creativity through designing being overwhelmed by the software this need not be the case. As a result the units work involving the use of CADCAM, were published on the Nuffield Primary Design & Technology website (http://www.primarydandt.org/news/cadcam-units-available-now,569,NNS.html). The enthusiastic response of the primary schools to this development has led Hayesfield School to continue this work with partner primary schools and to develop four further units of work to incorporate CADCAM.

As a result of government support for the use of ICT in teaching, one important piece of technology that is now widely available in most primary schools is the interactive white board (IWB). However the provision of this hardware does not of itself guarantee that teachers will be able to use it effectively. Promethean, a major supplier of interactive white boards and software to support their use in the classroom, was keen to develop software that would support primary Design & Technology. Through their work with Dave Mackerel, the graphic designer, Promethean became aware of the Nuffield Primary Design & Technology Project and began discussions with David Barlex as to the way the IWB might be used to support the teaching of primary Design & Technology. In 2009 an agreement was reached to develop IWB software to support six Nuffield primary Design & Technology units of work: How scary should a calendar be? What should be stuck to your fridge? What music would you like to make? Will this story surprise you? How will your beast move its mouth? Should your creature be fierce or friendly? This software will be made available from the Promethean website (www.prometheanplanet.com) at minimum cost. Scott & Asoko (2006) identified four classes of communicative approaches all of which have their place in teaching:

> Interactive/dialogic: teacher and pupils consider a range of ideas
>
> Non-interactive/dialogic: teacher reviews different points of view
> Interactive/authoritative: teacher focuses on one specific point of view and leads pupils through a question and answer routine with the aim of establishing and consolidating that point of view
>
> Non-interactive/authoritative: teacher presents a specific point of view. (Scott & Asoko, 2006, p. 14)

Julia Glass, the developer of the IWB software for the Nuffield units of work is an experienced teacher and has already shown in the pilot materials that it is possible to build all these different communicative approaches into the software.

Throughout 2009 the primary curriculum had been under a government instigated review led by Sir Jim Rose. He reported in 2009 (Department for Children, Schools and Families, 2009b) and, as might be expected, the review highlights the use of ICT in all subjects and in the case of Design & Technology makes specific mention of CADCAM. However there was concern within the Design & Technology community that the association of Design & Technology with Science in an area of learning to be called 'Scientific and Technological Understanding' would be misleading and underplay the role of designing. An on-line survey carried out by the Design & Technology Association indicated that of 507 respondents 90.1% thought that Understanding Science, Technology and Design was a preferable title. The Design & Technology Association in collaboration with the National Association of Advisers and Inspectors of Design & Technology provided an extremely thorough response to the Rose Review (2009b) suggesting a range of alternatives and amendments, making amongst other points that the connections between Science and Design & Technology as presented in the review were problematic in supporting an erroneous and limiting 'Technology as applied Science' view.

From tentative beginnings in 1990, through years of despondency in which the stranglehold of the literacy and numeracy strategies on the primary school curriculum almost eliminated Design & Technology from children's learning experiences, we moved to a position where there was renewed interest and opportunity for the contribution that Design & Technology can make to children's development.

The Nuffield Primary Design & Technology Project, working in collaboration with a variety of partners, had been instrumental in moving practice and perceptions of Design & Technology to the point where teachers and head teachers could take advantage of this new climate of opinion. The Rose Review, however well intentioned, did present problems for Design & Technology. However, the new coalition administration that is now in office has rejected the Rose Review proposals but has at the time of writing yet to make known its views on the nature of the Design & Technology curriculum it wishes to see in primary schools.

REFERENCES

Barlex, D. (1995). *Nuffield Design and Technology*. Harlow: Longman.

Barlex, D. (1999). Primary Design and Technology – A lost cause or everything to play for? In C.Benson & W. Till (Eds.), *Second International Primary Design and Technology conference 1999* (pp. 2–9). Birmingham, UK: CRIPT at UCE Birmingham.

Barlex, D. (2001). Using a website to support a community of practice. In M. J. de Vries & P. Mottershead (Eds.), *New media in technology education. Proceedings of the PATT 11 conference*. Retrieved from http://www.iteaconnect.org/PATT11/PATT11.pdf

Barlex, D. (2003). Developing and celebrating good practice in Primary Design & Technology. In C. Benson, M. Martin & W. Till (Eds.), *Fourth International Primary Design and Technology conference 2003* (pp. 2–4). Birmingham, UK: CRIPT at UCE Birmingham.

Barlex, D. (2004). Design decisions in Nuffield Design & Technology. In M. J. de Vries & P. Mottershead (Eds.), *Pupils' decision making in technology research, curriculum development and assessment, Proceedings of the PATT 14 conference*. Retrieved from http://www.iteaconnect.org/PATT14/PATT14.pdf

Barlex, D. (2007). Creativity in school Design & Technology in England: A discussion of influences. *International Journal of Technology and Design Education, 17*, 149–162.

Barlex, D., & Edwards, P. (1999). Technology in Scottish primary schools: Curriculum development between the Nuffield Design & Technology project and the Scottish CCC's Technology education development programme. In N. P. Juster (Ed.), *The continuum of design education, Proceedings of the 21st SEED annual design conference and 6th national conference on product design education* (pp. 15–24). Bury St Edmunds and London, UK: Professional Engineering Publishing Limited.

Barlex, D., & Miles-Pearson, S. (2008) Introducing CADCAM into primary schools, Part 1 of a case study – Developing the curriculum. In E. W. L. Norman & D. Spendlove (Eds.), *Designing the curriculum – making it work, the Design & Technology Association International Research Conference 2008* (pp. 5–10). Wellesbourne: Design & Technology Association.

Barlex, D., & Miles-Pearson, S. (2009). Assessing the impact of peer-peer mentoring on pupils' engagement with CADCAM. In C. Benson, P. Bailey, S. Lawson, J. Lunt & W. Till (Eds.), *Seventh International Primary Design and Technology conference 2009* (pp. 25–29). Birmingham, UK: CRIPT at Birmingham City University.

Becta. (2009). *Schools update* (Vol. 2).

Black, P., & Harlen, W. (1993). *Nuffield primary science teachers' handbook*. London: Collins Educational.

Department for Children, Schools and Families. (2009a). *Harnessing technology for next generation learning, children, schools and families implementation plan 2009–2012*. Nottingham: Department for Children, Schools and Families.

Department for Children, Schools and Families. (2009b). *Independent review of the primary curriculum: Final report (Rose Review)*. Nottingham: Department for Children, Schools and Families.

Department for Education and Employment. (1997). *The implementation of the national literacy strategy*. London: Department for Education and Employment.

Department for Education and Employment. (1998). *The national literacy strategy*. London: Department for Education and Employment.

Department for Education and Employment. (1999). *The national numeracy strategy*. London: Department for Education and Employment.

Department for Education and Skills. (2004a). *Primary national strategy, excellence and enjoyment: Learning and teaching in the primary years. Learning to learn: Progression in key aspects of learning*. London: Department for Education and Skills.

Department for Education and Skills. (2004b). *Primary national strategy, excellence and enjoyment: Learning and teaching in the primary years. Learning to learn: Key aspects of learning across the primary curriculum*. London: Department for Education and Skills.

Harris, M., & Wilson, V. (2003). *Designs on the curriculum? A review of the literature on the impact of design & technology in schools in England*. Glasgow, Scotland: Scottish Centre for Research in Education.

Mueller, A., & Welch, M. (2006). Classroom-based professional development: Teachers' reflections on learning alongside students. *Alberta Journal of Educational Research, 52*, 143–157.

Murphy, P. (1999). *Independent evaluation of the Nuffield Primary Design & Technology project*. unpublished.

Naylor, S. (1996). *Comment at invitation seminar at the Nuffield Foundation*.

Office for Standards in Education. (1995). *Subject report for Primary Design & Technology*. London, England: HMSO.

Office for Standards in Education. (1996). *Subject report for Primary Design & Technology*. London, England: HMSO.

Qualifications and Curriculum Authority. (1998). *Design and Technology: A scheme of work for key stages 1 and 2*. London: Qualifications and Curriculum Authority.

Rutland, M., Rogers, M., Hope, G., Prajapat, B., Haffenden, D., Seidel, M., et al. (2006). Student teachers' impressions of Primary Design & Technology in English schools: A pilot study. In E. W. L. Norman, D. Spendlove & G. Owen Jackson, (Eds.), *The Design & Technology Association International Research Conference 2006* (pp. 97–110). Wellesbourne: The Design & Technology Association.

Scott, P. H., & Asoko, H. (2006). Talk in Science Classrooms. In V. Wood-Robinson (Ed.), ASE *guide to secondary science education*. Hatfield, UK: Association for Science Education (ASE).

Scottish Consultative Committee on the Curriculum. (1996). *Technology education in Scottish schools: A statement of position from the Scottish CCC*. Dundee, Scotland: Scottish Consultative Committee on the Curriculum.

Welch, M., Barlex, D., Christie, C., Mueller, A., Munby, H., Chin, P., et al. (2000a). Developing an approach to assessment for the elementary Science and Technology curriculum of Ontario. In P. H. Roberts & E. W. L. Norman (Eds.), *International Design and Technology educational research 2000* (pp. 34–39). Loughborough, UK: Loughborough University.

Welch, M., Barlex, D., Christie, C., Mueller, A., Munby, H., Chin, P., et al. (2000b). Teaching Elementary Science and Technology in Ontario. In P. H. Roberts & E. W. L. Norman (Eds.) *International Design and Technology educational research 2000* (pp. 180–187). Loughborough, UK: Loughborough University.

Welch, M., Barlex, D., & Mueller, A. (2001). Curriculum materials writing: An opportunity for innovative professional development. In P. H. Roberts & E. W. L. Norman (Eds.), *International Design and Technology educational research 2001* (pp. 147–152). Loughborough, UK: Loughborough University.

David Barlex
Formerly of the School of Sport and Education
Brunel University
England

CLARE BENSON AND TARA TRELEVEN

12. DESIGNERLY THINKING IN THE FOUNDATION STAGE

INTRODUCTION

Whilst Design and Technology (D&T) was introduced into the primary curriculum (children aged 5–11 years) in England in 1990, there were no specific subjects in the first English Early Years curriculum for children aged 3–5 years (SCAA, 1996). Six areas of learning were identified and whilst D&T content could be identified in all areas, the focus for the subject was within Knowledge and Understanding of the World. As the curriculum has been updated the 6 areas remain (DfEE/QCA, 2000 and DfES, 2007) including Knowledge and Understanding of the World. Young children are to be given opportunities to explore materials, investigate products and how they work, build and construct with a wide range of materials, use a range of tools safely, and select tools and techniques that they need to shape, assemble and join materials in order to make their products. However there was little Continuing Professional Development (CPD) available to support the implementation of this aspect of the curriculum and certainly no national programme of CPD. From a small scale research study (Benson, 2001) for the Qualifications and Curriculum Authority (QCA) carried out in 49 Early Years settings[1], it was apparent that teachers lacked confidence and had little understanding of the nature of D&T. Activities focused on making, with little regard to designing. The teachers indicated that they were unsure as to how they could incorporate investigating and evaluating products into an Early Years curriculum – important activities to help children to look critically at the designed and made world around them. Teaching resources were available to support the development of knowledge and understanding of materials, mechanisms and making, but little to support designing including product evaluation – focusing children's attention on the design of a product and developing their designerly thinking skills.

The concept of designerly thinking used by, for example Baynes (1994), linked imaginative play and designing skills and provided a useful starting point for further work on this aspect of the curriculum. Following on from the initial research (Benson, 2001) a major project was funded by the Department for Education and Skills (DfES) that focused on encouraging teachers to develop children's designerly thinking skills in their own Foundation Stage settings (Benson, 2003). A key aspect of the project was to provide resources that could be used to show how to help the development of designerly thinking through exploring and evaluating a range of designed and made products - the Early Years Materials Kit (TTS, 2003).

C. Benson and J. Lunt (eds.), International Handbook of Primary Technology Education: Reviewing the Past Twenty Years. 137–150.

Talking about and evaluating existing products, including disassembling and reassembling items, is often the important first step in a school project and is the basis of investigating, disassembling and evaluating activities (IDEAs) introduced later on in the primary school in the National Curriculum. However, when looking at the opportunities for the development of evaluative skills in younger children, there is an under representation of talking about the designed and made world in the Foundation Stage curriculum guidance for teachers (DfEE/QCA, 2000). This is in contrast to the French Early Years curriculum (Senesi, 1998) that specifically advocates the assembling and disassembling of designed and made products.

It was the DfES designerly thinking project and research findings from this that provided the inspiration and focus for research carried out by Treleven (2004) and led to the research outlined in this chapter. Previously Benson (2003) and Treleven (2004) had presented the children with a range of products, using questioning as a key tool in developing their designerly thinking skills. Now Treleven wanted to try a different approach and put each product into a context as she introduced them to the children. The following is the report of a small scale case study research project carried out by Treleven in a London Nursery.

THE RESEARCH PROJECT

In assessing the current experiences of the children to be involved in the project, I saw the potential to develop these by working in the Nursery setting with a collection of designed and made products taken from the Early Years Materials Kit (TTS, 2003). Thus, I developed a series of activities designed to promote the importance of talking and making decisions about the products, but in a different way to the previous studies (Benson, 2003; Treleven, 2004). The objects were presented within a context to the children and the children were then asked to solve a problem or find a solution. I felt that the children might be more able to engage with the task set in an authentic context.

The aims of the project were to support the findings of the original project (Benson, 2003) and to show potential benefits of using a designerly thinking approach in a Foundation Stage curriculum. The skills that develop when children are involved in product evaluation activities would be identified. The investigation would focus on whether when tasks are meaningful to the children, they are able to use designerly thinking skills to evaluate different designed and made products. In order to do this, the study began by drawing upon different aspects of children's thinking, with particular reference to creativity and critical thinking.

LITERATURE REVIEW

The work of Piaget has had a huge impact on how we look at children's thinking today (Tassoni & Hucker, 2000). Piaget was primarily interested in how children make sense of the world around them. His work on children's developmental stages suggests that children are only capable of abstract thought from age twelve onwards. In the current study, according to Piaget, the children would be working within the pre-operational stage and would therefore, be unable to engage in logical thought. They would not be able to look at anything other than from their own viewpoint, as Piaget claims that at this stage, children's egocentricity prevents them from doing so.

The work of Piaget has been criticised for being too restrictive (McGarrigle and Donaldson, 1974; Hughes, 1996; Shayer & Adey, 1981). Other theorists, such as Bruner (1983) provide a wider explanation of children's learning and explore other influences on children, such as the role of the adult and the child's previous experiences. Edwards & Knight (1994) conclude that although Piaget's work does have limitations, it is a useful starting point for educationalists looking at children's learning as it highlights the need to understand the way that children think.

More recently, Fisher (2005) has discussed the importance of both creativity and critical thinking in children's development. He suggests that although the two types of thinking develop in different ways, most situations or problems require a person to engage in both types of thinking. But what is creativity? Many authors have concluded that creativity is difficult to define (Moyles, 1989; Sternberg, 2001; Spendlove, 2005; Craft, 2007), while others have linked it closely with divergent thinking (Guilford, 1957; Dansky & Silverman, 1973; De Bono, 1987; Fisher, 2005). Siraj-Blatchford & MacLeod - Brudenell (2003) conclude that although there is insufficient evidence to show that creativity can be taught, it can certainly be encouraged through providing an appropriate environment. They describe a case study in which 5 year olds were asked to design wolf-proof homes in relation to the story of The Three Little Pigs. In this instance, the teacher praised original, creative ideas but Siraj-Blatchford & MacLeod - Brudenell (2003) suggest that often, rather than encouraging creative thought, practitioners dismiss children's ideas as they do not fit existing plans or materials. In relation to D&T, Siraj-Blatchford & MacLeod - Brudenell (2003) believe creative thinking is an essential element of the design process and it is vital that creativity is encouraged. Fisher (2005) also highlights the importance of creativity and describes it as looking at a problem from a new angle so as to think up an original solution – one that is unique and innovative. This is vital in D&T where designers strive to develop and produce original and creative products.

In order to think of a different and/or better solution to a problem, critical thinking is needed, whereby the problem is analysed in a logical way to reach a solution (Fisher, 2005). In D&T, a critical thinker will challenge the work of others and begin to think about how improvements can be made to that work. When discussing critical thinking, Bloom's taxonomy has provided us with an interesting and useful view of thinking processes. Bloom (1956) identifies 6 levels of thinking, the lowest being knowledge and the highest level being evaluation. The evaluation stage is that which Bloom equates to critical thinking. It is at this stage that judgements and assessments are made and then acted on. For those working in D&T, this stage is a vital one in judging whether a product is successful or not, and if not, then identifying the reasons why.

Bloom's taxonomy has been adapted to relate to thinking skills in young children. Fisher (2005) applies the taxonomy when developing questions in response to the story of Goldilocks and the Three Bears and also to the topic of weather.

Parker Rees (1997) shows that when given the right context, young children are often able to engage in higher order, evaluative thinking. When considering evaluative thinking in relation to designed and made products, Siraj-Blatchford & Macleod - Brudenell (2003) highlight how even very young children are learning to evaluate products early on in their everyday lives. They may have a preferred cup, for example, based on attributes such as what it looks like, how it is useful to them or if it is easy to hold. Rogers and Stables (2001) provides two examples where young children are able to evaluate the design of familiar products such as pegs and lunch boxes.

Designerly thinking is the evaluating of designed and made products based on their design. It encompasses both critical thinking and creativity. In this new project, in linking designerly thought with the skills needed to solve problems, the children were being given an open ended invitation to offer their opinions and to evaluate their own ideas and the ideas of others (including the adult). They were asked to offer their own solutions - there being no right or wrong answer. They were presented with a number of scenarios or situations that required them to draw on their own experiences to suggest original ideas and possible alternatives.

METHODOLOGY

Action research is practical, flexible research aimed at improving educational practice (Costello, 2003). The fact that I chose to carry out action research in the setting where I worked meant that I could use the study to further develop my own practice in this way. Action research is more than just an isolated study. It is cyclical in nature, whereby the individual carrying out the research is systematically reflecting on and improving practice (Denscombe, 1998; Dick, 2002). The ultimate aim for the study was to improve practice within the area of designerly thinking in the Nursery, and improving practice is a key factor of action research (Cohen, Manion & Morrison, 2000). Although the study consisted of a small sample of 6 children, the conclusions made will effect how D&T is then developed further in the Nursery. From the findings, I will consider the implications for the planning in the Nursery and make changes accordingly. In keeping with a data-driven approach (Dick, 2000), although D&T, and in particular designerly thinking, was identified as the initial focus for the research, the nature of the study ensured there was flexibility for other important issues to be considered if they emerged. This was to prevent the study from having too narrow a focus that may occur with theory driven practice (Costello, 2003). As the focus was changing practice within just one setting, this was 'technical research' as identified by Zuber-Skerritt (1996). This process of action research followed an initial cycle of 10 months, with the in-school research being carried out in the second half of the Autumn term, 2005.

A series of 6 activities was devised, specifically aimed at developing designerly thinking. Each activity was based around one of the items in the Early Years Material Kit (TTS, 2003) which the children had not seen before. Bentley & Campbell (1990) believe a crucial element of D&T involves child centred problem solving activities. I embraced this approach, while incorporating aspects of previous work carried out on designerly thinking (Benson, 2003; Treleven, 2004; Benson, 2005). Each activity had a problem and a resolution - it is the 'problem' part of the activity that is directly related to the design of the product – for the children were encouraged to think about the product itself, the user of the product, its need, and how it is used. Throughout each activity the children were encouraged to use all their senses as they discussed the product. The following is a brief description of each activity:

Slipper Activity (Activity 1)

The children are told that a slipper has been found on the way to the classroom. They are asked to give ideas as to who it may belong to and why (the problem). They are asked what they think should be done with the slipper now and why (the resolution).

Truck Activity (Activity 2)

It is explained to the children that I have bought this truck for my little brother's birthday and I am not sure if he will like it – what do the children think (the problem)? Why do they think he will like it/not like it? What other things in the Nursery might he like to play with? At a suitable moment tell the children that they have been very helpful and I think that my brother will like it and I will let them know next time I see them (the resolution).

Cup Activity (Activity 3)

The children are shown a plastic beaker - when I go out I always get thirsty but if I put my cup (indicate the beaker) in my bag, the water always spills out – my mum gave me this cup – get out the upside down cup – do you think it will work any better? Do you have any other suggestions for what I can do? (the problem) Then depending on the responses of the children – get some water and test the cup to find out that it does not spill (resolution) or encourage the children to think of an original alternative solution.

Book Activity (Activity 4)

Explain to the children that when it was my brother's birthday he was given a book. Tell them that my brother thought the children in the Nursery might like it – why do you think he said that (the problem)? Ask the children to think about why they might like it, what is different about it – how does it work? Then say to the children that I will read it to the others at story time to see if they like it or not (resolution).

Glasses Activity (Activity 5)

Explain to the children that on the way to Nursery I found these (indicate the safety glasses) in the corridor. Ask the children whose glasses they think they might be and what are they used for (the problem)? Try to reach a consensus for whose they might be and then explain that I will go and ask that person and let them know the next time I see them (the resolution).

Toy Activity (Activity 6)

Explain to the children that my granny found this in her loft (indicate the toy) and said she used to play with it but did not have time to tell me what she did with it. Ask the children if they have any ideas? Do they think it was a fun toy? What could you do to make it more fun? Finish by showing the children some other toys from the past in a book and talking about them (the resolution).

QUESTIONING

There is a great deal of research concerned with the effect of questioning on children's performance. Despite being part of the group, I decided questioning would be appropriate in moving along the children's thinking. The benefits of asking open ended questions as opposed to closed 'one word answer' questions are widely discussed by many authors (Siraj-Blatchford, J & Siraj-Blatchford, I, 1995; Craft, 1997; Fisher, 2005). In using open ended questions my aim was to allow the children 'psychological freedom' (Rogers, 1961), freeing them from worries that they must reach the 'right' answer, a worry that Fisher (2005) suggests restricts children's motivation to offer their own ideas.

While using questioning to scaffold children's learning provides important support for the child (Smith, 1994), Moyles (1989) identifies the importance of listening and responding appropriately to children's explanations. This was a key factor in the current study in assessing what the children understood and what they were developing in terms of different types of thinking.

Restricting children's responses by asking ill thought out questions may limit their ability to think critically (Fisher, 2005). It is vital therefore, to ask questions that stimulate and extend children's thinking. Bloom's taxonomy places importance on asking questions that will elicit synthesis and evaluative responses if children's thinking is to be developed (Fisher, 2005). Therefore, in order to ensure that I had a range of questions, I divided possible questions into the different levels of thinking. This ensured that during each activity, there were opportunities for the children to respond using different levels of thought. I also considered the 'productive questions' recorded by Benson (2004) that were directly related to young children involved in an activity based on designed and made products. These questions focused on the product, its user and its purpose and they had a significant influence on the types of questions I decided to use.

FINDINGS

In order to gain an idea of any general trends in the findings, I carried out a count analysis on the children's responses. This is identified by Gillham (2000) as a useful summative tool. One disadvantage that emerged with this small group research was that if children were absent, there were gaps in responses for different children. This meant it was extremely difficult to follow the children's responses individually through the tasks. Therefore, it was more practical to take the responses to each activity and analyse them as a group. As discussed by Gillham (2000), when carrying out case study research you can begin by having a broad expectation of what may be indicated in the findings. However, as the case study is not restrained by experimental conditions, the findings may not necessarily follow an expected route. As is the nature of data driven action research, any issues that arise during the research need to be considered (Dick, 2002), and personal experience was an issue that although not having been identified as an initial focus, during an initial read through of the transcripts it was shown to be a reoccurring theme throughout the responses. Although the transcriptions are an accurate, reliable method of recording (Hopkins, 2002), it is my own interpretation of responses through my notes and subsequent evaluations that are more subjective. However, this must be taken into account as the following discussion is based upon these. Firstly, I categorised the responses into what I considered to be the correct category based on the following assumptions:

- Personal experience – responses that related directly to aspects of the children's lives outside of the Nursery.
- Critical thinking – in order for a response to be considered an example of critical thought, the children's responses would be considered comparison and justification responses when based on the higher order thinking skills identified by Bloom (1956).
- Creative thinking – On reviewing various views about the nature of creativity (De Bono, 1987; Moyles, 1989; Siraj-Blatchford & MacLeod - Brudenell, 2003), I concluded that the definition given by Fisher (2005) relating creativity to the ability to generate new ideas and explore alternatives would be the basis of what I deem to be creativity in the current study. Thus, a response fitted into this category if it was an example of the children using knowledge they already have and applying it to something new.
- Designerly thinking – reflects developing children's awareness of the designed and made world, looking at aspects such as features of a product, purpose and user (Benson, 2003). The children's responses were classified in this category if they related to one of those areas.

Using these four categories when examining the transcripts, I identified the following key findings:

- the children were enthusiastic and interested in the activities they participated in;
- most of the evidence gathered was verbal but significant non-verbal behaviour was also recorded;
- personal experience had a significant effect on the responses of the children;
- the activities provided numerous examples of children engaging in critical thinking, including higher order thinking as identified by Bloom (1956);
- the activities provided an ideal starting point for encouraging independent thinking and creative thought;
- although the number of designerly thinking responses did not increase with each activity, 44% of verbal comments related directly to designerly thinking – meaning they referred to a product, a user or the purpose of a product.

In the following section, the above findings will be discussed in more detail, ending with a particular focus on the findings relating to designerly thinking - this being the initial focus of the study.

DISCUSSION

Children's Interest in the Activities

The findings of the current study reflect those of the Benson (2003) research in that the children were motivated by the products they were shown. The children were interested and eager to come to the table when I arrived each week. This was reflected in their comments which included '*When are we going to help you today?* (Child 1) and '*Can we go over there (indicating where we sat last time) now?*' (Child 2) - both before the start of the truck activity. These comments show that the activities stimulated interest and enthusiasm among the group. Using the phrase from Woods (1990), the children's attention was 'captured' as opposed to being 'recruited'. That is, they wanted to be involved in the activity, rather than being told that they must participate in the activity, as often happens once children start Key Stage 1 (Edwards & Knight 1994).

Verbal/Non Verbal Evidence

Children will respond more if an activity is of interest to them (Craft, 1997). However, some children choose not to respond in a verbal manner but show interest through their behaviour (Moyles, 1989). This was particularly significant for child 5 in the study; although he said very little, he carried out a number of activities in silence. An example of this is when in the cup activity, child 5 left the table, while his actions showed he remained aware of what was happening in the group. He chose to leave the table to find his own materials to make a water container and only rejoined the group when he needed assistance. He had worked independently and developed his own ideas, but as he did so without interacting with anyone else, he was responding well to the activity, just not in a verbal way.

Personal Experience

As discussed by Fleer, Jane & Robbins (2004), the home environment has a big impact on the designerly thinking skills of young children. As they highlight, many young children have a wealth of experience with designed and made products before they start school and will continue building on this experience within their home environment. These individual, personal experiences will have an effect on how they respond to stimuli once they reach school. Wells (1988) and Roden (1999) both talk of personal experience as being a crucial factor in the way that young children will approach a problem.

The current study supports the view that a child's own experiences will have an effect on how they respond within the classroom. The children referred to family members: '*Sometimes my mum wears glasses.*' (child 2 glasses activity), and objects at home: '*I have a plastic cup, a fairy cup, it has fairies on.*' (child 6, cup activity). The children had knowledge about how to behave in certain situations: '*I think you need to wrap it up and give it to him.*' (child 3, truck activity). Child 3 has clearly had some personal experience of wrapping birthday presents in preparation for giving them to someone else.

It was also not only direct experience that affected decisions made by the children, Child 3 drew on the story of Cinderella as a solution to my 'slipper' problem: '*Knock on everyone's door 'Is that your slipper?' If they say no, then go to the next one. They say yes*'. The parallel to the fairy tale was quite clear; thus child 3 had drawn on this aspect of a previously heard story and applied it to the current problem, being able to select a solution that might work. She was clearly using her prior knowledge to help her make sense of the current situation. That she was able to do so quite confidently is a reflection of the independent, confident learners that the Foundation Stage framework encourages (DfEE/QCA, 2000).

Critical Thinking

Parker Rees (1997) suggests that young children are capable of higher level thinking, despite sometimes finding it difficult answering questions requiring lower level thought, as defined by Bloom (1956). In the current study, there was evidence to support this finding. One example was the truck activity, where at first the children seemed to find it difficult to explain why my brother would like a plastic truck. Although they had all agreed that he would like it, there was no response to the question '*Why do you think he will like it?*' This might initially have suggested that the children were in fact unable to engage in thinking beyond application level (they used what they already knew about trucks to make a decision). However, by the end of the truck activity, two of the children were engaging in higher level thinking as they could justify decisions they had made. I asked them if there was anything in the Nursery my brother might like and all the children were able to choose something and child 1 and child 2 were able to tell me why he would like it – child 1 '*He might like this*', 'Why?' '*Is he a boy?*' 'Yes' '*cause boys like dragons*'. Child 2 suggests he will like a hat as '*...he can wear it outside*'. With only a little adult intervention, both child 1 and child 2 began extending their own thinking – by linking what I had asked them to do, with reasons for their personal choices. The above examples of logical thought imply that even young children can engage in more complex thinking than Piaget suggests.

Creative Thinking

Fisher (2005) describes creative thinking as the ability to generate new ideas and explore alternatives. Young children are creative thinkers and given an object may not use it as it was originally intended (Edwards & Knight, 1994). Their ability to think creatively and identify possible alternatives should be encouraged and developed within the classroom (Siraj-Blatchford & MacLeod - Brudenell, 2003; Fisher, 2005).

Despite the fact that the activities were not developed with the aim of actively promoting creativity, there was clear evidence that the children were able to develop new ideas and alternatives independently. One example relates to the toy activity, where the children were shown an unfamiliar wooden toy they had not seen before. The children began to think of possible ideas for what it could be, which led them to attribute their own use for the product. Child 2 states, '*You make it fly. It looks like a kite*', suggesting it may be a kite, and child 5 also agrees with this idea saying, '*I can do it. It can fly up to the sky*'. Both children agree that the toy is to be used for air travel, perhaps basing this on past experience as the top of the toy is similar to a propeller. Thus the children are using information they already know – that propellers are things on helicopters which fly and are creatively adapting it to the new situation.

The children were also able to generate and develop alternatives independently. Young children will express their creativity through their free flow play (Moyles, 1989; Bruce, 1991) and this was reflected in this project. Completely self initiated, the children decided to experiment with different combinations of 'bottles' and 'lids' in order to make a successful water container that did not leak water. They were involved, enthusiastic and were developing their own learning. Independently they came up with a number of possible solutions to my original problem, thus were thinking creatively (Fisher, 2005). My initial input complete, the children took the activity further, while I was able to observe these confident, independent 'designers'. This is a clear example of the children being '*interested, excited and motivated to learn*' (DfEE/QCA 2000).

Designerly Thinking

Designerly thinking is a type of thinking based on evaluating a product's design. Nearly half of all the comments made by the children were judged to be related to designerly thinking, a positive finding that suggests that the activities were appropriate and did in fact encourage the children to think in a designerly way. There was no apparent pattern as to whether the children focused more on user, product or purpose. For some activities the comments were evenly spread between the 3 areas, or in the case of the book activity, the designerly comments were exclusively about the product.

Despite not having much in their Nursery experience relating to designed and made products – the children displayed a wealth of design knowledge. When first looking at a product, the children were confident in identifying the ones I had to show them, with the exception of the wooden toy. They were able to point out the physical attributes, such as the materials the products were made of, and the colour and size. '*It's pink.*' (child 1, slipper activity) '*wood, no plastic*' (child 1, truck activity). This supports the findings of Anning (1993) that colour, shape and material are often the focus for questions early years' practitioners give to children.

When moving on to the purpose of the product, the children were able to come up with sensible explanations relating to the purpose of the product. Slippers were 'rough' so '*it makes everybody not fall*' (child 2, slipper activity) and they put '*bendy stuff in them to make them (glasses) bend*' (child 6, glasses activity). The truck '*is a building thing where you pick things up*' (child 1, truck activity). These were instances that showed that the children were thinking about what an item was going to be used for and were able to make judgements accordingly.

The user of the product was the most open ended aspect of the questioning as questions relating to the object itself and what it was used for were more obvious to the children as they had the object in front of them. This might not be so for the user, so any judgements made relating to the user of the product were completely independent. The 'problem' was often related directly to user - '*Who do you think it belongs to?*' (slipper activity) and were designed to encourage the children to come up with their own ideas.

The children were clearly able to consider the user of the products shown to them and the range of possible users varied widely. Sometimes the children's choices were limited to the sex of the user – '*a girl*' (Child 2, slipper activity) '*a boy*' (child 5, glasses activity) but there are many examples that show this was not always the case and the children were able to provide actual users for the products. In the glasses activity, child 1 said that the glasses may belong to '*Bob the builder, 'cause I watched him cut with these on*'. This quote shows that child 1 has related the glasses to a familiar character and is able to name him as a possible user. Family members, '*my grandma*' (child 6), and members of the community, '*dentist*' (child 5) and '*doctor*' (child 3), were also considered by the children in the glasses activity. Child 6 was able to identify herself as a possible user of products: '*I have a plastic cup, a fairy cup, it has fairies on*' (child 6, cup activity). '*When it's sunny I wear sunglasses.*' (child 6, glasses activity). In both examples child 6 was talking about things that relate directly to her. She was identifying herself as a possible user for both the glasses and the cup – showing that in these instances – personal experience relates directly to designerly thinking.

To conclude therefore, although the findings suggest that the children were not acquiring a brand new skill, it appears that the activities were accessing and developing skills that the children had already begun to acquire. Certainly the responses given by the children in this project suggest that given a starting point, the children are able to provide evidence of what is quite clearly designerly thought. There are examples to show that young children were already familiar with questions relating to the designed and made world, for example looking at colour and shape, but that they are equally adept at considering both the purpose and user of a product.

Implications for Personal Future Practice

The implications for my own setting are that the activities need to be planned into the Nursery long term and medium term plans, so that as different groups of children move through the Nursery, all get the chance to participate in designerly thinking activities. For the children who have participated in the current study, I now need to work with the Nursery staff to decide how to move these children on and the best way of doing that. I will need to make decisions with the staff as how best to use the information that has been discussed.

Implications for Foundation Stage Practitioners

For designerly thinking activities to become a part of everyday teaching in the Foundation Stage, practitioners would need to feel confident in developing designerly thinking skills. This would possibly raise staff training issues regarding how to question and respond to children in the best possible way. It is most likely, based on previous research (Anning, 2003: Benson, 2003) that the majority of Foundation Stage practitioners are unaccustomed to this type of activity.

Implications for the Foundation Stage Curriculum

The Foundation Stage curriculum aims to be a cross curricular one (DfEE/QCA 2000) and in the Foundation Stage the influence of D&T can be found across all areas of the curriculum. Despite this however, the current Foundation Stage curriculum guidance does not place emphasis on how D&T can be carried out at this stage of children's education (Benson, 2005). Whilst it is explicit in the Early Learning goals, there is little reference to product evaluation skills or designerly thinking in the Guidance materials (DfEE/QCA 2000). For designerly thinking activities to be carried out regularly in all Foundation settings, there would need to be recognition of the importance of designerly thinking in the Foundation Stage Curriculum. These findings support a case that designerly thinking activities should be given such acknowledgement. Some children receive these experiences at home (Fleer, Jane & Robbins, 2004) but many have to rely on the activities they are provided with at school (Anning, 2003).

CONCLUSION

This study highlights how designerly thinking based activities can make a positive contribution to children's early experiences of evaluating designed and made products. The children have been encouraged to think and respond to objects and were able to confidently contribute their own ideas, making their own decisions that they could then justify. They were able to talk about attributes of different products, explore who might use the product and make suggestions relating to the use of the product, giving appropriate explanations. The findings certainly suggest that these types of activities are the building blocks on which later evaluative work can be based and should be embraced and recognised as vital starting points for children in early years' settings and beyond. However, the use of these authentic tasks provides an important context within which children's thinking can be developed for use across the whole curriculum and as a useful life skill as they are faced with situations in which critical and evaluative thinking are integral to decision making.

NOTES

[1] Early Years Setting – this is any educational place for 3–5 year olds. It can include a nursery, a play-group, or a child minder.

REFERENCES

Anning, A. (1993). Technological capability in primary classrooms. In *International Design and Technology educational research 1993*. Loughborough, UK: Loughborough University.

Anning, A. (2003). Young children meaning making through drawing. *Journal of Design and Technology Education, 9*(3), 138–144.
Baynes, K. (1994). *Designerly play*. Loughborough: Loughborough University of Technology.
Benson, C. (2001). *Research into the early years framework for design and technology*. London: QCA.
Benson, C. (2003). Developing 'designerly' thinking in the Foundation Stage. In C. Benson, M. Martin & W. Till (Eds.), *Fourth International Primary Design and Technology conference 2003* (pp. 5–7). Birmingham, UK: CRIPT at University of Central England.
Benson, C. (2004). Creativity: Caught or taught? *Journal of Design and Technology Education, 9*(3), 138–145.
Benson, C. (2005). Developing designerly thinking in the Foundation Stage - Perceived impact upon teachers' practice and children's learning. In C. Benson, S. Lawson & W. Till (Eds.), *Fifth International Primary Design and Technology conference 2005* (pp. 15–18). Birmingham, UK: CRIPT at University of Central England.
Bentley, M., & Campbell, J. (1990). Primary Design and Technology: An introduction. In M. Bentley, J. Campbell, A. Lewis & M. Sullivan (Eds.), *Primary design and technology in practice*. Harlow: Longmans.
Bloom, B. S. (1956). *Taxonomy of educational objectives, Vol. 1: The cognitive domain*. New York: McKay.
Bruce, T. (1991). *Time to play in early childhood education*. London: Hodder and Stoughton.
Bruner, J. (1983). *Child talk*. New York: Norton.
Cohen, L., Manion, L., & Morrison, K. (2000). *Research methods in education* (5th ed.). London: Routeldge Falmer Press.
Costello, P. J. M. (2003). *Action research*. London: Continuum.
Craft, A. (1997). *Can you teach creativity?* Nottingham: Education Now.
Craft, A. (2007). Changes in the landscape of creativity in education. In A. Wilson (Ed.), *Creativity in primary education* (2nd ed.). Exeter: Learning Matters.
Dansky, J., & Silverman, I. (1973). Effects of play on associative fluency in preschool aged children. *Developmental Psychology, 9*, 38–43.
De Bono, E. (1987). *Letters to thinkers*. London: Harrap.
Denscombe, M. (1998). *The good research guide*. Buckingham: Open University Press.
Department for Education and Employment (DfEE)/QCA. (2000). *Curriculum guidance for the foundation stage*. London: Author.
Department for Education and Skills (DfES)/QCA. (2007). *The early years foundation stage*. Nottingham: Author.
Dick, B. (2000). A beginner's guide to action research. In P. J. M. Costello (Ed.), (2003). *Action research*. London: Continuum.
Dick, B. (2002). *Action research: Action and research*. Retrieved June 22, 2006, from www.scu.edu.au/schools/gcm/ar/arp/aandr.html
Edwards, A., & Knight, P. (1994). *Effective early years education: Teaching young children*. Maidenhead: Open University Press.
Fisher, R. (2005). *Teaching children to think*. Cheltenham: Nelson Thorne.
Fleer, M., Jane, B., & Robbins, J. (2004). Designerly thinking: Locating technology education within the early years' curriculum. *Early Childhood Folio*, (8), 29–33.
Gillham, B. (2000). *Case study research methods*. London: Continuum.
Guilford, J. P. (1957, March). Creative abilities in the arts. *Psychological review, 64*(2), 110–118.
Hopkins, D. (2002). *A teacher's guide to classroom research*. Maidenhead: Open University Press.
Hughes, M. (1996). *Teaching and learning in changing times*. London: RoutledgeFarmer.
McGarrigle, J., & Donaldson, M. (1974). Conservation Accidents. *Cognition, 3*(4), 1974–1975, 341–350.
Moyles, J. (1989). *Just playing? The role and status of play in early childhood education*. Maidenhead: Open University Press.
Parker-Rees, R. (1997). Making sense and made sense: Design and Technology and the playful construction of meaning in the early years. *Early Years, 18*(1), 5–8.
Tassoni, P., & Hucker, K. (2000). *Planning play and the early years*. Oxford: Heinemann.

QCA/DfES. (2000). *Curriculum guidance for the foundation stage.* London: Author.

Roden, C. (1999). How children's problem solving strategies develop at Key Stage 1. *Journal of Design and Technology Education, 4*(1), 21–7.

Rogers, C. (1961). *On becoming a person.* Boston: Houghton Mifflin.

Rogers, M., & Stables, K. (2001). Providing evidence of capability in literacy and Design and Technology in both year 2 and year 6 children: Alternative framework for assessment. In C. Benson, M. Martin, & W. Till (Eds.), *Third International Primary Design and Technology conference 2001* (pp. 165–169). Birmingham, UK: CRIPT at University of Central England.

School Curriculum and Assessment Authority (SCAA). (1996). *Desirable outcomes for children's learning.* London: DfEE/SCAA.

Senesi, P. (1998). Technological knowledge, concepts and attitudes in nursery school. In J. S. Smit & E. W. L. Norman (Eds.), *International conference on Design and Technology educational research and curriculum development 1998* (pp. 27–31). Loughborough, UK: Loughborough University.

Shayer, M., & Adey, P. (1981). *Towards a science of science teaching.* London: Heinemann.

Siraj-Blatchford, J., & Siraj-Blatchford, I. (1995). *Educating the whole child.* Buckingham: Open University Press.

Siraj-Blatchford, J., & MacLeod-Brudenell, I. (2003). *Supporting science, design and technology in the early years.* Maidenhead: Open University Press.

Smith, P. (1994). Play and the uses of play. In J. Moyles (Ed.), *The excellence of play.* Buckingham: Open University Press.

Spendlove, D. (2005) Creativity in education. *Design and Technology Education: An International Journal, 10*(2), 9–19.

Sternberg, R. J. (2001). What is the common thread of creativity? *American Psychologist, 56,* 360–362.

Tassoni, P., & Hucker, K. (2000). *Planning play and the early years.* London: David Fulton Publishers.

Treleven, T. (2004). *Investigating the provision of design and technology in selected early years settings.* Unpublished M.A. Ed. research project. Birmingham: University of Central England in Birmingham.

TTS. (2003). *Technology teaching systems home page.* Retrieved October 20, 2003, from http:www.tts-group.co.uk

Wells, G. (1986). *The meaning makers – children learning language and using language to learn.* London: Hodder and Stoughton.

Woods, D. (1990). *How children think and learn.* Oxford: Basil Blackwell.

Zuber-Skerritt, O. (1996). *New directions in action research.* London: Falmer Press.

Professor Clare Benson
Director, CRIPT
Faculty of Education, Law and Social Sciences
Birmingham City University
England

Tara Treleven
Primary School Teacher
South London
England

PASCALE BRANDT-POMARES

13. TECHNOLOGICAL EDUCATION

The Issue of Information Retrieval Via the Internet

INTRODUCTION

This chapter focuses on what pupils know about the information they retrieve on the Internet and about what is at stake in the learning process that teaching and especially technology education have to recognise. Information is considered as a major element in personality construction giving the activity of information retrieval on the Internet a specific status in the access to knowledge.

The perception pupils have about the information retrieved on the Internet is examined by means of a questionnaire. This questionnaire gives prominence to the existing confusion about the quality of the information that can be consulted on the Internet and especially on the Wikipedia website. This phenomenon justifies the necessity of investigating information retrieval within technology education.

CONTEXT AND THEORETICAL APPROACH OF THIS STUDY

It is frequently said that children who have grown up with the Internet have no trouble mastering the use of computer systems. And yet, the difficulties quite young pupils (around 11 years old) are often confronted with while doing information retrieval on apparently simple subjects can be very surprising. Before reporting on the research project, it is important to investigate further the subject especially since computer activity is a relatively frequent activity undertaken by young people. Initially, we will review how access to information plays a part in personality construction, particularly if it is via the Internet and with this perspective in mind, we will consider how the situation is in the education system, from the point of view of both information retrieval as education technology and as a subject taught in technology education.

Information on the Internet and Personality Construction

It is obvious that access to information as a source of learning and knowledge constitutes a major part of personality development. The psychological instruments (Vygotski, 1985) to which the Internet gives access can only contribute to the personality construction process. According to Simondon (2005), information lies at the core of the individuation process linking information, communication and formation. Admittedly, the fact that the information is available on a declaration basis,

C. Benson and J. Lunt (eds.), International Handbook of Primary Technology Education: Reviewing the Past Twenty Years. 151–166.

does not guarantee the learning of the concept nor of the underlying notion, but the thinking process necessary for comprehension and the language communicating the information are indistinguishable. In any case whether the information is right or wrong can modify, complete, or increase the knowledge of a student. From this point of view, the significance of the information is essential as it contributes to cognitive development. Whether it is widespread or not is not to be taken into consideration in this process. This missing concept has prevailed over the Wikipedia project where control by many people does not necessarily guarantee the nature of its intent.

Information on, and from, the Internet: the Example - Wikipedia Website

In 2001, Jimmy Wales and Larry Sanger using wiki technology created the Wikipedia website (*http://www.wikipedia.org/*) which belongs to the Wikimedia foundation. This website was first conceived in English and then, very quickly a French version became available. In the beginning, the designers saw what this technology could provide on the Internet - a free universal, multilingual encyclopaedia that was written collaboratively. This notion is displayed on the home page of the Wikipedia website:

"the free encyclopaedia that anyone can edit"

The cofounder Jimmy Wales had planned that Wikipedia could reach a quality level at least equivalent to that of the Encyclopaedia Britannica. But the original aim of creating a freely distributed encyclopaedia that anyone can improve makes it the basic strength of Wikipedia but also its great weakness. Articles in Wikipedia are written in a collaborative way, which means that the contributions can come from any person who wants to create or modify web pages under one control - its own self regulation. The intention, which is in itself praiseworthy, can in fact allow any assertion to be published, and as long as nobody else decides to modify it, it can be a total nonsense. The anonymous nature and the lack of control provide the conditions necessary for a quickly evolving website but it does not allow any reliable guarantee on exact direction and meanings. Despite the "pseudo" supervision by the virtual community, mistakes can very well slip by and only a specialist would recognise the errors, even though Wikipedia warns the users by announcing on the website that one of the characteristics of Wikipedia is to be based on mistrust: all wikipedians are encouraged to be careful and critical about the quality of other participants' contributions[1].

No media can escape the problem of information control and the only known defence of traditional editing is the one the cofounder Larry Sanger took into account when launching the Citizendium project in September 2006 (http://en.citizendium.org/wiki/Main_Page). This project is similar to Wikipedia, the difference being calling on experts to guide the public when writing articles. These experts check the articles, as their aim is to avoid mistakes that are not systematically controlled on the Wikipedia website. It is therefore possible to say that quality editing prevails over speed editing.

With Wikipedia, we are really able to see the importance that pupils give to the quality of the information retrieved via the Internet because there is no editorial line valid on many sites, and not only on Wikipedia but furthermore on all personal sites and blogs. Their preliminary knowledge does not always give them the means to separate the wheat from the chaff at this key stage of human development: childhood and adolescence. It is therefore not surprising that the education system needs to have a particular interest in this matter.

What are the Issues Relating to the Internet and the Education System?

The distance between access to and learning of knowledge gives a crucial aspect to the matter of education. Access to knowledge can only constitute a first step in the learning process, but the initial access ensures a first meeting with the knowledge itself but not necessarily with its acquisition. Therefore, access to different media through which information is transmitted cannot be imposed. If documents uploaded onto the Internet are to be submitted to the critique of all, then this requires an education in critical evaluation and becomes a fundamental educational matter. It comes down to giving each and everyone the means to judge the quality of documents transmitted through the Internet.

Numerous injunctions from the French National Ministry of Education lead towards the integration of ICT in the overall teaching of every subject. In general, this policy emerges through impact initiatives or initiatives related to technology education. On a European level, most initiatives fall within this framework[2]. Each in their own way brings an answer to the problem of ICT integration to education (La Borderie & Perriault , 2002). The Educnet[3] website, launched in 1998, gathers for example reference texts on the matter, examples of teaching practices, and lists of resources. In some way, it represents the showcase of governmental educative measures and is available through the Ministry website: www.éducation.fr. The B2i (Computer and Internet Certificate) and the Educnet projects are part of initiatives organised or favoured by the French National Ministry of Education.

The educaunet program. Educaunet is part of a European initiative supported in France by the Clémi[4]. Within the Ministry, the Clémi is a centre in charge of conceiving and developing educational programs concerning the media. The aim of this program is to develop education as a means of defence against the risks of accessing 'wrong' information on the Internet. Solutions integrated into computer systems are available and can be transferred to the computers, the selection of which can be watched (filtering software, authorised access, browser security system, etc.). This 'human-machine' coupling (Deforge, 1985) is not the one that has been selected by the Educaunet program. The opposite approach has been chosen. It focuses on favouring education by warning pupil users of the possible risks, while teaching them how to protect themselves whenever possible. It hopes to avoid the trauma of shocking pictures but also to allow them to seize the originality of this kind of communication where you have trouble identifying the persons you are dealing with, to become self-sufficient, critical and responsible, and able to appreciate the resources of the Internet while skilfully escaping its pitfalls[5].

The éducaunet program is mainly centred on prevention and can only function if young people are being supported. The list of risks as well as the list of answers is a long one (Chenevez, 2001). The reasoning used in the éducaunet program seems interesting as it shows the connection established between education and consulting websites, from the skills that the pupils are missing to be able to use the Internet critically as well as a cautious adult, who is, to some extent, able to protect him/herself against harmful, improper or illicit contents, fraudulent, deceiving, and false practices or manipulative behaviours which can hide amongst the unquestioned resources of the network, and are not always easy to locate.

The B2i: computer and internet certificate. While other training institutions (some of the Greta[6], IUT, Universities or private organisations) or other countries (for example Italy, Austria, Ireland, Norway, Sweden) have already adopted the PCI (Computer Driving Licence) originating from associations[7]; the National Education in France has chosen to set up a Computer and Internet Certificate, the B2i. It was created in 2000 in order to validate the skills acquired by pupils in primary school (level 1) and in secondary school (level 2) and the abilities mastered in the computer field of ICT. This certificate is not a qualification but an attestation. A Computer and Internet Certificate for teaching, the C2i2e was planned for teachers.

> Level 1 validates the following:
>
> The pupil can use the information and communication technologies available in school in a self-sufficient and well reasoned way; to read and produce documents; to retrieve information that is useful; and to communicate through electronic mail. To be able to do so, the pupil has to have command of the first basis of computer culture in its technological and citizen dimensions.
>
> Level 2 validates the following:
>
> The pupil has command over all the skills covered in level 1 of the certificate. Besides, he/she is able to control usual computer tools in order to produce, communicate, get informed and organize his/her own documents. He/she, in particular, is able to organize complex documents consisting of tables, formulas and links with other documents. In order to proceed, he/she has to know the elements of computer culture directly useful to him/her (specific vocabulary, essential technical characteristics, and methods for data processing through computer systems). He/she can perceive the limits relative to the use of nominative information as well as the limits determined by the respect of intellectual property.

The teacher in charge of his/her class in primary school is responsible for the B2i level 1, while in secondary school (although recommendations suggest that any teacher can undertake the work) in reality, the technology teacher often takes care of it, as certain computer technology units of the technology education program are the same as the skills acquired in the B2i. However, using the Internet is also part of accumulation, classification, and dissemination of information activities.

Internet: a subject taught in library studies. In secondary school, the work of librarian teaching staff in the library and document centre (CDI) is mainly of an educational nature and has to be conducted in close collaboration with the teachers. Their actions contribute towards increasing the use of books and in a more general fashion, towards information sources. To this end, they favour the introduction of pupils to reading graphic and audiovisual documents and the use of the computer, in collaboration with the teachers within the framework of the programs. But, now-here is it stated a detailed description, with a teaching programme, as to what the librarian staff really has to teach.

Information retrieval in various subjects. Even in school, information retrieval allows activities to be set up rather easily without a clear learning process. It certainly presents numerous challenges. However, many sites, amongst which 'educnet' can be found, offer to take into account requirements to evaluate the information which tends to show that this assessment is not obvious but requires a special learning process. In order to know if information can be trusted, this method relies on a set of questions. While following this method, the first question to ask is: 'Who?' This question focuses directly on the source of the information. To be able to identify this source represents a major element concerning the assessment on the reliability of the information. When the author is identified (whether a person or a legal entity), it becomes possible to think about his/her competences. The TLD (Top Level Domains) give us information on the editor: .org, .net indicates an association or a non profitable organisation. The .com TLD concerns websites dedicated to the Internet network itself.

The second question this method suggests is: What? It concerns information accuracy. In order to answer this question, we need to check if the information found is just a collection of facts or whether it is attested and well argued, and whether it is bringing the information closer to the kind of audience the site is aimed at (for example, specialist; initiated; any kind of audience). In fact information found on a website visited by specialists and elaborated by specialists that has links to other websites where we will find this information, is likely to be more accurate than information that is published by an individual, even if the latter information is as valuable as any other.

The third question: 'Where?' relates to the origin of the information. With reference to legislation for instance, it seems appropriate to choose first, information provided by a website located in a geographical area connected to the required information. Generally, the website address is useful, as it brings valuable information as to the origin of the website. However, it is useful to know that the country code (for example TLD like .fr or .uk) is not necessarily the code of the country where the person has published the website, but the TLD linked to the server.

The fourth question relates to time: 'When?' It is necessary to know how frequently the information is updated. Of course different kinds of information require different frequencies of updates. It is therefore necessary to check the date when the article was written and, if it is the case that the article could be outdated, it would be appropriate to look for a more recent one.

For the fifth question: 'How?' It is necessary to investigate how the information is put at the disposal of the user, how the document is structured and if the information is written or backed up by figures. If we are dealing with written information, then we must look under what kind of form this information is offered: whether it is on an assertion basis, to assess something or to be controversial in order to start a debate.

Finally, the purpose of this method is to get us to question the reasons why the website offers this information. This is the question: 'Why?' What is the aim of this website? Does the author provide indicators about the purpose of the information he/she is publishing on the network (such as passion, personnel training, altruism, proselytism)? Is the information free? Is there advertising on the website? If so, is it connected to the information you are looking for? Is the advertising clearly separated from the content of the documents?

This method suggests a certain number of questions we have to ask ourselves, in order to assess properly the credibility of the information found, whatever the subject studied.

Information retrieval and technological education. The fact that information retrieval on the Internet is useful in all subjects taught makes it difficult to identify a vertical continuum within technological teaching. The fact that there are constant fluctuations concerning the place of information technologies in the teaching programmes in secondary school bears witness to this problem. Even though there are many opportunities to undertake information retrieval, it is somehow difficult to know how it is taught exactly and if the right idea of the knowledge about information retrieval is passed down by the teachers.

Technological analysis of information retrieval activity. Kolmayer (1998) considered that the information retrieval situation is a problematic task in which the cyclic aspect of information retrieval (Dinet, Rouet, & Passerault, 1998) in the database has been observed by different writers; in particular in the cognitive model of Guthrie (1988) consisting of 5 phases (formation of objectives, selection, information extraction, integration, and recycling) and the evaluation-selection-treatment process from Rouet & Tricot (1998). One of the most significant elements of this cyclic aspect consists of the modification of the objectives currently used (Marchioni, 1992 ; Osmont, 1992 ; Villame, 1994).The cognitive processes of planning, control, and regulation (Rouet & Tricot, 1998) that are brought into operation during the activity of information retrieval in formalised databases, remain true with the use of the Internet. But, pupil activity differs from expert users' activity in a variety of aspects (Brandt-Pomares, 2003).

Information retrieval techniques and therefore the technological knowledge relating to information retrieval on the Internet is a matter of using data processing equipment, such as a computer, browser software, and the Internet network. A particular analysis was made of this process and it has enabled the elaboration of expert knowledge (Brandt-Pomares, 2003) linked to the instrumental origin of the tool (Rabardel, 1995) and therefore to the acquisition of research mode techniques in implementation schemes, tool choices, selection in the result lists, data base on which the research is based, of hypermedia browsing, of website notion and network referencing. Thirteen year old pupils frequently use the Internet, despite having evaluated the tool only through experimentation and having limited perception as to the potentialities of that tool. They underestimate what the tool can do (Norman, 1988, 1993; Leplat, 2000). It justifies the fact their practices must be enriched and widened in relation to keywords (Blondel, Schowb & Kempf 2001), multiple requests (Hôlscher & Strube, 1999), and regulation processes (Brandt-Pomares, 2003) as students turn to these less than other users during research retrieval via the Internet.

Analysis of pupils' activity. Retschitzki & Gurtner (1996) underline the powerful motivation seen in most children when they spend time on computer activities. This is easily verified in the classroom when observing how speedily pupils leave their desks to settle in front of the computer screen. When 13 years old pupils are placed in an information retrieval situation, the link between what they find and what they are looking for, is based on an evaluation of the nature and the relevance of the information they have access to. The information is not only linked to the implementation of the tool, if a number of elements are intrinsic to the artefact, others are not depending on it, for instance, the wording of a website address is dependant on the Internet organisation (official websites, personal websites, trademark websites, etc.). This wording can give indications of the sources of the retrieved information. But the different sources do not seem a determining factor in pupil practice. It would seem that only the existence of the information gives it a probative strength. The natural tendency to believe what is asserted (Goffard & Goffard, 1998) belongs to the credulity of childhood. Children first believe the propositions that are made are true, before they can step back and consider them, something which is favoured by education, as children do not spontaneously question the nature and sources of information. The fact that pupils consider the information seen on the Internet to be true, leads us to think that it is difficult for them to discriminate between right and wrong information. This is a worrying fact as anybody can create their own internet website and publish it after writing any information - true or false. Some websites can give free access to any kind of information even if it is illegal. Besides which, anybody can modify the content of some websites or articles, as we have seen with Wikipedia which is not the most unreliable website there is. Information retrieval is very much linked to the actual subject of the retrieval, to the informative nature of what is retrieved. Regarding this, we are able to underline that the efficiency of the retrieval made by pupils, depends greatly on their initial knowledge (Rouet & Tricot, 1998).

It is therefore important to investigate the hypothesis that 11 years old cannot see the difference between various information sources and that they hold information published on the Wikipedia website to be true, when using the Internet to retrieve information.

EMPIRICAL RESEARCH

A survey of 47 pupils in two classes of 11 years old has been conducted: the aim being to verify the hypothesis that:
- pupils go on the Internet to undertake information retrieval;
- pupils do not make any distinction between different sources of information;
- pupils believe that information published on the Wikipedia website is true.

Creation of the Questionnaire

The questionnaire contained 13 questions. The aims of these questions are detailed below:
- Question 1: Do you know when you are connected to the Internet?

This question allows us to know if the pupil knows at which exact moment, he/she is on the Internet or if he/she is on a local system (CD-rom, hard disk, local network) and to know if he/she is able to identify the information source he/she is consulting.
- Question 2: You have to do some research to trace the history of boats in order to give a presentation in your technology class. How are you going to proceed? Give number 1 for the means you will use first, number 2 for the second way, number 3 for the third way etc.

This question will allow us to know which research methods the pupils are going to prioritise.
- Question 3: Give a score from 0 to 10 if you think the information you have found will be right in any case.

This question will allow us to know if the pupil gives more value to one source of information rather than another.
- Question 4: Do you think that what is written on the Internet is verified, and if so by whom?

Here we want to examine whether the pupil thinks that the information available on the Internet is verified and if so by whom.
- Question 5: Do you distinguish between something you read in a book and something read on the Internet? If you do, what difference/s do you identify?

This question will allow us to examine if pupils give more credit to books and if they see a difference between what is written in a book (which is then not easily modified) and what is written on the Internet.
- Question 6: At the end of your presentation, will you be able to create a website or a blog on boat history?

This question will allow us to examine whether the pupil is conscious of the fact that he can himself publish a website on a subject he has little knowledge about. If the pupil answers "yes", he/she should then know that what is said on the Internet is not necessarily written by experts and that some of the information is wrong. However, there is a risk that the pupil will interpret this question in another way "will you be able to" as it is a rather wide concept.

- Question 7: Have you heard of Wikipedia? If you have, can you write an article on Wikipedia? Can you modify an article on Wikipedia?

The aim of this question is to give us information on what the pupils know about Wikipedia.

- Question 8: Do you know who can create a website?
- Question 9: Do you think a company can create a website?
- Question 10: Do you think an association can create a website?
- Question 11: Do you think anybody can create a website?

Questions 8 to 11 will allow us to know if the pupils are aware who is publishing the information on the Internet.

- Questions 12: While looking at the document, tell me how many hulls a catamaran has and tell me (if you can) in what year the catamaran was invented.

How many hulls ? Year of invention?

- Question 13: Give a score from 0 to10 (0 meaning I am not sure at all about the information I found to 10 I am absolutely sure about the information I found). If you want to explain why you gave this score you can do it below.

The document to be consulted in question 12 is a screen copy of the Wikipedia website concerning the definition of the catamaran.

These two questions will allow us to see up to which point the pupils believe what is said on the Internet and if they stop their information retrieval as soon as they have found the answer to their questions.

Analysis of Answers to the Questionnaire

In answer to the first question, 42 pupils have indicated that they knew when they were connected to the Internet. On the other hand, out of those 42 pupils, only half were able to give an answer that indicated that they know when they are actually on the Internet. In fact, only 21 pupils really know when they are connected to the Internet (cf. Table 1).

Table 1. Answers to question 1

Question 1: Do you know when you are connected on the Net?	
Yes	No
21	26

Answers to question 2 are gathered in Table 2.

According to the pupils' classification, the results have been graded in the chart and each result has been multiplied by a value according to this classification (6 points for the 1st method, 5 points for the 2nd method, 4 points for the 3rd method, 3 points for the 4th method, 2 points for the 5th method, 1 point for the 6th method).

Table 2. Answers to question 2

Question 2: You have to do some research to trace the history of boats in order to give a presentation in your technology class. How are you going to proceed? Give number 1 for the method you will use first, number 2 for the second way, number 3 for the third way etc.						
Position	*1st*	*2nd*	*3rd*	*4th*	*5th*	*6th*
Web site	15	21	9	1	1	
Wikipedia	14	13	12	4	3	1
Books	9	7	13	10	5	3
Parents	4	5	8	17	10	3
Friend	5	1	5	10	13	13
Blog				5	15	27

The results are:

Internet websites: 236
Wikipedia: 216
Books: 184
Parents: 155
Friends: 124
Blogs: 72.

The results show that pupils mainly use the Internet for their research work. Table 2 and the previous results underline the fact that pupils favour information retrieval via the Internet (Internet websites then Wikipedia) rather than research in books (books or encyclopaedia). Next we find information given by parents and friends used and lastly the pupils' research on blogs.

Results to question 3 are gathered in table 3.

Table 3. Answers to question 3

Question 3: Give a score from 0 to 10 if you think the information you have found will be right in any case											
Scores	*0*	*1*	*2*	*3*	*4*	*5*	*6*	*7*	*8*	*9*	*10*
Books					2	2	1	3	10	9	20
Internet	1	1	2	1	1	1	1	8	8	11	12
Wikipedia	1		1	1	1	6	5	3	5	12	12
Parents	1		1	2	1	10	4	11	9	2	6
Friend				3	5	10	7	6	4	9	3
Blog	16	6	10	4	1	3	2	2	1	1	1

The average scores given by this chart are:

Books: 8.6
Internet websites: 7.7
Wikipedia: 7.6

Parents: 6.7
Friend: 6.5
Blog: 2.3.

These results prove that pupils are really sure of the information they found, when it was found in a book rather than on the Internet. They are also rather confident in the answers given by their parents or their friends while they have little confidence in the results of researches from blogs.

The results to question 4 (cf. Table 4) allow us to see that 18 pupils think that what is said on the Internet is verified. In the second part of the question, when asked "who verifies the information published on the Internet?" 6 answers show that pupils do not know who verifies the content of the web pages. 4 answers concern the 'owner' of the website. In other answers we find: teachers, parents, parental guidance, computer specialists or the police. We notice a certain confusion in the pupils' minds.

Table 4. Answers to question 4

Question 4: Do you think that what is written on the Internet has been verified?	
Yes	No
18	29
If it has, by whom?	
I don't know	6
The website owner	4
Parental guidance	2
Parents	2
Teachers	2
Webmasters	1
The police	1

It is possible to group the answers from Table 5. We can see that a little fewer than half the pupils note a difference between what can be read in a book and what can be read on the Internet. Among those 22 pupils differentiating between what is written on the Internet and what is written in books, 12 seem to think in the same way as one who wrote the following answer:

> "What is said in a book is necessarily right, and you can't be sure of the result of what is on the Internet."

It must be said that 3 pupils answered "information found in books and on the Internet are not the same" and two of them answered "that there are less explanations in a book than on the Internet".

To the question n°6, at the end of your presentation, will you be able to create a website or a blog on boat history? 25 pupils think that they will not be able to create a website. But as we had anticipated, this question can have resulted in this kind of answer for reasons we are not really able to distinguish. Actually some pupils seem to have answered that they could not publish on the Internet because of lack of appropriate skills. The pupils have thought that they did not know enough on the history of boats to create a website on this subject which is not what we were trying to find out. It is therefore difficult to interpret their answers to this question. We note once more the intricacy between technical questions of publishing on the Internet and the nature of the information to be published itself.

Table 5. Answers to question 5

Question 5: Can you distinguish between something read in a book and something on the Internet?	
Yes	23
No	24
If you answered yes, what difference/s do you think there are?	
In the book, it will be more right	12
Not the same information	4
There are less explanations in books	2
Books can't lie	2
Books are more serious	1
The writer has been through enough effort to write the book	1

Responding to the 7th question (Table 6), 25 pupils knew of the Wikipedia website. On the other hand, 15 children amongst them (meaning a majority) stated they could not modify or create an article on this website. The analysis of the answers to questions 8, 9, 10 and 11 (Do you know who can create a website? Do you think a company can create a website? Do you think an association can create a website? Do you think anybody can create a website?) allows us to find out that 33 pupils are able to express that they know who can create a website. On the other hand, 46 pupils say that a company or an association can create a website, but only 29 pupils are positive that anybody can create a website. Finally, a large number of pupils (18) are left who seem to believe that individuals cannot create a website.

In answer to the 12th question, Table 7 shows that 28 pupils answer that catamarans have two or three hulls (from what they could read on the screen print of the Wikipedia web page), 9 said that catamarans have three hulls and 10 said catamarans had 2 hulls. Concerning the year when the catamaran was invented, all the pupils answered that it was invented in 1700, whereas the text indicated that the English pirate and adventurer William Dampier was the first one to describe a catamaran around 1690.

Concerning question 13, the answers of the 47 pupils show that they consider the information they found to be right. They are almost sure regarding the number of hulls and completely sure regarding the invention date. As it happens, the total number of answers comes to an average of 8.4 (from 0 "I am not sure at all about this information" to 10 "I am completely sure about this information") while 25 pupils give a score of 10. Regarding the invention date, 27 pupils give a score of 10 and the total average of the answers is 8.3.

Table 6. Answers to question 7

Question 7: Do you know Wikipedia?	
Yes	25
No	22
Can you create or modify an article on Wikipedia?	
Yes	10
No	15

Table 7. Answers to question 12

Question 12: How many hulls are there on a catamaran?	
2 hulls	10
3 hulls	9
2 or 3 hulls	28

Synthesis of Results

The pupils' answers to the questionnaire have allowed us to know first of all (Question 1) that not all pupils are able to understand the difference between documents consulted directly on line on the Internet or documents consulted in other media, such as cd-rom, local network, and local copy.

Observing questions 2 and 3, it is obvious that pupils favour research on the Internet more than research in books or encyclopaedias. Even so, and that makes the results rather reassuring, they give more credit to a result found in an encyclopaedia than to a result found on the Internet. This proves that they can see a difference, even though it is difficult to define the nature of this difference. Perhaps a false link is made between credibility and the effort required to retrieve the information. It is rather strange to note that numerous pupils (29/47) think that what is written on the Internet is not verified, while pupils trust in the majority of the information given on the Internet (cf. Table 3, score 7.7/10). This means that a certain number of pupils know that the information is not verified but despite knowing this, still believe in it. We have previously noticed that it was very difficult to make use of answers to question 6. However, this question, or rather the way it was answered, attests once more to the intricacy between the nature of published information and the technology used to publish it.

Considering Wikipedia, on average, pupils will trust this website, despite a number of pupils knowing that information can be modified or created on Wikipedia (10/47 cf. results to question 7), (cf. average score of 7.6/10 obtained from Table 3, average score of 8.3 and 8.4 obtained on Question 13.)

CONCLUSION

Although we had a limited number of questionnaires, the results analysis allows us to write that the hypothesis stating that pupils use the Internet to do information retrieval is verified, as 32 pupils out of 47 use the Internet (Wikipedia or other websites) in first or second position when asked which means they would use first to prepare a presentation (cf. answer to question 2).

Our second hypothesis concerned the fact that pupils would not be able to distinguish between information sources. Relying on the results of questions 3 to 5, we can say that this hypothesis is not really verified, which is rather a reassuring fact even though it does not relate to all the pupils. As a matter of fact, we notice that pupils give more importance to information that is not provided by the Internet. They also know (a large majority) that information on the Internet is not always verified. Moreover, pupils tell us that they do not find the same information in books or in encyclopaedias but grant more importance to the content of books. On the other hand, concerning the third hypothesis saying "pupils think information published on the Wikipedia website is true", we can say this hypothesis is verified. As a fact, when looking closely at the results related to question 3 or to question 13, we are able to say that pupils trust the results coming from this website despite the way in which editing occurs.

In fact, pupils have a correct intuition that all information is more or less the same, but are not sufficiently equipped to find out from the information they have access to, via the Internet, the one that they can identify as more trustworthy than any other. When looking at the results of the questionnaires, it seems important to warn pupils about the risks they are taking while retrieving information on the Internet and in particular on the Wikipedia Website. The quantity and variety of information available on this website does not allow them to realise that the people who have written the articles published on this website, are not necessarily expert and competent in the subject but it does not make them question the quality of the information published on this website.

Debate about Technology Education

Involving pupils in real activity is necessary to improve teaching. If technology education has a role to play relating to the use of the Internet, including information retrieval, it has to be structured around a real activity, in which pupils can learn about information retrieval and teachers can teach the key learning objectives. The fact that information retrieval on the Internet is useful in all subjects makes it difficult to identify a vertical continuum within technology education. Even though activities involving information retrieval are frequently practised, it is always difficult to know exactly how to teach it, if it involves a real kind of teaching, and the exact knowledge that should be taught by teachers. Nevertheless, at a specific time in schooling, technology education should contribute to the teaching of information retrieval in order to improve the efficiency of pupil research.

NOTES

[1] http://fr.wikipedia.org/wiki/Wikip%C3%A9dia:R%C3%A8gles

[2] European Schoolnet: European program gathering this type of initiative can be consulted at http://www.eun.org

[3] http:/www.educnet.education.fr

[4] Training to understand and use media in the classroom is part of these priority missions http://www.clemi.org/formation.html. It gives advice and follow up in class projects. It trains the education staff. It is a conciliation and mediation institution. It produces educational tools http://www.clemi.org/formation.html

[5] EDUCAUNET, critical education about the Internet and the risks linked to its use, consult http://www.educaunet.org/versions/francais.html.

[6] GRETA is a group of secondary schools within the National Education system in 6000 locations across France.

[7] The PCI is an international independent standard acknowledged by the European Union. Created by the CEPIS (The Council of European Professional Informatics Societies [http://www.cepis.org]) it is held up by the EDCL foundation (European Computer Driving Licence Foundation [http://www.ecdl.com/main/index.php]). PCI website: http://www.pci.tm.fr/sitepcie/html/instit_education.htm

REFERENCES

Blondel, F. M., Schwob. M., & Kempf, O. (2001, octobre). *Pratiques de recherche d'informations sur internet en sciences physiques: difficultés et compétences, Actes des deuxièmes rencontres scientifiques de l'ARDIST* (pp. 97–106). Marseille, Skôlé, Numéro Hors Série.

Brandt-Pomares, P. (2003). *Les nouvelles technologies de l'information et de la communication dans les enseignements technologiques; de l'organisation des savoirs aux conditions d'étude: didactique de la consultation d'information*, thèse, Université de Provence.

Chenevez, O. (2001, Septembre). Internet, est-ce que c'est dangereux ? L'Odyssée des réseaux, *Les cahiers pédagogiques, 396.*

Deforge, Y. (1985). *Technologie et génétique de l'objet industriel.* Paris: Maloine, Université de Compiègne.

Dinet, J., Rouet, J.-F., Passerault, J.-M. (1998, 15–17 oct). Les «nouveaux outils» de recherche documentaire sont-ils compatibles avec les stratégies cognitives des élèves?. In J.-F. Rouet & B. De La Passadière (Eds.), *Hypermédias et apprentissages, Actes du quatrième colloque* (pp. 149–161). Poitiers, Paris: INRP EPI (Technologies nouvelles et éducation).

Goffard, M., & Goffard, S. (1998). *Les activités de documentation en physique et en chimie*. Paris: Colin.

Guthrie, J. T. (1988). Locating information in documents: Examination of cognitive model. *Reading Research Quarterly*, *23*, 178–199.

Hölscher, C., & Strube, G. (1999). Web search behavior of Internet experts and newbies. In *WWW9 proceedings*. Retrieved from http://www9.org/w9cdrom/81/81.html

Kolmayer, E. (1998, 15–17 Oct). Démarche d'interrogation documentaire et navigation. In J.-F. Rouet & B. De La Passadière (Eds.), *Hypermédias et apprentissages, Actes du quatrième colloque* (pp. 121–134). Poitiers, Paris: INRP EPI(technologies nouvelles et éducation).

Leplat, J. (2000). *L'analyse psychologique de l'activité en ergonomie. Aperçu sur son évolution, ses modèles et ses méthodes* (p. 164). Toulouse: Octares.

Marchioni, G. (1992). Interfaces for end-user information seeking. *Journal of the American Society for Information Science*, *43*, 156–163.

Norman, D. A. (1988). *The psychology of every things.* New York: Doubleday, Basic Books.

Norman, D. A. (1993). *Things that make us smart : Defending human attributes in the age of the machine.* Reading, MA: Addison-Wesley.

Osmont, B. (1992). *Itinéraires cognitifs et structuration du lexique. Études d'interrogation de banques de données* (p. 448). Thèse, Université de Paris 8, Département des Sciences du Langage.

Perriault, J., & La Borderie, R. (2002). Dir., *Education et nouvelles technologies: Théorie et pratiques*, Nathan, 128 Éducation.

Rabardel, P. (1995). *Les hommes et les technologies* (p. 239). Approche cognitive des instruments contemporains. Paris: Armand Colin.

Retschitzki, J., & Gurtner, J.-L. (1996). *L'enfant et l'ordinateur*. Bruxelles: Mardaga.
Rouet, J. F., & Tricot A. (1998). Chercher de l'information dans un hypertexte: vers un modèle des processus cognitifs. In A. Tricot & J.-F. Rouet (Eds.), *Les hypermédias. Approches cognitives et ergonomiques* (PP. 57–74). Paris: Hermes.
Simondon, G. (2005). *L'individuation à la lumière des notions de forme et d'information*. Grenoble: Jérôme Million.
Villame, T. (1994). *Modélisation des activités de recherche d'information dans les bases de données et conception d'une aide informatique*, Thèse, Université de Paris 13, Laboratoire communication et travail.
Vygotski, L. (1985). *Pensée et langage* (1ère éd., p. 416) (Terrains). Paris: Messidor/Éditions sociales.

Pascale Brandt-Pomares
Gestepro UMR ADEF
IUFM University of Provence
France

ALAN CROSS

14. IN SEARCH OF A PEDAGOGY FOR PRIMARY DESIGN AND TECHNOLOGY

INTRODUCTION

During a recent visit to a primary school a teacher commented about the many changes in Design and Technology over her twenty one year career. She described how the subject had changed and of course its title. She claimed that, "the name of the subject does not matter, the children still make buggies!" An experienced non-specialist teacher, she had seen Design and Technology in English primary schools develop from the non-compulsory craft, Design and Technology (CDT) and home economics (HE) taught in a proportion of schools to a statutory element of a National Curriculum (DfEE/QCA, 1999) for all children of five to eleven years, entitled Design and Technology. It is interesting to reflect that while many of the Design and Technology tasks presently given to primary pupils might be indistinguishable from those made in the 1970s there have been shifts, for example in teacher acceptance, in teacher knowledge and in the emphasis of what is taught under this heading of Design and Technology. These shifts and developments have the potential to inform others implementing Technology education around the world.

Internationally primary school Design and Technology is a developing subject. Its subject status is, however, variable around the world; neither is there agreement about what to call it (Keirl, 2006a). What is taught in Design and Technology in England equates roughly to what, around the world, is more commonly called Technology education (Keirl, 2006b). The exclusion or inclusion of Design and Technology in the curriculum and its nature reflects much about the place and value of aspects of technology and design within a culture. Human beings have always made much of technologies and periodically develop them - for example, new biotechnologies and nanotechnology. Preschool and primary aged pupils around the world do things, design things, make things and develop things. From tree houses to sand pits and spaceships to other imagined worlds, they adapt and invent. Educationalists and others seek to harness this drive to design and construct and develop it for the good of learners and for the communities in which they live. Developments in England mirror dilemmas experienced elsewhere and present a curriculum design scenario which has the potential to inform those interested in Design and Technology education in schools.

TWENTY YEARS IN ENGLAND

As a teacher and teacher trainer of pre- and post- qualification students during this period, I have been motivated to explore the way Design and Technology is taught

C. Benson and J. Lunt (eds.), International Handbook of Primary Technology Education: Reviewing the Past Twenty Years. 167–180.

by primary teachers, most of whom have had modest or no training for the subject. There has been very little research into primary Design and Technology (Kimbell et al., 1996) and in particular investigation of teaching methods. Explanations for this might relate to the perceived status of the subject and its relative priority. Perhaps it is viewed as unproblematic; it may be seen as necessitating simple demonstration to pupils of, for example, the safe use of a tool followed by opportunity for them to practise the skill.

The research reported here explored the teaching of Design and Technology and the related areas of pupil autonomy and teacher direction in primary school Design and Technology lessons. If pupil autonomy is, as is suggested, a helpful focus for those considering primary Technology education, it might be that consideration of the place of pupil autonomy in the Design and Technology experience of children around the world is a powerful device for giving direction and perhaps raising educational expectations.

The past twenty years has seen Design and Technology develop from its previous incarnation, Craft, Design and Technology (CDT) into a subject created in 1989 (DES/WO, 1989). CDT had brought together subjects dealing with resistant materials such as wood and metal under an umbrella which emphasised aspects of design (HMI, 1987). The 1990 National Curriculum (DES, 1989) saw the transformation of CDT into Design and Technology with the inclusion of food and textiles technology. It was not long however before there were reservations about the nature and size of the statutory requirements and so the mid 1990s saw Design and Technology metamorphose from its original form to a somewhat amended and reduced National Curriculum subject (DFE/WO, 1995; DfEE/QCA, 1999). The changes above may have contributed to problems encountered by the subject but other factors have limited the development of Design and Technology including government's continued emphasis on so called 'standards' in English and mathematics (Ward, 2002). This may be particularly so in primary education where the limited personal knowledge and confidence of non-specialists (DATA, 2003; HMI, 2005) and lack of curricular time (Ward, 2002) have been highly influential.

Teaching methods in primary education have been a focus of continual though limited debate in England. Little apparent interest in the study of teaching led Simon (1981) to ask, "Why no pedagogy in England?" and Alexander (2004) to review this question concluding that the situation was not improved. He pointed to the Primary Strategy (DfES, 2003) which he felt at best had produced a pseudo pedagogy. Such influences have produced the most recent shifts in the use of teaching methods in English primary schooling in a rather prescriptive approach to the teaching of English and mathematics (DfEE, 1998; DfEE 1999; DfES, 2003). These strategies have been produced by expert groups rather than resulting directly from research into the efficacy of different teaching methods. They have been prescriptive although later this was disguised by a language of flexibility (Alexander, 2004). The relative employment and success of different teaching methods in Design and Technology remains a somewhat unexplored area.

From a state in the 1980s where primary enthusiasts taught CDT we now have a situation where Design and Technology is a compulsory subject taught to all children aged 5–11. Statutory requirements have resulted in some uniformity. This might be seen by some as an advantage but there are disadvantages. Whilst uniformity plays to the political rhetoric of entitlement it can have a limiting affect on creativity.

Some intimation about teaching can be drawn from three key governmental documents relating to Design and Technology. Firstly the National Curriculum (DfEE/QCA, 1999) stipulates that 'teachers teach' and that pupils 'should be taught to' each of which is followed by a series of broad subject related objectives like those below.

> ...that teachers should teach five to seven year old pupils to:
> a) select tools, techniques and materials for making their product from a range suggested by the teacher;

In the section on evaluation the document states that five to seven year olds should be taught to:

> b) identify what they could have done differently or how they could improve their work in the future. (DfEE/QCA, 1999, 92–95)

The terminology here is mixed. There is intimation, for example, that pupils should talk about their ideas in Design and Technology but at the same time, be taught to measure and identify improvements. The verbs imply a role for the teacher, for example, pupils will 'be taught to...' and at another point pupils will '...explore sensory qualities...'. The document indicates that the role of the teacher is a purveyor of knowledge. Such language may contribute to tensions in a subject which has at its heart a need for creativity (Nicholl, 2004). Concern in the early nineties about lack of advice for primary teachers led to the publication of the second key document, the Non Statutory Guidance: Design and Technology (NCC, 1990). Unfortunately this document did not provide significant guidance about teaching methods and it has not been replicated or updated.

Primary teachers seeking advice about how to teach Design and Technology in the last ten years have tended to refer to the third key governmental document, the non-statutory national scheme of work for primary Design and Technology (QCA/DfEE, 2001). Here they will find statements such as the following learning outcomes for five and six year olds in Unit 1A Moving Pictures:

> Most pupils will: ... have used tools safely to make a moving picture that incorporates a simple lever...
>
> and
>
> Some children will have progressed further and will: ...have developed their own ideas from the initial starting points ...

The scheme's (QCA/DfEE, 2001) emphasis on the experience of pupils implies a high degree of pupil activity and thus indicates something about the teacher's role. The teacher appears to be initiator, instructor and guide. It does not provide however, as do not other widely used published materials (e.g. DATA, 1999), clear advice on teaching methods. Is it better to instruct or demonstrate? What are the features of a well framed pupil task? How can a teacher promote increased pupil autonomy?

Thus the past twenty years has seen Design and Technology establishing itself in primary education and yet there has been little in the way of development or training about how the subject might be taught.

TEACHING BEHAVIOURS

Teaching can be viewed as a natural human activity (McNamara, 1994) although few would deny it is a complex activity (Stenhouse, 1975; Gipps et al., 2000). Dialogue about teaching has in the past been somewhat limited in England (Simon, 1981). This is exemplified and compounded by the considerable variety in the use of and confusion about terms such as 'teaching', 'teaching style', 'teaching methods', 'pedagogy', 'didactics' and 'instruction'. Stenhouse (1975) considered 'teaching' to encompass any strategies utilised by a school to promote learning. Gipps et al. (2000) usefully defined teaching as: 'a presentation in various ways of adult-decided knowledge, skills and understanding'. One form of presentation, instruction, is an important part of teaching as it includes the giving of commands or teaching a 'correct and non-negotiable way of doing or going about something' (Gipps et al., 2000, p. 39). Stenhouse saw limitations of instruction as an exclusive approach:

> Teaching is not merely instruction, but the systematic promotion of learning by whatever means. (Stenhouse, 1975, p. 24)

Alexander's distinction between components of pedagogy including teaching methods and classroom organisation provide at least a basis for considering teacher behaviour (Alexander, 1992). He divided what he called the observable practice of teachers under four subheadings shown in Figure 1.

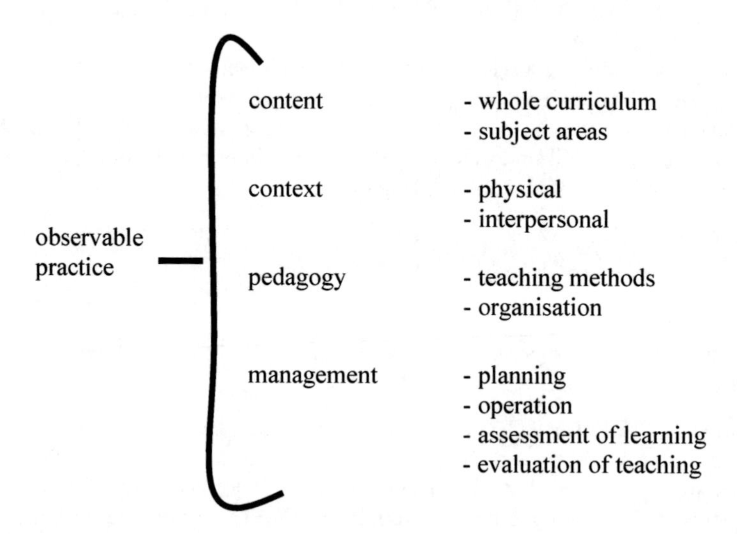

Figure 1. Part of Alexander's conceptual framework of educational practice (Alexander, 1992, p. 198).

Our thinking about teaching is influenced by our understanding of the nature of learning and knowledge. Traditional approaches to teaching see it as a process of transmission, one in which the teacher, the expert, passes on knowledge to the learner. The learner in this model is seen as passive. Such a positivist view of teaching may therefore assume that if the teacher accounts fully for each learner and for each concept to be taught then learning is guaranteed. Such a view ignores the involvement of others in the activity, most notably the pupil and his or her peers. Social constructivism (Adams, 2006) offers a perspective which seeks to account for teacher, learner and social setting. However, as Gipps (1992) acknowledges, constructivist approaches offer little guidance in terms of a clear pedagogy and therefore teaching behaviour.

In order to encourage a dialogue about teaching Design and Technology, a distinction was drawn between three sets of factors which teachers could vary in Design and Technology lessons (Cross, 2000). Examples were taken from lesson observations in the research reported in this chapter and placed in the table below (Figure 2) to illustrate how these categories might distinguish teaching behaviours. An additional distinction is included, based on Alexander's (1992) framework illustrated in Figure 1 which distinguishes between the organisational aspects or variables of a lesson and teacher behaviours which can also be seen as variable. The usefulness of these categories is open to question as they are unlikely to be comprehensive or sufficiently sophisticated. However a framework or taxonomy of some kind which is directly relatable to Design and Technology may be of use in assisting, for example, non specialists. The greatest potential contribution may be as a reflective tool for teacher self-review. Figure 2 illustrates an initial step towards such a framework which might encourage teachers to distinguish between the various features or variables in a lesson.

	Organisational Variables e.g. space allocation grouping of pupils time allocation resource allocation	
	Teaching Variables	
teacher directly engaged	Use of media which enable teaching	the pupil activity/task
e.g. explanation/clarification discussion questions instruction providing tasks listening	e.g. artefacts chalkboard construction kit computer tools context e.g. a story materials e.g. wood	e.g. recording designing making testing evaluating communicating

Figure 2. A possible framework for the consideration of aspects of teaching in Design and Technology (Cross, 2000).

In order for teachers, in particular non-specialists, to cater for this relatively new practical subject and the different learning styles of pupils, consideration of teaching methods in Design and Technology may offer considerable benefits. Whilst useful in some ways, those employing any reductionist approach must be alert to the dangers associated with attempts to atomise teaching. Teaching can be seen as far greater than the sum of its competencies. For Brophy & Evertson (1976) teaching was an orchestration of many factors. The inadequacy of any such framework is revealed when, for example, the social context of the classroom is considered. Davies & Elmer (2001) and others (e.g. Hennessy & Murphy, 1999; Rowell, 2002) emphasise the need to go beyond the utterances and behaviours of the teacher when considering the teaching of Design and Technology. The significance of the social context, including verbal and non verbal interactions between the lesson participants is increasingly being recognised. Teacher behaviours are part of, and highly influential in, this context. They are powerful forms of communication. The expectation is that greater understanding of their detail in Design and Technology lessons will assist all those interested in the teaching and learning of Design and Technology.

The significance of social interaction was observed in the lesson observations and has been recognised by others (Hennessy & Murphy, 1999; Rowell, 2002). Hennessy & Murphy link this to the nature of Design and Technology, recognising the potential for the use of open-ended activities which have, they suggest, the potential to increase independence or autonomy. In the small sample reported here there was a clear difference in the amount of pupil interaction. For example, in the first lesson considered below, which was less open and more tightly controlled by the teacher, there was less pupil interaction.

PUPIL AUTONOMY AND TEACHER DIRECTION IN PRIMARY DESIGN AND TECHNOLOGY

The research summarised in this chapter sought to examine teaching methods as part of pedagogy. Aspects which became significant during the research were pupil autonomy and teacher direction. Pupil autonomy proved in the analysis and discussion to be a useful lens to consider primary Design and Technology education. Boud's (1987) definition of autonomy appeared meaningful for Design and Technology:

> …to develop in individuals the ability to make their own decisions about what they think and do. (p. 18)

Thus autonomy might be considered to be a capacity for self-governance within Design and Technology. Such autonomy does not necessarily equate to independence though the terms are sometimes used interchangeably. Kimbell (1982) was referring to facets of autonomy when he recognised the challenge this creates for teachers:

> The child will move in small steps from almost total dependence on the teacher to almost total independence… The function of the teacher… is to steer children towards the goal of independent thought and action, along the tortuous path of guided or supported freedom. (Kimbell, 1982, p. 16)

My interest in Design and Technology pedagogy has been, at least partly, an attempt to illuminate that tortuous path.

As we explore ideas about pupil autonomy we might consider whether primary aged pupils have the capacity for autonomy. Putting aside questions about whether any of us ever achieve true autonomy we might usefully ask whether by the age of eleven children are likely to become autonomous in subjects such as Design and Technology. Children are unlikely by the age of eleven to become autonomous individuals in many areas such as their own care and welfare. Is it reasonable however to expect a developing autonomy in Design and Technology and a schooling which actively promotes this?

Candy (1987) challenged ideas of what might be expected of an autonomous person:

> We don't want people who can find resources for themselves, manage their own time … …rather we want learners who know and understand enough to distinguish plausible from implausible knowledge claims or convincing from unconvincing evidence. (p. 60)

Design and Technology teachers might reply that they want both. Perhaps Candy is assisting here with a distinction between independence (gathering resources etc.) and autonomy. Boud (1987) commented on what might be called degrees of autonomy. He felt that as teachers we should be less concerned about the magnitude of autonomy than in the direction of change towards self reliance. It might therefore be possible to construct a Design and Technology curriculum which would aim to move children in this direction.

How this is manifested in lessons may be seen in the extent to which pupils are aware of their situation, the context, their understanding, skills, their capability and how these are employed in Design and Technology tasks. Truly autonomous individuals in this subject would have at least a developing understanding of Design and Technology, a bank of knowledge and a set of skills. They would have a growing awareness of their limitations and might seek expertise, help and advice appropriately. Such individuals could operate effectively with others and, if required, alone.

The way in which pupils move towards autonomy in Design and Technology varies. When learning to cut wood with a saw many move to a level of independence or autonomy in a fairly predictable way. For example, an initial demonstration will be followed by numerous attempts to cut wood with increasing accuracy, safety and ease. Further opportunities to observe others sawing, demonstrating use of different saws, advice about tool care and safety, hints and corrections from a teacher lead towards autonomy. However, when a skill such as designing is considered, the path of progression is less clear and is influenced by many factors including whether the children are working alone, the familiarity of the context, the familiarity of materials and tools and whether they have had previous design experience.

The role of the teacher and the decisions he or she makes, for example about framing tasks and the degree of teacher support and direction, are likely to have a very powerful influence on learning in Design and Technology. This raises questions relating to teacher intervention. Is the best way to become autonomous in Design and Technology to be left to get on with things yourself? Does this question illustrate the confusion between autonomy and aspects of independence?

THE RESEARCH

The research summarised below focused on aspects of pedagogy in Design and Technology (Cross, 2000). This began in the late 1990s as an examination of case studies - firstly through interviews with primary school teachers and later lesson observations. Two of these case studies will assist here in demonstrating the contrast seen in lessons. The results illustrated a range of teaching methods and their effects, one example of which is the impact on pupil autonomy in Design and Technology lessons.

The Interviews

The seven teachers interviewed were all Design and Technology coordinators within their schools. They were selected following recommendation from local authority school inspectors. It was assumed that they would be best placed to have some knowledge of the teaching and learning of Design and Technology in their schools. Whilst this was true with respect to aspects such as resources and planning, it was much less the case when aspects of pedagogy were discussed. All seven were much happier talking about their own teaching and so this was how the interviews progressed. Initial comments from teachers illustrated different approaches and something of the potential influence of the teacher:

> …but I tell them what the product will be… (and)… they have some freedom but there are always constraints.
> I tell them what the product will be, I give them the criteria, lead the research and focus their attention.
> I just give them the brief, for example make an animal, their imaginations run free!

Teachers have the opportunity in their task design to construct a very structured framework to which pupils respond within tight parameters, or to provide a much more open brief. The teachers here talked about variety within their own teaching of Design and Technology, for example on some occasions they would give the pupils more options and what they often referred to as 'freedom'. Other factors they varied were pupil choice about resources and the extent to which a task or questions were open or closed. The teachers interviewed referred to: time, materials, grouping and task design. They assumed a high degree of control over aspects such as content, context selection, timing, grouping and selection of materials and tools. Often they claimed that aspects such as time were out of their control; they cited the length of lessons as constraining their options. Some complained about specific topics in the national scheme of work (QCA/DfEE, 2001) sometimes citing replacement topics. Autonomy was they said, most often given to the pupils in terms of choices children might make within materials available, tools and decoration of products such as choosing colours and designs for the side of a boat or vehicle.

The teachers appeared to prefer the term 'freedom' but saw negative as well as positive consequences. All agreed that freedom for pupils was a good thing; however, one teacher complained that giving the children what she called "complete freedom" had in the past led to "poor results". These results were seen in terms of the product produced. Comments from two other teachers indicated some negativity towards the notion of autonomy:

> Teacher A: They think that they can make anything!
> Teacher B: I find that they cannot constructively create with materials, they are very wasteful.

Play within Design and Technology was raised as a possible indicator of autonomous action. Most of the teachers felt that there was a place for play but could not give examples of play extending beyond trial and error in Design and Technology lessons. One expressed real concern about the idea of play in a Design and Technology lesson:

> Teacher: I'd feel unhappy about play, just playing!

Lesson Observations

The interviews were followed by seven lesson observations, the most striking feature of which was their differences. The lessons described below, with different age groups illustrate some of these differences. The first focused on the design and construction of tree decorations prior to Christmas and the second was a lesson about the design and construction of treasure chests used by pirates. The first Design and Technology lesson was highly directed as illustrated by the teacher's introduction to the lesson:

> The sheet says design your Christmas tree decoration. You've got to draw a picture for your design. Now to give you some idea (holds up a snowman template) you might want to do a snowman.

The teacher explained that they should draw a picture as a design, colouring the picture to match the colour of felt material they planned to use. She went to considerable trouble to explain that the decorations should be of moderate size and simple in design.

> …as you've got to sew it, if it's too small, it could be fiddly and so you need a simple design.

Another example was shown of a simple shape which would present few problems with sewing. The teacher went on to talk about other aspects of the making process.

> We are not going to sew all the way around, we are only going to sew part of the way around (partly circles template with finger) because we are going to fill it.

The teachers interviewed identified a wide range of methods used in the teaching of Design and Technology. These included specific teacher activity such as explaining, describing, demonstrating and leading discussion. There was also a broader use of strategies by the teacher, for example, grouping the pupils, providing leadership with regard to the context of the Design and Technology, tailoring timing within a lesson and varying the choice of tools and materials. The variation in these features appears to allow teachers to direct the attention and activity of pupils.

This lesson was typified by careful explanation including examples and the teacher checking with pupils to determine that they understood. The pupils appeared happy and enjoyed the opportunity to work with a range of materials including fabric and needles. The teacher was pleased at the end of the lesson as all the children had chosen a shape for their decoration. All had chosen to use the templates provided. They had selected fabric and thread colour combinations and had begun to sew. The teacher felt that she had given them some choice. She admitted that this was limited but justified this referring to previous experience which had seen pupils "waste time" constructing very small decorations or ones so complex that stitching was almost impossible. Time appeared to be a very powerful factor here as five forty minute timetable slots were available and this activity was viewed as one where pupils might not complete their task in the time available.

The second case study involved a longer lesson conducted over one morning with six and seven year olds. The teacher first reminded the twenty-eight pupils of two books she had recently read to them about pirates and then, with the use of the books, she drew the attention of the class to the pirates' need for a treasure box. The fifteen minute introduction emphasised the characteristics of and need for a treasure box.

> Teacher: and he's sitting on his treasure chest can you see? It's not a box; it's not a straightforward box like a cuboid, is it?
> Pupil: No
> Teacher: What's it got on it? (pause 3 seconds)
> Pupil: A lock
> Teacher: It's got a lock on it, why do you think it has got a lock on it?
> Pupil: So that nobody can get into it.

The pupils were then given a task involving the deconstruction of a grocery carton in order to examine its construction and the net employed.

> Teacher: The first thing that you are going to do is work with a friend on your table; you can choose and undo a box...

This was followed by reconstruction of the box with the addition of parts in order that it became a model of a pirate's treasure box.

> Teacher: ... sit down and think carefully about what materials you are going to use, how you are going to do it and what you need your treasure box to look like.

The teacher drew the pupils' attention to a large collection of construction materials including a range of cards, papers and adhesives.

> Teacher: You can choose the way that you are going to join it together.

The contribution of the children was largely confined to responding to teacher questions, although some opportunity was given for offering suggestions and posing questions. This was directed and brief. A child suggesting that weight might be important received little acknowledgement.

Having moved to the deconstruction of the box and subsequently to design and making, a change was noted. More time was available for experimentation and discussion. The pupils made choices, for example, about the type and colour of paper, the shape and position of handles, and the operation of the lid. However, the teacher retained direction by regularly reinforcing the characteristics she had identified for treasure box construction. She did this by stopping the whole class to remind them, by speaking with groups and on a one to one basis with pupils.

These two lessons illustrate different approaches. The first was directive and focused on teaching basic skills of stitch craft. The teacher asked the pupils to design a decoration, but little time or emphasis was given to the element of design. When selecting a motif for the decoration the pupils were highly influenced by the examples shown by the teacher. The second lesson was highly structured and directive but included, in different sections of the lesson, greater variation in the level of teacher influence.

The two lessons observed involved different degrees of teacher direction. In the first lesson, the teacher was quite directive about the shape of the decorations. In the second lesson where the pirates' box was modelled by the pupils, the teacher moved the pupils towards greater autonomy within a highly structured lesson. Such an example may be contrary to a view which would see pupil autonomy increase only as teacher direction decreased. In this case the teacher structured the session highly, appearing to utilise the degree of her direction as a variable factor of her teaching. The result was opportunity for increased autonomy in the pupils. What was perhaps most interesting was the shift towards pupil autonomy which occurred during the lesson. The initial discussion between the teacher and the pupils was highly directed by the teacher, as was the first practical activity. Until this point the pupils were led by teacher questioning, instruction and illustration. The pupils were then asked to select materials and later to generate ideas for handles and locks for the boxes. However, guidance was still provided:

> Teacher: Think about the size you'd like... ...choose what way you're going to join it together....

This example indicates that the degree of pupil autonomy is not necessarily determined in a straightforward way.

PART OF THE FUTURE – A DISCOURSE?

How teachers teach Design and Technology remains a relatively unexplored area. In England, the subject has suffered from a number of changes made in rapid succession (DES/WO, 1989; DFE/WO, 1995; DfEE/QCA, 1999). The teaching workforce in primary schools lack specialist knowledge and show at best only a little increased confidence. What is apparent is that by discussing this teaching with primary teachers and observing their behaviour in Design and Technology lessons interesting issues arise, such as the use of language, approaches to discussion with the children, teacher direction and the promotion of pupil autonomy in Design and Technology lessons. There remains considerable scope for the articulation of teaching or Alexander's (1992) 'observable practice' in the subject and for an exploration of the relevance of theory and ideas which have been proposed about teaching and learning. The scope for research also includes the extent of the knowledge required by teachers and in particular the pedagogical knowledge required. Are there similarities between the teaching of other subjects and Design and Technology? Are teaching methods imported to Design and Technology? Are some ideas exported? An example may be constructivist approaches which have been widely employed by teachers of science. What is the potential contribution to Design and Technology? Is the unique nature of Design and Technology represented in its pedagogy? The importance of pupil autonomy (rather than mere independence), might be stressed within the pupil experience. This might be more important in a subject like Design and Technology which is about human needs and human responses to those needs.

For any country or educational system wishing to implement Design and Technology as a subject in the primary curriculum there are important questions to address about how the subject is taught and how teachers will be prepared to teach the subject well. For the wider educational community there are further considerations about the nature of subjects such as Design and Technology and the value of describing teaching behaviours. Any such consideration should be prepared to accept complexity.

There is limited agreement around the world about this subject. Its title almost always employs the term technology, but interpretations of the term vary and the extent to which design is part of it or is related to it. Perhaps this variation in interpretation should be seen as a strength rather than a weakness?

It might be beneficial for educationalists around the world to agree about features of this experience: that there is content, that there are products; there are skills to be learned; and a process with which to engage. Are we agreed about the relative balance of these elements in Design and Technology education? Could we as Design and Technology educationalists articulate our intended outcome for Design and Technology in terms of movement towards autonomy? If we value autonomy for pupils how do we regard autonomy for the profession? If we are to work with a national curriculum do we need to recognise that the way it is constructed and articulated has a significant impact on the choices teachers make about teaching methods? If teachers are merely instructed to teach is there a danger that they will opt for an overly didactic approach? It may be that this research and discussion illustrates a need in Design and Technology education to take its pedagogy more seriously. Alexander (2004) defines pedagogy as the act of teaching and the associated discourse. Perhaps such a discourse would help us discover more in the next twenty years.

REFERENCES

Alexander, R. J. (1992). *Policy and practice in primary education*. London: Routledge.

Alexander, R. J. (2004). Still no pedagogy? Principle, pragmatism and compliance in primary education. *Cambridge Journal of Education*, *34*(1). Retrieved from http://www.robinalexander.org.uk/docs/Camb_Jnl_article_04.pdf

Adams. P. (2006). Exploring social constructivism: Theories and practicalities. *Education 3–13*, *34*(3), 243–257.
Boud, D. (Ed.). (1987). *Developing student autonomy in learning*. London: Kogan/Page.
Brophy, J. E., & Evertson, C. M. (1976). *Learning from teaching: A developmental perspective*. Boston: Allyn and Bacon.
Candy, P. (1987). On the attainment of subject matter autonomy. In D. Boud (Ed.), *Developing student autonomy in learning*. London: Kogan/Page.
Conway, J. (1997). *Educational technology's effect on models of instruction*. Retrieved from http://copland.edel.edu/~jconway/EDST666.htm
Cross, A. (2000). *Pedagogy and curricular subjects*. Unpublished thesis, University of Manchester.
Davies, T., & Elmer, R. (2001). Learning in Design and Technology: The impact of social and cultural influences on modelling. *International Journal of Technology and Design Education, 11*, 163–180.
Department of Education and Science and the Welsh Office (DES/WO). (1989). *Design and technology for ages 5–16*. London: HMSO.
Department of Education and Science and the Welsh Office (DES/WO). (1990). *Technology in the national curriculum*. London: HMSO.
Department for Education and the Welsh Office (DFE/WO). (1995). *Design and technology in the national curriculum*. London: HMSO.
Department for Education and Employment (DfEE). (1998). *The national literacy strategy*. London: Author.
Department for Education and Employment (DfEE). (1999). *The national numeracy strategy*. London: Author.
Department for Education and Employment/Qualifications and Curriculum Authority (DfEE/QCA). (1999). *The national curriculum: Handbook for primary teachers in England.* London: Author.
Department for Education and Skills (DfES). (2003). *Excellence and enjoyment: A strategy for primary schools*. London: Author.
Design and Technology Association (DATA). (1999). *DATA helpsheets for the DfEE/SEU/QCA exemplar scheme of work for design and technology in primary schools*. Wellesbourne: Author.
Design and Technology Association (DATA). (2003). *Survey of provision for design and technology in schools 2002/2003*. Wellesbourne: Author.
Gipps, C. (1992). *What we know about effective primary teaching*. London: Institute of Education.
Gipps, C., McCallum, B., & Hargreaves, E. (2000). *What makes a good primary school teacher? Expert classroom strategies*. London: Routledge/Falmer.
Hennessy, S., & Murphy, P. (1999). The potential for collaborative problem solving in Design and Technology. *International Journal of Technology and Design Education, 9*, 1–36.
HMI. (1987). *Craft, Design and Technology from 5 to 16*. London: HMSO.
HMI. (2005). *Ofsted subject reports 2003/04: Design and technology in primary schools*. London: Ofsted.
Keirl, S. (2006a). On being international and the phenomena of difference.... *Design and Technology Education: An International Journal, 11*(2), 3–6.
Keirl, S. (2006b). Design and Technology education: Whose design, whose education and why? *Design and Technology Education: An International Journal, 11*(2), 20–30.
Kimbell, R. (1982). *Design education: The foundation years*. London: Routledge and Keegan Paul.
Kimbell, R., Stables, K., & Green, R, (1996). *Understanding practice in Design and Technology*. Buckingham: Open University Press.
McNamara, D. (1994). *Classroom pedagogy and primary practice*. London: Routledge.
National Curriculum Council (NCC). (1990). *Non statutory guidance: Design and Technology*. York: NCC.
Nicholl, B. (2004). Teaching and learning creativity. In E. Norman, D. Spendlove, P. Grover & A. Mitchell (Eds.), *DATA international research conference 2004: Creativity and innovation* (pp. 51–59). Wellesbourne: Design and Technology Association.
Qualifications and Curriculum Authority/Department for Education and Employment (QCA/DfEE). (2001). *Scheme of work for Key stages 1 and 2: Design and technology.* London: Author.
Rowell, P. (2002). Interactions in shared technology activity: A study of participation. *International Journal of Technology and Design Education, 12*, 1–22.

Simon, B. (1981). Why no pedagogy in England? In B. Simon & W. Taylor (Eds.), *Education in the eighties*. London: Batesford Academic and Educational Ltd.
Stenhouse, L. (1975). *An introduction to curriculum research and development*. London: Heinemann.
Ward, H. (2002, November 1). The forgotten joys of making and doing. *Times Educational Supplement*.

Alan Cross
Department of Education
University Of Manchester
England

JOHN R DAKERS

15. THE RISE OF TECHNOLOGICAL LITERACY IN PRIMARY EDUCATION

INTRODUCTION

Whilst I acknowledge that the subject did not exist 500 years ago, try nevertheless, to imagine a design and technology classroom at that time. The practical skills taught at school during this period were restricted to the crafts or trades associated with writing, reading and record keeping as required by clerks for example. Skills for trades involving bodily effort, such as for cooks, tailors, masons, wheelwrights or blacksmiths, were not taught at school but by apprenticeship (Orme, 2006). Design was not a curricular subject during this period. Technologies, at that time, were what we might now term 'primitive' and involved learning not only how the technology functioned, but the procedural skills associated with its operation. If we were to move forward in time to the 1800s, a time traveller from the 1500s would recognise that whilst significant refinements had been made to technologies, they would nevertheless have little difficulty understanding their functionality and would, therefore, easily adapt to their use. The function of a technology, particularly complex technologies like machines, for example, was very limited compared to today. Those who used technology learned how to operate the technology in question but also learned how the technology operated. They could see inside the black box as it were. Moreover, technologies were limited in number. A carpenter during this period would have at his (*sic*) disposal a lot less tools and machines (technologies) than his (*sic*) counterpart today.

> Fast forward another hundred years, however, and it is a different story. By the end of the nineteenth century, a panoply of new technologies had appeared that were qualitatively different from earlier technologies: steamboats and ironclad ships, the telegraph and telephone, the transcontinental railroad, the phonograph, the internal combustion engine, gasoline and other petrochemicals, aspirin and a wealth of other drugs, the automobile, and the machine gun. The world of 1900 was much more dependent upon these machines and tools, which posed challenges that were entirely new. A competent, contributing member of society had to understand and use an increasing number of technological devices (Pearson & Young, 2002, p. 48).

The technologies that mediate our lives today are very much more complex again. Even in the first half of the 20th Century, technologies were mostly understandable in the context of the uses: horse drawn carts and then automobiles, even up to the

C. Benson and J. Lunt (eds.), International Handbook of Primary Technology Education: Reviewing the Past Twenty Years. 181–193.

1970s, were relatively simple enough to be both used and maintained by the user. However, to understand the way modern technologies function as distinct from how they are operated today, including automobiles, for example, requires specialist knowledge. Whilst it would be impossible for any individual to understand how all modern technologies functioned, modern (domestic) technologies are designed to be operated by their users without requiring them to understand how the technology functions. When your automobile breaks down you consult an automobile specialist; when your central heating boiler ceases to operate you consult a heating engineer; and when your television breaks down, or is out of date, you may buy a new one.

This suggests that technology education today cannot be expected to teach children how all modern technologies function, and this is reflected in classroom practice. Primary technology education across the industrial nations, in terms of teaching and learning how technologies function, focuses on pre-modern technologies such as mechanisms, structures, technical drawing by hand, food and textiles and simple electronic processes. When modern technologies are studied, the focus is centred upon their operation. Computer aided design and computer aided control, to offer but two examples, are more concerned with teaching the student how to operate the computer and its programme, as a tool to aid in the design and operation of an artefact. Neither the student nor the teacher is expected to understand how the computer and its programme actually function. This is as it should be. A number of countries, including England and the USA, have incorporated the notion of 'design' as a major component of the technology curriculum.

However, in order to better understand the modern technological world we occupy, whether as a fabricator, designer or consumer, technology curricula from around the world have begun to place a significant emphasis on issues relating to values and attitudes regarding technology. Issues relating to ethics, sustainability, environmental impact, social impact and moral impact have entered the various technology education rationales.

Over the past ten years this has manifested itself conceptually as 'technological literacy'. It might be argued that the concept of technological literacy in a classroom context had its genesis in 1996. It was the International Technology Education Association in the USA who published a new set of standards for technology education. These standards evolved into the 'Standards for Technological Literacy' which were subsequently published in 2002. Several associated publications relating to the concept of technological literacy continue to be produced by ITEA and are available at their website http://www.iteaconnect.org/. Moreover, a growing body of research and research literature has evolved in this context; just try an internet search for 'technological literacy' and see what you find.

To this end, I have long advocated for the incorporation of technological literacy as an essential component in the classroom delivery of Technology education. (I use the term Technology education throughout this chapter as a generic term: I recognise that in England, for example, the subject is known as Design and Technology, whereas in Scotland, it is still known as Technical education and Home Economics in a number of schools). Part of the problem, however, has been reaching a consensus, or even a common understanding as to what the term 'technological literacy' actually means from the perspective of a classroom teacher (and others for that matter).

I have written scholarly papers and books about the concept of technological literacy and its relation to technology education (see for example, Dakers, 2005; 2006a; 2006b; 2007a; 2007b), but when asked by teachers to explain what it is and how to incorporate it in the classroom, I usually reply that it is contingent upon the utilisation of both procedural and conceptual knowledge in technology education and that I know it when I see it! Not entirely helpful I hear you say, and you would be right! To that end, I hope to explain, in more practical terms, how teachers might engage students in the process of developing technological literacy by teaching within a more philosophically orientated pedagogical framework – without panicking! Moreover, I will illustrate this with a non-prescriptive case study example that will enable primary colleagues to continue to teach established design and making skills along with the development of knowledge and understanding related to technology, whilst enabling pupils to develop their technological literacy.

UNPACKING THE CONCEPT OF EDUCATION IN GENERAL AND TECHNOLOGY EDUCATION IN PARTICULAR

The terms 'educate' and 'pedagogy' both have very similar meanings. The term educate is related to the Latin *educere* which translates as 'bringing out' or 'leading out'. In ancient Greece, a *pedagogue* was a slave who took his masters' child to school, hence, 'to lead'. I mention this in order to highlight that the Latin and Greek concepts relating to education and pedagogy, have more to do with bringing something (knowledge) out, as against putting something (knowledge) in. It was Socrates who argued that all knowledge was already inside us, in other words we are all born with a sort of hard drive, which might be called a brain, which is pre-programmed with all knowledge. It is the job of the teacher to draw out of the pupil useful knowledge and guide them how to use it sensibly and virtuously. Conversely, a more recent philosopher than Socrates called John Locke (1632–1704), postulated that human beings are born with no inherent mental content, that the mind is *tabula rasa*, a blank slate in which knowledge is built up from external sources gradually and is based upon interaction with the external world. These two world-views form the essence of what is known as the *nature* versus *nurture* debate.

I would ask you to reflect here for a moment and consider where you place yourself on this continuum. This will have an impact on the way you teach technology education. If, for example, you follow Socrates, you will be inclined to support the nature thesis. This will lead you to seeing your job as being more related to guiding and questioning in order to reveal that knowledge which is already present within the child: the Socratic method as it were. If, on the other hand, you prefer Locke's thesis then you will be inclined to believe that your job is to pass on, or 'put in' existing pre-established knowledge for the child to absorb, learn and develop. This will require that you have already acquired that body of knowledge resulting from your own expertise and experience in the world. This is often referred to as the transmission model of teaching where prescribed skills and pre-existing facts are transmitted from experts to novices. The transmission model, I would argue, is the one that is most prevalent and the one that most people think about in relation to the art or science of teaching. This is understandable to some extent. The thinking behind school systems around the world tends to work from the general to the particular; from holistic to specialization. Teachers in the secondary sector are considered to be specialists or even experts in their subject area and teach only that subject. It is thus understandable, given that they consider themselves to be 'subject experts', that they will default to a transmission model of teaching where they, as experts, pass on their knowledge and skills to the pupils who are considered to be novices. Current assessment arrangements tend only to help exacerbate this situation, but that is another debate beyond the scope of this chapter. Primary teachers' level of expertise in particular subject domains varies around the world. In Scotland, for example, primary teachers are not considered nor required to be 'experts' or 'specialists' in technology education, although they are expected to teach it. In England, there are various routes that primary teachers can take to have technology as one of their 'specialisations'. The point that I wish to make is, however, that if a transmission model of teaching is adopted, in whatever sector, it assumes that knowledge relating to technology is universal, it exists 'out there' so to speak and is fixed and immutable. The same can be said for the model that Socrates advocates, but just in reverse; it is 'in there' so to speak.

Mathematics, with some justification, can make the claim that it is universal and immutable. In other words, the same procedures and rules apply universally around the world; 2 + 2 = 4 or a quadratic equation is always the same whether calculated in Saudi Arabia, Ethiopia, Wales or even under water[1]. The same claim cannot be made for all knowledge relating to technology and consequently technology education. Significant issues that are neither universal nor algorithmic in technology education include ethics, values, social consequences initiated through technological development and design to name but four (although design is not always a feature of some technology curricula). These issues are aligned to the thinking mentioned earlier about values and attitudes.

If some issues relating to technology are not always absolute they must then be abstract or conceptual and so open to interpretation. This is where these aspects of technology education, taught as a transmission model, get into a little trouble. It is not either nature or nurture, using the hands or the mind, teaching practical skills or teaching academic concepts. It is a fusion of all these and is informed by context.

AN ALTERNATIVE TO THE EITHER/OR PARADIGM

It was eighteenth century Europe that witnessed a quantum leap from what was considered to be the tyranny of the medieval period, where the authority and teachings of Church and State was transmitted as universal and absolute fact which was not open to challenge, to that of a new age of reason and scientific revolution which began to question the received wisdom of the day. This period was known as the Age of Enlightenment. One influential thinker to emerge from this period was the German philosopher Immanuel Kant (1724–1804). He argued that non-universal issues, such as those relating to technology education mentioned above, should not reside in some external higher authority such as the Church or State (or school or expert), but should be the responsibility of the individual's own judgement which would be informed by reason. Kant used the Latin term *sapere aude* meaning 'dare to know' or 'have the courage to use your own reason'. Steven Law (2006) cites one of Kant's most quoted characterisations of Enlightenment as follows:

> [Enlightenment is the] emergence of man from his self-imposed infancy. Infancy is the inability to use one's reason without the guidance of another. It is self-imposed, when it depends on a deficiency, not of reason, but of the resolve and courage to use it without external guidance. Thus the watchword of enlightenment is: *Sapere aude!* Have the courage to use one's own reason! (Law, 2006:6–7: italics in original).

This has interesting and profound implications for teaching a subject like technology education, particularly where design takes a central role. Can we truly expect primary school children to apply reason to aspects of design and technology without the guidance of another? My answer (along with a significant number of others as we shall see later) is a qualified yes. However, my qualification is predicated upon pedagogy and not upon content. Or, borrowing from Fun Boy Three and Bananarama: 'T'Ain't What You Do (It's The Way That You Do It)'! (MCA Music Ltd., 1939).

PHILOSOPHY FOR CHILDREN

It was in the late 1960s that a US professor called Matthew Lipman started the Philosophy for Children movement. His motivation for doing this is accounted for in the Stanford Encyclopaedia of Philosophy as follows:

> As the Vietnam War escalated in the mid-1960s, so did heated arguments about the wisdom and morality of the war and society's ills in general. Matthew Lipman became dismayed at the quality of argumentation employed by presumably well-educated citizens. Convinced that the teaching of logic should begin long before college, he tried to figure out a way to do this that would stimulate the interest of 10–11 year olds. Leaving Columbia University for Montclair State College, he launched his efforts with his first children's novel, *Harry Stottlemeier's Discovery* (1974). Lipman's concerns about the level of critical thinking in society in general, and the schools in particular, were not his alone. By the 1970's the hue and cry for teaching critical thinking in the schools was, if not clear, at least loud; and it has continued largely unabated to the present (SEP, 2007).

This interest in philosophy as a form of pedagogy emphasises the importance of questioning, enquiry and communication. This resonates strongly with the idea of social constructivism as espoused by Vygostky (1978). Moreover, the movement has grown considerably. Influenced by Lipman, SAPERE (taken from Kant's mantra *sapere aude - dare to discern*) was set up in the United Kingdom in 1992 with the express aim of 'developing better reasoning, more reflective consideration of values and the development of communities of enquiry at all levels of education and in a wide variety of contexts.' (SAPERE, 2007a).

Significantly, this is not considered to be exclusive to post primary education, quite the reverse. A number of publications and training courses are now available for the primary sector. These include classroom text books by authors such as Philip Cam from Australia (1993; 1994; 1995), Robert Fisher from England (1996; 1997; 1999), and Paul Cleghorn from Scotland (2004). Moreover, a growing body of empirical research in the primary sector is now available. This research indicates that by creating a community of enquiry or a philosophical approach to teaching results in significant increases in pupil attainment in all subject areas. One such recent study took place in a number of primary schools in the Scottish Council of Clackmannanshire (Law, 2006; SAPERE, 2007b). This study was subject to a set of very rigorously controlled experimental studies carried out by Professor Keith Topping from the University of Dundee, the results of which are summarised as follows:

- A whole population of children gained on average 6 standard points on a measure of cognitive abilities after 16 months of weekly enquiry.
- Pupils increased their level of participation in classroom discussion by half as much again following 6 months of weekly enquiry.
- Pupils doubled their occurrence of supporting their views with reasons over a 6 month period.
- Teachers doubled their use of open-ended questions over a 6 month period.
- When pupils left primary school they did not have any further enquiry opportunities yet their improved cognitive abilities were still sustained two years into secondary school.
- Pupils and teachers perceived significant gains in communication, confidence, concentration, participation and social behaviour following 6 months of enquiry (SAPERE, 2007b).

Another example is given by Law (2006). The Buranda State School near Brisbane in Australia incorporated a philosophy for children programme using materials developed by the philosopher Philip Cam (see above). Law cites a report on the success of the intervention.

> For the last four years, students at Buranda have achieved outstanding academic results. This had not been the case prior to the teaching of Philosophy. In the systemic Year 3/5/7 tests (previously yr 6 tests), our students performed below the state mean in most areas in 1996. Following the introduction of philosophy in 1997, the results of our students improved significantly and have been maintained or improved upon since that time (Buranda State School Showcase 2003 Submission Form) (Law, 2006, p. 37).

The classroom application of these philosophies for children's programmes has, as its basis, the use of stories to develop thinking. Fisher (1998) argues that the use of stories provides us with metaphors for life. 'Human life', he says, 'can be regarded as a story, a narrative structured in which everyone has a part…To understand the narrative structure of stories, or of human lives, requires more than the exercise of human reason, it requires what Egan calls 'the other half of the child', namely, imagination.' (p. 96). However, Haynes (2002) qualifies the use of stories in this context as having to be selected carefully in order 'to express ambiguity, to produce puzzlement, or to evoke a deep response' (p. 22). This style of pedagogy turns notions of what constitutes teaching technology on their head and appears to be counterintuitive. Certainty gives way to ambiguity, order becomes confusion and the surface learning of prescribed facts transforms into deeper forms of enquiry. Technology education becomes 'fuzzy' as it becomes more concerned with uncertainty and risk and

> [d]esign tasks [become] typically multi-dimensional, messy and value laden (i.e. wicked). (Kimbell & Perry, 2001, p. 6).

Teaching technology education using a philosophically orientated pedagogical framework involves open-ended enquiry and dialogue; it requires that teachers listen to and respect pupil voice. Haynes offers us a succinct checklist outlining the necessary structures that need to be put in place in order to create a community of enquiry:

Children need to be confident that:

- their accounts of their experiences and their opinions will be treated with respect and accepted as valid;
- teachers will be faithful to the detail of their contributions;
- they will be given time to speak;
- they will not be mocked or humiliated;
- they can be tentative, playful or exploratory in their thinking;
- they can change their mind if they want to;
- minority views will be supported;
- challenges to the status quo, when raised as part of the process of enquiry, will not be punished.

Teachers need to make the effort to:

- be open-minded and encourage open-mindedness;
- be willing to reconsider established ideas and to view facts, ideas and theories as provisional;
- be supportive when there is a struggle to articulate new ideas;
- allow proper time for each person's contribution;
- be skilled in recognising connections between ideas;
- hold back on their own interests in any enquiry;
- check for possible misunderstandings;
- be flexible, intuitive and responsive to the dynamics of a discussion. (Haynes, 2002, p. 66)

As I hope I have demonstrated, there is strong supporting empirical evidence to show that this works. Secondly, this type of pedagogy articulates almost exactly with the findings of Black *et al.* (2003) and the Wiliam & Black studies (2006) on the use of formative assessment in classroom practice. Third, and of equal importance, you no longer need to be an expert technologist or an expert designer, you just need to be an expert in creating a community of enquiry.

A POSSIBLE SCENARIO FOR CREATING A COMMUNITY OF ENQUIRY IN TECHNOLOGY EDUCATION

I offer this example as a suggestion. It is not prescriptive and is both open to various interpretations and subject to alteration. Moreover, rather than write a story I will instead suggest several stories which constellate around the same theme, that of castaway(s). The stories that spring immediately to mind are *Robinson Crusoe*, *Swiss Family Robinson, The Coral Island, The Lord of the Flies* and *The Admirable Chrichton*. Recent movies with a similar theme include *The Beach* starring Leonardo DiCaprio and *Castaway* starring Tom Hanks. The television series *Lost* offers yet another more contemporary source, albeit having less orientation towards the traditional castaway genre in favour of a somewhat fantastic plot. Other shows that have some relevance are the reality TV programmes *Survivor* and *I'm a Celebrity, Get Me Out of Here*!

I have not chosen this theme randomly; I have borrowed from Karl Marx's book *Das Kapital* (2000). Without wishing to delve through the complexities of Marxist theory, I do wish to show how the theme of Robinson Crusoe, as used by Marx, can be used in relation to technology education.

In Marx's way of thinking, there are two principle values associated with an object; its 'use value' and its 'exchange value'. The former represents the utility of the object, so a chair, for example, is used for sitting upon and a frying pan is used for frying food. Exchange value, on the other hand, refers to the expression of the value an object has in relation to another object. So in barter, a chair might have a value of three frying pans or, using money, it might have a value of fifty pounds sterling. What Marx wanted to argue was that Robinson Crusoe, as a castaway having very little belongings other than those salvaged from his ship, was more interested in use value than exchange value, or in a modern context where, for example, designer labels have more symbolic value than use value. This story sets the scene for children to discuss a number of issues such as, why a designer set of trainers, or a designer T-shirt is more desirable to them than one which is not 'designer' labelled for example. It also sets the scene for making children more aware of the technologically mediated world they inhabit by casting them, metaphorically, into an environment which has none of the 'taken for granted' technological devices to which they have become so accustomed. It helps to create a viable scenario which can serve to reveal for children the hidden, taken for granted, technologically mediated world in which they live. The scenario, moreover, allows for the design and fabrication of a multitude of 'primitive' technologies as might normally be used, but in the context of being cast away.

CASTAWAY AS A SCHEME OF WORK FOR TECHNOLOGY EDUCATION

It should be noted first that this theme offers a range of cross-curricular linking potential. However, in relation specifically to technology education the class can discuss the concept of the castaway through the characters in the movies or TV programmes mentioned earlier or some other relevant storyline.

Using the castaway story and the concept of use value, the teacher can begin to elicit from the class their vision of a technological lifeworld where they will need to 'design' and fabricate, not only a physical infrastructure capable of sustaining life, but a cultural system based upon their own personal histories and experiences. It would be important to establish with the class at the beginning, whether the scenario is one where the individual child is castaway, as in *Robinson Crusoe*, or whether a group are castaway, as in *Swiss Family Robinson* or *Lost* for example. The class would then have to establish the islands eco-system. A starting point might be looking up information on the Juan Fernández Islands. These were the actual islands where the real Alexander Selkirk, the inspiration behind *Robinson Crusoe*, was actually cast away. Much of the information about these island ecologies can be used to set the scene so to speak, and can be easily accessed via the internet or in the library.

Another area for philosophical reflection might involve an enquiry into whether or not the children think that they would be able to fabricate the necessary tools required for constructing artefacts such as shelters or rafts. A given scenario might be that no tools were available from the wreck so how would they adapt? What might they use as a hammer or a saw and what might they use to join raw materials together? (Tom Hanks, in the movie Castaway, found several novel tool uses for a pair of ice skates, including one to help remove his decayed tooth. He also used a form of grass to fabricate rope in order to construct a raft). Discussions surrounding the novel use(s) of unlikely materials can stimulate much creative thinking in terms of design. These discussions can also serve to reveal the pupils' current understanding of the material world in which they live. (Tom Hanks' character's experience in the world, for example, enabled him to make the connection between the hardness and sharpness of the metal ice skate blades, by translating that knowledge into a novel use of ice skates as a tool to break open coconuts or to serve as a form of axe).

Open ended discussions where the teacher has input, but not as an expert castaway or technologist might take the form of exploring what their shelters or rafts or pots might look like? Would they use designs and fabrication techniques that they had knowledge about or would they learn 'on the job' so to speak? Would their priorities in designing and fabricating a shelter for example, be directed towards functionality or aesthetics, or put another way, use value or symbolic value? This sub-theme further allows both the pupils and the teacher to explore the nature nurture debate. Would they (or a character from the book, movie or TV show), adopt a Lockean *tabula rasa* model (as mentioned earlier) by adapting to their external environment or will they draw upon their previous life experiences and translate those skills accordingly? This is an interesting philosophical question for the pupils. Do they need to learn how to use tools first, in a classroom for example, before they can use them in daily life, or do they learn to use tools as the needs arise, on the job so to speak?

A design project to consider might be to design a shelter on a desert island which must protect them from the rain, wind and sun. They can only use natural materials found on the island. It would be worth discussing what these might be. Suppose they have retrieved a set of tools from the wreck, what might these be and what use would they be?

What would the priorities be for the castaways? Would they make artefacts such as shelters, plates and beds, for each other or individually? This begins a debate on use value – exchange value. LOST (a television show shown internationally) serves as a good example for this. One character in the series collects and hoards resources from the plane wreck and in so doing, creates a power base. Discussions about individuality versus team work and ultimately politics can emerge out of these scenarios (and many other things). These are all opportunities which can serve to both enable the child's imagination to be set free, as well as facilitating a form of critical pedagogy around issues relating to technology. Technology, it will be revealed, is not neutral; it is actually value laden and political in nature.

An initial or opening discussion on this theme might include exploring what technologies the children miss by being cast away on a desert island. These technologies are so taken for granted that this discussion will help to expose how reliant we all are upon technology. Some things might include all kinds of buildings and their associated infrastructures like plumbing, electricity, telephones, televisions and computers. There will be no roads or bridges; in fact there will be no transportation system at all other than by foot. Our domesticated pets will have vanished as will farm animals and the surrounding environment will be considered as being alien. In other words, there is nothing on the island that has been shaped or developed by human intervention. Plants, trees, bushes and flowers are not grown in nurseries; they are, like the animals, wild. The children should be encouraged to look at the technologically mediated world they occupy at present and contrast that with how their lives would be affected by being cast away on a desert island.

Many design scenarios can evolve from the castaway theme from the simple, such as design and make a useful tool to help the castaway community, to the more complex, such as design a sustainable village for the castaway community. Some priorities to think about might include:

- Shelter - What should it provide and protect against?
- Water - Humans can only last up to three days without water and get weaker as time goes on. Where might you look to source water? What technologies might you fabricate to make clean water?
- Fire - How might you start a fire without matches or a lighter? For what reasons would you need a fire?
- Food - What food would you need and how would you collect it? How would you prepare it? How would you present it? If it was an animal or a fish, how would you hunt it, trap it or catch it?
- First aid - Suppose you had a rudimentary first aid kit from the wreck, what might you need to treat? What plants can be used medically and for what purpose?
- Navigation - How would you know where you were on the island? How might you read the sun?

These scenarios can be incorporated into a whole class holistic pedagogy which is not exclusively related to design and technology education, but can cover a number of areas in the curriculum. Moreover, it affords the teacher the opportunity to be flexible and creative in the learning experience. The transmission model discussed above can be used to teach practical skills such as the correct use of prescribed tools; however this pedagogy cannot facilitate learning which is normative in nature. A more open ended and flexible approach is required for this. For example, the teacher can, and should, demonstrate the correct and safe way to use craft knives to cut card for example, but cannot dictate a right or wrong solution for any political or value laden judgement the castaway community might instigate. Modern visions of education are, for Slattery (2006):

> ... characterized by the Tylerian Rationale, behavioral lesson plans, context free-objectives, competitive and external evaluation, accountability politics, dualistic models that separate teacher and student, meaning and content, subjective persons and objective knowledge, body and spirit, learning and environment, and models of linear progress through value-neutral information transmission. (pp. 213–214)

The theme of castaway offers a way to reconcile these dualistic paradigms by incorporating technological literacy into the design and technology curriculum. If as educators, we believe that technological literacy can be best understood, (1) in terms of a set of social activities (enacted on a desert island or elsewhere) which are mediated, appropriately or inappropriately by technology whether modern or primitive; (2) that there are different technological literacies associated with different domains of life (living in London or as a castaway for example); (3) that technological activities are patterned by social institutions and power relationships, and some technologies are more dominant, visible and influential than others; and that (4) technological activities change and new ones are frequently acquired through processes of informal and formal learning and sense making.

> In the simplest sense [technological activities] are what people do with [technology]. However, [critical aspects of technological activities] are not observable units of behaviour since they involve values, attitudes, feelings and social relationships. This includes people's awareness of [technology], [their] constructions of [technology] and [their] discourses of [technology], how people talk about and make sense of [technology] (Barton & Hamilton, 2000, p. 8).

Technological literacy essentially seeks to explore the social shaping which occurs as consequence of the interface between human beings and their active involvement with technologies. The castaway theme, in association with the adoption of a philosophically orientated pedagogical framework, can help to bring the social aspects of technology into relief by transporting children, who currently live in the highly technologically mediated twenty first century and who have little critical understanding of the impact that modern technology has on their lives, into a world without technologies in which they will become responsible for creating technologies that they find meaningful, purposeful and essential. This theme and pedagogical framework creates a learning environment which goes beyond the simple transferring of facts and skills from teachers to children. Rather it invites children to think critically about subject matter, doctrines, the learning process itself, and their [castaway] society. The teacher poses problems derived from the children's imagined castaway life which will involve social issues, design issues and technical issues and these will be considered in a mutually created dialogue.

NOTES

[1] Whilst I acknowledge that arguments questioning the universality of mathematical principles do exist in chaos theory for example, for the sake of brevity I shall assume that the basic nature of arithmetic, algebra, geometry and calculus, as taught in schools around the world, have universal structures.

REFERENCES

Barton, D., & Hamilton, M. (2000). Literacy practices in situated literacies: Reading and writing in context. In D. Barton, M. Hamilton & R. Ivanic (Eds.), *Situated literacies: Reading and writing in context*. London: Routledge.

Black, P., Harrison, C., Lee, C., Marshall, B., Wiliam, D. (2003). *Assessment for learning: Putting it into practice*. Buckingham: Open University Press.

Cam, P. (1993). *Thinking stories 1 - teacher resource/activity book*. Sydney: Hale & Iremonger. (Republished in German 1996 and Latvian 1997).

Cam, P. (1994). *Thinking stories 2 - teacher resource/activity book*. Sydney: Hale & Iremonger.

Cam, P. (1995). *Thinking together philosophical inquiry for the classroom*. Sydney: Hale & Iremonger/ PETA.

Cleghorn, P. (2004). *Thinking through philosophy series*. Blackburn: Educational Printing Services Ltd.

Dakers. J. (2005). Technology education as solo activity or socially constructed learning. *International Journal of Technology and Design Education, 15*(1), 73–89.

Dakers, J. (Ed.). (2006a). *Defining technological literacy: Towards an epistemological framework*. New York: Palgrave MacMillan.

Dakers, J. (2006b). Technology Education in Scotland. An investigation of the past twenty years. In M. J. de Vries & I. Mottier (Eds.), *International technology education studies: The state of the art* (pp. 331–346). Rotterdam: Sense Publishers.

Dakers, J. (2007a). Incorporating technological literacy into classroom practice: Conceptualizing technology lessons. In M. J. de Vries, R. Custer, J. Dakers & G. Martin, (Eds.), *Exemplary practice in technology education*. Rotterdam: Sense Publishers.

Dakers, J. (2007b). Vocationalism – Friend or foe to technology education. In D. Barlex (Ed.), *Design and technology: The next generation* (pp. 90–107). London: The Nuffield Foundation in association with TEP.

Fisher, R. (1996). *Stories for thinking*. Oxford: Nash Pollock Publishing.

Fisher, R. (1997). *Games for thinking (Stories for thinking)*. Oxford: Nash Pollock Publishing.

Fisher, R. (1998). *Teaching thinking: Philosophical enquiry in the classroom*. London: Continuum.

Fisher, R. (1999). *First stories for thinking*. Oxford: Nash Pollock Publishing.

Haynes, J. (2002). *Children as philosophers: Learning through enquiry and dialogue in the primary classroom*. London: RoutledgeFalmer.

Law, S. (2006). *The war for children's minds*. Abingdon: Routledge.

Lipman, M. (1974). *Harry Stottlemeier's discovery*. Upper Montclair, NJ: Institute for the Advancement of Philosophy for Children.

Kimbell, R., & Perry, D. (2001). *Design and Technology in a knowledge economy*. London: Engineering Council.

Marx, K. (2000). *Das Kapital*. Washington: Regery Publishing Inc.

Orme, N. (2006). *Medieval schools: From Roman Britain to Renaissance England*. China: Yale University Press.

Pearson, G., & Young, T. A. (2002). *Technically speaking: Why all Americans need to know more about technology*. Online book retrieved from http://www.nap.edu/catalog.php?record id+10250#toc

SAPERE. (2007a). Retrieved from http://sapere.org.uk/what-is-sapere/

SAPERE. (2007b). Retrieved from http://sapere.org.uk/2005/08/04/research-project/

SEP. (2007). *Stanford encyclopedia of philosophy*. Retrieved from http://plato.stanford.edu/entries/children/

Slattery, P. (2006). *Curriculum development in the postmodern era*. New York: Routledge.

Vygotsky, L. S. (1978). *Mind in society. The development of higher psychological processes*. Massachusetts, MA: Harvard University Press.

Wiliam, D., & Black, P. (2006). *Inside the black box: Raising standards through classroom assessment*. London: NFER Nelson Publishing.

John R Dakers
University of Delft
Netherlands

WENDY FOX-TURNBULL

16. AUTOPHOTOGRAPHY

A Means of Stimulated Recall for Investigating Technology Education

INTRODUCTION

This chapter discusses the use of 'stimulated recall' as a tool for investigating Technology in the primary classroom. Stimulated recall usually uses video and audio recordings of the participants in action, which are later shown as a prompt to reflect and comment upon. Typically, recordings are taken by researchers which are then used in the interviewing process (Moreland & Cowie, 2007; Slough, 2001). This study differs slightly in that the students were provided with disposable cameras to take the photographs of their practice themselves. They could request someone else to take a photo for them if they wanted to feature in the picture themselves. The photos were then used to stimulate discussion with the researcher about their learning. The process of the participants taking and selecting their own photographs is termed 'autophotography' (Moreland & Cowie, 2007).

Technology involves students working collaboratively in the development and production of a technological outcome - a product or system that meets a previously identified technological need (Ministry of Education, 2007). As students work together teachers facilitate learning and guide students through their technological practice. Teachers are often deeply involved in discussion and problem solving with individuals or small groups while others work quite independently. Involving the students in recording and recalling their own technological practice allows teachers to assess students' technological processes as well as receiving further insight into their technological outcomes as they develop, together with their reasons for making particular design decisions.

STIMULATED RECALL

Stimulated recall can be viewed as a subset of introspective research methods which accesses participants' reflections on mental processes and has its origins in philosophy and psychology (Mackey & Gass, 2005).

> Stimulated Recall is a family of introspective research procedures through which cognitive processes can be investigated by inviting subjects to recall when prompted by a video sequence, their concurrent thinking during that event. (Mackey, cited in Lyle, 2002, p. 34)

Slough (2001) credits Benjamin Bloom with the first description of stimulated recall in 1953 which he described as a method for retrieving memories. Many studies have

C. Benson and J. Lunt (eds.), International Handbook of Primary Technology Education: Reviewing the Past Twenty Years. 195–209.

used stimulated recall to study classroom practice and interaction (Beers et al., 2006; Plaut, 2006; Sime, 2006; Slough, 2001). Both audio and video recording have frequently been used (Plaut, 2006; Seung & Schallert, 2004; Slough, 2001). Moreland & Cowie (2007) employed stimulated recall by using 'autophotography', i.e. photographs taken by children and then used as prompts in semi-structured interviews. Stimulated recall interviews are used to gain qualitative insight into the actual working memory processes (Beers et al., 2006). Plaut (2006) used stimulated recall to investigate students' and teachers' constructs of 'confusion' in their study of transferring teacher expertise to student teachers. Slough used stimulated recall with interviews, videotaping, observation and field notes, thus providing a comprehensive range of data. Beers and colleagues (Beers et al., 2006) published a study into how information and communication technologies (ICT) tools augment learning in a variety of tasks (Seung & Schallert, 2004).

Stimulated recall protocols should include opening interviews with background questions and open-ended prompts to give the researcher information on participants' understanding (Plaut, 2006; Slough, 2001). Mackey & Gass (2005) suggest that when using stimulated recall extreme care must be taken, given issues of memory, retrieval, timing and instructions. The following recommendations are made to avoid the pitfalls associated with these issues.

- Give clear guidelines to each participant (Schepens, Aelterman, & Van Keer, 2007).
- Carry out the stimulated recall interviews as soon as possible after the actual incident (Mackey & Gass, 2005; Schepens et al., 2007; Seung & Schallert, 2004).
- Audio-tape each stimulated recall interview (there are incidences of participants using observation field notes) (Seung & Schallert, 2004) and transcribe participant conversations (Moreland & Cowie, 2007; Schepens et al., 2007).
- Participants should be minimally trained to enable them to carry out the procedure but they should not be cued to extra and unnecessary knowledge (Lyle, 2002; Mackey & Gass, 2005).
- Stimulus should be as strong as possible (Lyle, 2002; Mackey & Gass, 2005).
- If participants are involved in the selection and control of the stimulus episodes there is less likelihood of researcher interference (Lyle, 2002; Mackey & Gass, 2005).

Advantages of Stimulated Recall

One advantage of stimulated recall is that it allows participants to explain their decision making (Mackey & Gass, 2005; Sime, 2006; Slough, 2001). The use of multimedia sources in recall sessions has the benefit of replaying and reintroducing cues that were present during the task (Sime, 2006; Slough, 2001). Stimulated recall also provides an opportunity for real life context. It is a valuable tool when accompanied with 'carefully constructed research designs' (Mackey & Gass, 2005; Sime, 2006), and if the recall session is organised as soon after the event as possible, participants are less likely to have to rely on memory alone (Lyle, 2002; Mackey & Gass, 2005; Sime, 2006). As a research tool stimulated recall requires a minimal training of participants in relation to research goals (Lyle, 2002; Mackey & Gass,

2005; Sime, 2006) and it also allows relatively unstructured responses from them (Lyle, 2002). Stimulated recall has considerable potential when studying cognitive strategies and other learning processes (Sime, 2006). Mackey & Gass (2005) suggest it is an effective way to gain the perspectives of learners, their interpretation of events, and their thinking at a particular point in time.

Limitations of Stimulated Recall

One limitation to stimulated recall is that recall procedures should occur as soon as possible after the task is completed. Once information is established in the long term memory it ceases to be recall or a direct report of the experience but rather reflection or a combination of experience and other related memories (Plaut, 2006; Sime, 2006; Slough, 2001). Another limitation is that participants may censor or distort their thoughts and ideas in order to present themselves more favourably (Seung & Schallert, 2004; Sime, 2006). Participants also have the opportunity of adding tacit knowledge and therefore possibly provide inaccurate reasons for their actions (Sime, 2006). Stimulated recall alone does not capture teacher/students or student/student interactions over time. Stimulated recall records participants' thinking, but not their actual behaviour (Plaut, 2006) because classroom interaction is very complex and often automated with information being difficult to access (Lyle, 2002). It is therefore suggested that stimulated recall be used in conjunction with other data gathering strategies such as observation, interviews, recorded conversations, and participants' work sample to triangulate the data gathered (Plaut, 2006; Seung & Schallert, 2004; Slough, 2001).

THE STUDY

This study was undertaken in a primary school within the mid-socioeconomic decile range in urban New Zealand. The aim of the study was to gain insights into children's learning in Technology through an analysis of children's conversations with their teachers and peers while participating in Technology education. The researcher investigated what insight could be gained about children's Technology learning and understanding through the analysis of children's conversations about their technological practice supported by autophotography as a recall tool.

Two classes participated, one Year 2 (six years old) and one Year 6 (10 years old). Over the period of a year, two Technology units were taught in each class. The units were designed and planned by the classroom teachers in conjunction with the researcher taking the needs of the school into consideration. During the planning stage the teachers and researcher used *The New Zealand Curriculum* (Compton & Harwood, 2007).

The purpose of the first round of research was to enable the researcher to gain a rapport with the students and teachers to increase the likelihood of rich conversations during the second round and to teach the children how to photograph their learning. All the children in both classes were taught how to take photographs using a digital camera and were asked to photograph their learning in specially planned lessons.

The identification of learning to photograph was very difficult and the children struggled with this task. One child, Tullan, put it very succinctly when he said to the researcher

> I cannot photograph my learning; it's in my head.

During the teaching of the second unit, six children were selected from each class as research participants and all children were given a camera to record their learning and activity in the classroom. This time they were instructed to photograph the things they thought might help them design and build their props and the important stages in designing and building them. They were able to ask another person to take their photos if they wished to feature in them. At the conclusion of the unit the participants were interviewed using their autophotographs to assist them in the recall of their practice with the aim of gaining insight into their thinking, understanding and decision making.

METHODOLOGY

A qualitative methodology was used within a sociocultural framework. Cohen, Manion & Morrison (2000); Fraenkel & Wallen (2006); and Lichtman (2006) all cite a number of characteristics or critical elements of qualitative research. Those relevant to this study include the following:

- People construct their own meanings (direct link to social constructivist theory).
- Meaning arises out of social situations and is handled through interpretive situations.
- Words and often direct quotes rather than numbers are used to illustrate a certain point. Thick description is desirable.
- Researchers go into the natural setting to observe and collect their data and use everything from pad and pen to sophisticated audio and video taping equipment to gather data. They are very concerned with context.
- There is no one right way to conduct qualitative research. For example there are several ways to interpret what is seen and heard. However, interpretation will hold more credibility if supported by well organised data.
- The researcher plays a pivotal role in the research. It is through the researcher's eyes that the data is collected. Bias is a problem however; it can be eliminated or controlled through triangulation.

This research drew on aspects from an ethnographic approach. The research occurred in the natural setting, with a clear focus on the actions and interactions of the children and their teachers. The researcher's role was clearly understood by all participants and she was clearly present in the classroom during data gathering, undertaking ongoing conversations with the children as they worked. The main phase of data gathering occurred in Round Two during the teaching of the second unit in each class. In this unit the children were developing props for their class item in the forthcoming school production. The unit was essentially the same at both levels. Data gathered included researcher observation; participant interviews; recorded and transcribed child/child and teacher/child conversations; child work samples; and stimulated recall child interviews using autophotography.

DATA ANALYSIS

The autophotographs of twelve participant students were used in conjunction with semi-structured interviews to help recall the technological practice undertaken. Interviews were recorded as the participants discussed their photographs with the researcher; these were later transcribed. All photos and transcripts were printed and matched. The researcher then searched the data for emerging themes and patterns in relation to insight and evidence of participant learning within the field of Technology practice in relation to the New Zealand Curriculum (Ministry of Education, 2007). Emerging themes were coded and included evidence of:

- links between the participants' decision-making and prior learning (knowledge gained earlier in the unit and prior to the unit);
- insight into technological practice in relation to the New Zealand Curriculum Technological Practice Strand (Ministry of Education, 2007) including brief development, planning for practice, and outcome development and evaluation;
- understanding or discussion around the technological process undertaken by the participants;
- insight into the advantages of working collaboratively.

The photos and matching discussion also highlighted a number of issues associated with stimulated recall as a research tool. Data triangulation occurred through the use of collected work samples, interviews with the classroom teachers and researcher observations.

Findings

In this section significant findings are discussed in each of the categories mentioned above giving insight into the use of stimulated recall as a research tool in order to gain understanding of learning in Technology education.

The autophotography and associated recall during the interviews provided clear evidence that children in both age groups were able to make links to prior learning. In Year 2 all six children identified significant learning by photographing work that related to the Taiwanese or to the fact that props established for the school production had associated specifications. Ryan was able to discuss the link between a real Taiwanese boat which they viewed on video and the need for a realistic Taiwanese boat prop.

> Ry: It's the part of the other boat, of the same boat but we painted it red and white instead of just red.
> R: Why was the boat painted red and white?
> Ry: Because that's the same colour as the umm, real boat.
> R: Where was the real boat?
> Ry: At Taiwan.
> R: In Taiwan. Why are we making things for Taiwan?
> Ry: Umm, because we're doing a production about Taiwan.

Anna was able to articulate that her group's prop needed to be durable, and seen by the audience. She also recognised that a prop helps with the show, again aspects taught early in the unit.

> An: We've also got some hot dog sticks in the tail so the tail wouldn't flop around.
> R: Oh, why were they there? You said to flop around, why didn't you want the tail flopping around?
> An: Because, then nobody would see the tail.

Ajay photographed the boat and also identified that it had to be 'seen' by the audience. These statements indicate that both Anna and Ajay drew on learning from a guest speaker from a local theatre who talked to the children about the need for props to be clearly visible, and from viewing a small stage play which used props and associated discussion about the props used. Following these two activities the children and the teacher established through co-construction a number of criteria for their props, among other things these included durability and visibility for the audience.

Conversations with the Year 6 children revealed that two of the children were able to put skills and knowledge they had learned at home from their parents to use during the development of their props. The students were asked to plan their final props to scale. Alex was able to employ a strategy used by his father, of drawing a ruler down one side of the planning page. Below is the comment Alex shared with the researcher about the photo he requested the researcher take of him (Figure 1). The conversation also indicated that Alex knew that plans had to have considerable detail.

> Al: We put like scale and yeah, just all that sort of stuff.
> R: How did you know to put all that on a plan?
> Al: Well just because plans have like scales and all that......because I've seen plans that my Dad makes and stuff.
> R: Does your Dad deal with plans quite a bit?
> Al: He designs..... rally cars and stuff.

He also used one of his father's terms, 'make-shift', when referring to the making of mock-up washers:

> Al: Yeah. I basically did the stand. Then that's a bad picture of it standing up again. And then, oh, yeah, that's (pointing to one of his photos) the practice screws and making a 'makeshift' one, one of the makeshift washers.

Figure 1. Alex drawing the 'scale' on one side of his planning paper.

Maddy photographed bracing that she had purchased for her radio. She knew about the bracing because her father had used it in the construction of a tree hut at home.

R: And you asked, when you were joining the bits of wood, you asked me to get some little ...
Ma: Umm, brace.
R: Bracing. How did you know about those?
Ma: From Dad making our tree hut.
R: Right.
Ma: He used them to make it strong.

Stimulated recall also allowed the researcher to identify the participants' insight into technological knowledge and practice gained during the unit (Beers et al., 2006). Both classes of children and their teachers were undertaking their second unit of work in Technology using the 2007 New Zealand curriculum (the first had been done with the same researcher as an earlier part of this study). All but one of the participants in Year 2 recalled the term 'mock-up' and were able to articulate its purpose. One participant could not recall the name but understood it was 'a practice'. Anna's conversation below indicates a clear understanding of the purpose of producing a mock-up.

An: And then that's our fish there. This is our umm, mock-up fish.
R: Tell me what a mock-up is.
An: Well, it's something that is going to look like your real fish because we haven't actually like done our real fish.
R: So why do you do a mock-up first?
An: That gives you an idea what your fish is going look like.

Ryan's conversation with the researcher indicated that he had a very good understanding of what a plan was and the purpose of developing a plan. It also shows that he had an understanding of annotations (Figure 2).

Ry: The plan. This is the plan, it would tell you what it looks like.
R: What else would it tell you?
Ry: How big and how long.
R: What is this? (Researcher points to mark on the photo)
Ry: That's part of the fish, it's the eye.
R: What's this word here?
Ry: Eye
R: Why have you got that word written there?
Ry: Well, we write 'eye' there and then we do a point to where the eye is.

In Year 6 the autophotography interviews revealed some quite sophisticated thinking in terms of technological practice and knowledge. Alex understood that one of the structured learning activities, 'Pros and Cons', helped his group establish criteria for their prop. He also indicated that he used a mock-up to express an idea to his group. Alex had to change his technological outcome halfway through to accommodate

Figure 2. Ryan's group's plan with 'annotated' eye.

changing needs of a client. Alex and his group were designing a 1930s style microphone for a pair of Olympic commentators who were to originally stand on stage. Alex's responsibility was the stand for the microphone. When almost finished, Alex was informed by the script writers that the commentators would be seated at a desk on stage. The interview revealed that he was able to use the knowledge he had learned from creating the first version, to create the second in a fraction of the time. It also revealed Alex's understanding that the designed outcome must be functional and that he was happy to take his time "getting the design right".

During her interview Millie happily acknowledged that her early design was not from the correct era - their prop needed to reflect the era 1900–1936 Olympic Games. This response indicated a clear understanding of the need to meet the established criteria for the project. When discussing the project she used the term 'specifications' correctly and recognised that identifying them was an important part of the design process. She also realised that specifications could be referred back to. Millie's recall also indicated that materials may have an impact on functionality and design and that a mock-up guided her practice.

Like Millie, Maddy's interview revealed that she recognised the importance of specifications and she used the term 'mock-up' correctly; her interview signified she was aware of its importance. She recognised that planning needed to have scale, detail and different views, and ideas do not always go to plan. Maddy was able to modify her ideas to fit cost and availability of materials.

Ma: That's cutting out the wire for the speaker and it wasn't, the speaker wasn't really big enough so we had to cut out another one.
R: Explain that to me a little more. Why didn't you just make this bigger?
Ma: Well we couldn't really because it was just a cut-off that Miss D [classroom teacher] had got for free because otherwise she would have had to pay for more.

Jiyong understood the purpose of the plan and the need for accuracy. He also used 'mock-up' in context and the researcher was able to determine that he recognised initial research influenced final design and that materials are sometimes selected for ease of use. Jiyong's interview also revealed that materials influenced the authentic appearance of the prop his team was developing:

> These were our speakers and yeah, we did that on wire and we got, that card and we hot glue gunned them, it looks more like a radio and it kind of brings the message. ...Yeah, it looked, like an old radio.

Insight was also offered into Jiyong's understanding of the need for his technological outcome to meet the established criteria:

> ... then we've got to put a bit of cardboard over the top of it to make it strong and durable.

Tullan photographed a list of criteria needed for his prop. This was what he said about that photograph:

> I was trying to remember ....all of the things that the props need to go by, like durable, safe, ergonomically designed and the era and stuff.

From this statement we can see he clearly understood the significance of criteria to his practice.

The stimulated recall interviews also gave insight into the students' participation in technological design process. The participants were given freedom to photograph what they wanted to while developing their intended outcome; however the researcher did remind all the children about their cameras on occasions. When interviewed with their autophotographs, all but one of the students offered some insight into aspects of technological process. In Year 2 Dylan was able to recall the papier maché making process, Ryan could give an overview of his total process, and Ethan accurately described in detail the process of making the wings for his group's fish. In Year 6 Alex recognised that experience from the first Technology unit aided his practice:

> It's ok to change designs as you go.

Millie stated that the mock-up was something that she could refer back to, to assist her design process. Maddy recognised that one of the features on her mock-up was not needed in her final technological outcome and Tullan's interview revealed that he understood the importance of planning for his practice and referred to it as a guide through the development phase.

> Tu: Yeah, the timeline....was important because I had to remember what to do and remember what I had to put into my props.
> R: Why was the timeline important?
> Tu: Umm, because otherwise you could have as long as you want and it, it, you might sort of forget about it....or you might sort of, I don't know how to explain it, but sort of a deadline where you sort of have to have it done.

Working collaboratively was a significant component of the participants' practice. One of Millie's photos showed her working with her two team mates and her comment indicated her appreciation of the process of putting together ideas from the group:

> And this (Figure 3) is us working on our final design with our, we're like putting forward all our ideas.

The interview also revealed that she appreciated the ideas of others.

> It [the idea] came from Dochrane because he wanted to add it, it goes in the side of it so it can stay. It looks really good.

Figure 3. Millie and team members working on their final design.

DISCUSSION

Stimulated recall has allowed the researcher to gain an insight into the participants' thinking in relation to their technological practice; their interpretation of events; and their thinking at a particular point in time. This would have been difficult to have gained through direct observation and interview alone (Mackey, 2005). During the stimulated recall interview the autophotographs prompted children to recall and discuss aspects of their total practice that might not have otherwise been apparent or visible. In this section the findings are discussed around the four themes mentioned above.

From the interviews it became apparent that participant technological outcomes were clearly influenced by the prior learning experiences planned by the teachers as part of the unit. The unit was collaboratively planned by the two participant teachers and the researcher. Before the children began the planning and the construction of their props, a number of activities were set up to engage them in developing their understanding of 1) the function and purpose of stage props; and 2) the culture and/or era of their production which needed to be reflected in their props. Many of the autophotographs taken were of these activities and children were able to recall the activity and the purpose and direction of the learning facilitated by it. In Year 2 a number of visual activities the children undertook were deemed significant. These activities included a film clip of flying fish, a short film of Taiwanese fishing in action and a visiting speaker with a range of props from a local theatre. The data shows that these activities influenced their later practice. Two Year 6 children built on knowledge from their fathers which contributed to development of their outcome. While this was not necessarily the reason they took the specific photographs, when interviewed, both Alex and Maddy suggested their practice was influenced by something their fathers either did or said.

Participants were asked to take photographs of anything they thought was important or which would help them build their props as well as the important stages of building their props. Autophotography and the subsequent interviews allowed the researcher to gain insight into the depth of the participants' understanding of technological practice (Mackey, 2005). This was identified through the selection of photographs taken and what the students said about each photograph. The photographs evidenced significant aspects of their technological practice. Information from the interviews shows that a majority of children engaged with and understood the term 'mock-up'. Most of the participants photographed as least one aspect of their mock-up design and when talking to the researcher they were able to correctly explain its role in their practice. Another aspect of technological practice evidenced in the data was the participants' understanding of and engagement with established criteria for their props. In Years 2 and 6 criteria were referred to in a number of ways. Some participants linked to a guest speaker; others linked to a video of a stage show they were shown; others referred to the actual criteria that were co-constructed with the teacher in both classes. The participants' understanding of the process and purpose of planning technological outcomes was another aspect of insight. Four Year 2 and four Year 6 participants specifically referred to their planning or their plan in the correct context - as a guide to inform the construction of their outcome, clearly indicating understanding of planning as a technological strategy (Sime, 2006).

For the purpose of this study the researcher differentiated between the participants' understanding of specific terms and understandings as discussed above and their ability to identify and articulate the design and construction processes they undertook. The data revealed that two Year 2 children were able to recall in detail aspects of their process, for example: creating reinforced wings for Ethan; the process of papier maché for Dylan; and Ryan was able to recall his total practice. In Year 6 four of the participants shared insight into the design process by discussing how their process changed as new information came to light either through failure analysis or interaction with peers and stakeholders. One child recognised that planning for his practice through task identification and an associated timeline was able to guide his practice and keep him on task. Three of the Year 6 children specifically mentioned the benefit of referring back to either specifications, planning or task timelines. This clearly supports the literature claims that stimulated recall allows participants to explain their decision making (Mackey, 2005; Sime, 2006; Slough, 2001).

The final significant theme discussed in this paper is the insight gained into the participants' opinions of working collaboratively. All children in both classes worked in groups of three to design and develop a prop for their class item in the school production. The researcher observed the participants struggling to work together collaboratively; they argued about their decision making; had some difficulty in making decisions; and frequently approached either their teacher or the researcher with problems they were experiencing while working together. Surprisingly a number of autophotographs showed participants working in groups, and when asked about working collaboratively, five Year 6 and five Year 2 participants were able to identify positive aspects of group work. Recognition of the benefits of working collaboratively is significant because working collaboratively is an authentic way of undertaking technological practice (Turnbull, 2002). The data supports the findings of Doise and Mugny (1984) in that children working in pairs solved problems at a more advanced level than those working by themselves.

Two issues compromised the use of stimulated recall with autophotography in this study. The children in both classes were given cameras to record their learning and practice. Disposable cameras were used (for practicality reasons and to prevent deletion of taken photographs) and therefore the interviews could not take place until after the conclusion of the unit and when the photographs had been developed. Ryan and Dylan, Year 2, had each forgotten what two of their autophotographs were and why they had taken them, even though the Year 2 children undertook their unit during a 'Technology week' and were interviewed the following week. Two of the participants in Year 6 had also forgotten why they took the photographs. Plaut (2006), Sime (2006) and Slough (2001) suggest that stimulated recall should occur as soon as possible after the task is completed. Given that the Year 6 unit occurred over a five week period with the interviews taking place a further week after that, it is hardly surprising that some participants had forgotten why they had taken a particular photograph. To avoid this in the future I would use digital cameras and interview each child at the end of each lesson or at the end of each day. However given the low number of forgotten photos and the fact that all participants gave insightful comments related to their technological practice and process, the researcher does not see this as an issue significantly affecting results.

Three of the children claimed that someone had 'hijacked' their camera meaning the photo was taken by someone else when they had not requested it. Seung & Schallert (2004) and Sime (2006) suggest participants may censor or distort their thoughts and ideas in order to present themselves more favourably. Whether 'hijacking' occurred or whether the participants just took a few 'off task' photos was difficult to determine; however the result was the same - a few irrelevant autophotographs. Originally the researcher planned for students to take a maximum of 15 autophotographs; however the only disposable cameras available were 22 exposures. Given that the participants had a finite number of photographs they were able to take, there was a risk that they might have run out of photos on their cameras before the end of their technological practice. However most participants appeared to finish their cameras approximately in conjunction with their practice and some had photographs left. Therefore a few 'off task' photographs would have had little impact on the selection of aspects of their practice that the participants photographed. The researcher does acknowledge that some students might have taken more photographs during and at the end of their practice if more had been available. This is another reason to suggest the use of digital cameras in future studies.

A TOOL FOR TEACHERS

Although this study used stimulated recall using autophotography primarily as a research tool, the researcher, a New Zealand registered primary teacher, was also able to gain insight into the potential of it as an assessment tool for Technology. Allowing students to select and photograph their own work, evidencing their technological practice and photographing significant aspects of their learning or activities they find helpful during the development of their technological outcomes, is empowering. It also allows teachers opportunity to access insight into student thinking and learning at a later date through conversation using the photographs to stimulate students' thinking. This is particularly significant given the practical hands-on approach to Technology, especially in the primary school. It is not uncommon for teachers to be focused on physical and safety aspects of running a practical session and thus miss opportunities for both formative and summative assessment. Acknowledging that time in any classroom is precious, teachers may need to use this tool in a more selective manner than outlined in this study. It is hardly practical for teachers to interview all children in their class or discuss all photographs taken. In the section below the researcher suggests four strategies for using stimulated recall – autophotography in a more selective manner.

Selection and critique of existing technological outcomes. In the early stages of a unit teachers could ask students to photograph a range of existing effective designs within a given context. The photographs could then be used to stimulate discussion with the teacher and/or their peers about the physical and functional features for one or more of the designs.

Ranking and rating for functionality. Students could also use the above mentioned photographs to rate or rank the designs against given attributes or simply rank the designs according to functionality. Students could be asked to select one and justify either the most or least successful design or design feature. Conversations with the students about their selected photograph will allow teachers insight into the students' understanding of attributes and their relationship to functionality.

Significant aspects of the design process. Students could be asked to photograph significant aspects of their practice. These aspects may be the components that are a teaching and learning focus for the unit. For example teachers could ask students to photograph aspects of their design planning within one unit, modelling in another and brief development in another. Again ranking the significance of the photographs or the selection of the most significant will give teachers insight into students' understanding in specifically targeted areas. Teachers could use the photographs to engage the students in focused conversation about specific aspects of their practice. Photographs could also be used to develop a portfolio of technological practice and are particularly useful for reluctant or non readers and writers.

Photographing their final outcomes. Photographing a final outcome from a variety of views or aspects allows students and their teachers to engage in conversation about the physical and functional aspects of designs at a time that is mutually suitable to both. In the event of authentic technological contexts, final outcomes are often presented to stakeholders upon completion. Using stimulated recall – autophotography allows students to recall how their thinking, research and design ideas have influenced their final outcome. These photographs could also be used to engage students in evaluative conversation in relation to outcome attributes and specifications.

CONCLUSION

Stimulated recall is a research method that allows the investigation of cognitive processes through inviting participants to recall their thinking during an event when prompted by a form of visual recall. In this study stimulated recall with autophotography facilitated participants' conversations about their own learning and technological practice.

In the first round all children in both classes were taught how to take photographs using a digital camera and were asked to photograph their learning in specifically planned lessons. The identification of learning to photograph was very difficult and the children struggled with this task. In the second round students were asked to photograph the things that they thought might help them design and build their props and the important stages in designing and building their prop. In the second round the students had complete ownership of the photos they took or asked to have taken.

Stimulated recall using autophotography has considerable potential as a research tool in Technology education. Autophotography facilitates participant ownership of the items recalled. This participant ownership of the photographs taken is one advantage of this method of stimulated recall over researcher taken video recording. The selection of photographs in itself gives insight into participant thinking about what is or is not significant for them. Four themes emerged from the data which include insight into: participants' use of prior learning and knowledge; participants' knowledge and understanding of technological practice; participants' understanding and recall of technological process; and participants' thinking and attitude towards working collaboratively. This tool has allowed the researcher insight into the participants' understanding of technological practice and process.

For teachers stimulated recall could become a valuable tool for assessing students' learning, both formatively and summatively. Allowing students to photograph their own technological practice empowers them to consider their practice critically and to share with their teachers aspects of their learning and thinking that might otherwise be missed in a normal busy classroom.

REFERENCES

Absolum, M. (2006). *Clarity in the classroom: Using formative assessment* (2007 ed.). Auckland: Hachette Livre NZ Ltd.

Beers, P. J., Boshuizen, H. P. A., Kirschner, P. A., Gijselaers, W., & Westendorp, J. (2006). Cognitive load measurements and stimulated recall interviews for studying the effects of information and communication technology. *Education Technology Research Development, 56*, 309–328.

Cohen, L, Manion, L., & Morrison, K. (2000). *Research methods in education* (5th ed.). London: Routledge Falmer.
Compton, V., & Harwood, C. (2007). *Design ideas for future technology programmes.* Wellington: Ministry of Education.
Compton, V., Harwood, C., & Northover, A. (2000). *Technology education assessment: National professional development* (Unpublished paper). Wellington: Milestone Four and LITE (Assessment) 1999 for the Ministry of Education.
Fraenkel, J., & Wallen, N. (2006). *How to design and evaluate research in education.* New York: McGraw Hill.
Lichtman, M. (2006). *Qualitative research in education.* Thousand Oaks, CA: Sage Publications.
Lyle, J. (2002). Stimulated recall: A report on its use in naturalistic research. *British Educational Research Journal, 29*(6), 861–878.
Mackey, A., & Gass, S. M. (2005). *Second language research: Methodology and design.* Mahwah, NJ: Lawrence Erlbaum Associates, Inc.
Ministry of Education. (2007). *The New Zealand curriculum.* Wellington: Learning Media.
Moreland, J., & Cowie, B. (2007). *Young children taking pictures of technology and science.* Waikato: University of Waikato.
Plaut, S. (2006). "I just don't get it": Teachers' and students' conceptions of confusion and implications for teaching and learning in the high school English classroom. *Curriculum Inquiry, 36*(4), 391–421.
Schepens, A., Aelterman, A., & Van Keer, H. (2007). Studying learning processes of student teachers with stimulated recall interviews through changes in interactive cognitions. *Teacher and Teacher Education, 23,* 457–472.
Seung, L. D., & Schallert, D. L. (2004). Emotions and classroom talk: Toward a model of the role of affect in students' experiences of classroom discussions. *Journal of Educational Psychology, 96*(4), 619–634.
Sime, D. (2006). What do learners make of teachers' gestures in the language classroom? *International Review of Applied Linguistics in Language Teaching, 44*(2), 211–230.
Slough, L. (2001). *Using stimulated recall in classroom observation and professional development.* Paper presented at the American Educational Research Association.
Turnbull, W. (2002). The place of authenticity in technology in the New Zealand curriculum. *International Journal of Technology and Design Education, 12,* 23–40.

Wendy Fox-Turnbull
University of Canterbury
New Zealand

KEITH GOOD AND ESA-MATTI JÄRVINEN

17. EXCITING ELECTRICS – THE STARTING POINT APPROACH TO DESIGN AND TECHNOLOGY IN ACTION

INTRODUCTION

Creativity is arguably central to Design and Technology and much has been published on this (Kimbell, 2001; Spendlove, 2003; Davies & Howe, 2004). It is sometimes associated with genius but there are other interpretations. Benson (2004) writes that while teachers may have a future Picasso or Freud in their class, it is more likely that they will have children who have:

> … an original idea or solution that is original to themselves and not necessarily totally original. (Benson, 2004, p. 138)

This is what Craft (2002) calls little 'c' creativity which is within the reach of all children. The study described in this paper was based on the premise that all children are capable of a degree of creativity in identifying design problems and generating solutions to them.

The approach featured in this study has been used increasingly in the researchers' work with children, students and serving teachers in England and Finland. The *spa* model has been used by Good (1988) and more extensively in his *Design Challenge* series of books (1999a,b,c,d, 2000).

The Qualifications and Curriculum Authority (QCA, 1998) scheme of work is currently used as the basis for Design and Technology in 'over 90% of primary schools in England' (DATA, 2005, p. 18). In this scheme, the outcomes in a class are directed to have the same purpose, for example, the children are told to 'design a photograph frame'. In Finland one can see hangovers from handicraft education where pupils make artefacts almost to a 'recipe'. However, during the 2004 revision of the Finnish compulsory curriculum, teaching Technology was introduced as a cross curricular theme where children develop [technological] ideas and evaluate them.

THE STARTING POINT APPROACH

The starting point approach (*spa*) has some specific features which distinguish it from approaches that are characterised by outcomes with a common purpose (O'Sullivan, 2005). In the *spa* children are first introduced to specific technology and its applications in society. Then they are taught to make their own working

C. Benson and J. Lunt (eds.), International Handbook of Primary Technology Education: Reviewing the Past Twenty Years. 211–219.

example of the starting point, gaining knowledge and skills in the process. This involves what English teachers recognise as teaching ‘focused practical tasks’. In the *spa* resulting practical work is the starting point for designing. During group brainstorming led by the teacher, children develop a wide variety of different ideas for using the starting point.

Unlike the usual approaches in England and Finland, some making precedes designing and children can design ‘what they like’, as long as it is based on the starting point. The children have to select their favourite idea to make and evaluate. The *spa* seems to reconcile the apparently conflicting demands of teaching specific skills and knowledge whilst encouraging individuals to be as creative as possible. The common starting point is intended to provide stimulus for the children and make diverse projects feasible for the teacher.

PURPOSE OF THE STUDY

The central purpose of the study was to find out if the children could do what the *spa* asked of them and if it helped them to develop projects with different purposes within the group.

METHODS OF INQUIRY

The theoretical framework of the study was qualitative in nature and based on interpretative skills and inductive analysis, whereby the researchers continually explored the relationship between data and emergent findings (Ritchie & Hampson, 1996). The chosen starting point was a pressure sensitive switch made from card and kitchen foil. The study included an open search for children’s emerging ideas for ways to make a pressure pad go on. The researchers also wanted to see whether the children could apply this starting point in innovative and creative ways in their own environment.

The English children taking part in the study were from urban schools and were attending the Children’s University at the University of Greenwich. There were 16 children in the group, aged 11. The Finnish children were from Karhukangas Primary School, a small rural school in Haapavesi Township. The head teacher Markus Tornberg, helped to set up and carry out the Finnish part of the study. All 11 children from classes 5–6 (11–12 year olds) participated in the study.

Studies in England and Finland were conducted following an agreed ‘script’ designed to epitomise the *spa*. Before starting, the children were given an overview of the session. It was seen as important that the children knew from the outset that they would be asked for ideas for using the pressure pad. This was so that subsequent activities could be used as stimulus and to give maximum time for ideas to emerge.

Phase 1

The basic concept of a switch was discussed. This was revision for the English children who had covered switches as part of their National Curriculum Science. The children were shown a large pressure pad and how it worked. The characteristics of pressure pads were discussed, for example that they are thin, take up little room, are tough, and are operated by pressure. It was hoped that focusing on these special qualities might lead to ideas that were prompted by them.

Phase 2

The children were given a copy of the basic pressure pad pages from one of Keith Good's *Design Challenge* series of books: *Exciting Electrics* (1999, p. 12–13). The Finnish children were given a translated version of the same pages. They were also provided with all the materials needed to make a working pressure pad and a circuit for it to control. Every child made their own pressure pad.

Phase 3

The children were asked to think of where pressure pads were used in everyday life and their ideas were recorded on a flip chart. This was intended to consolidate the concept of a pressure pad and allow one idea to prompt others. The researchers then encouraged the children to brainstorm as many ways as possible to make the pressure pad switch go on (i.e. close the circuit).

Phase 4

During the final brainstorming, children were encouraged to generate lots of new ideas for using a pressure pad. Again, the flipchart was used for recording purposes. These ideas were intend to stimulate design and make projects of their choice. The research was focused on the following questions:
- Could children identify the existing uses of pressure pads in the world around them?
- Could children generate ways to turn on pressure pads in different ways?
- Could children find possible uses for their pressure pads?

ANALYSIS AND RESULTS

The researchers assumed the role of participating observers. This procedure enabled them to be 'inside' the study, true to the nature of qualitative research (Erickson, 1986). Data collected included brainstorming recorded on a flipchart, the children's notes and drawings, video recordings of the brainstorming sessions as well as photographs of the children's practical outcomes. Verbatim transcriptions were derived from the video recordings. During the analysis process, irrelevant data, such as children talking outside the project were excluded (Miles & Hubermann, 1994). All the collected data was submitted to analysis.

During the first round of analysis, the data indicated that the children were creating ideas for their own projects. This prompted the researchers to carry out further viewing of the data in order to specify those emerging features.

During the analysis process, the researchers shared observations during a series of meetings in Finland and England. Data examples presented in this article were analysed by both researchers individually and also in collaborative discussion (Ritchie & Hampson, 1996). Finally, the researchers reached the stage where they considered that they had investigated the data sufficiently from the viewpoint of the research problem. From this point the researchers proceeded to present results.

The inductive interpretative analysis process used in this study enabled the results to be presented as empirical assertions, with supportive data (Erickson, 1986, p. 45). Examples are referred to within the commentary in order to clarify the analysis process (see Järvinen & Twyford, 2000).

Empirical Assertion 1: The Children are Able to Find Existing Uses for Pressure Pads in the World Around them

The Finnish children came up with the following examples:
- scales (weighing fruit, etc. in the supermarket)
- car radios
- shop tills
- control panel for milking machine and feeding control in barn
- motor workshop – used to control engine hoist
- digital cameras
- cash point machines.

The English children came up with:
- cash machines
- light switch
- mobile phone
- TV remote control.

Commentary. The above examples demonstrate that the contributing children were able to find existing uses for pressure pads in the world around them and that the basic idea of a pressure pad was understood. Interestingly, the child who referred to control panels of milking and feeding devices identified quite recent applications of pressure pads in modern barns (his parents were farmers). The child also referred to pressure pads used to control an engine hoist in the workshop of their farm. Children's understanding of the technology in their surroundings was increased. The 'black box' technologies of control panels, weighing scales and other everyday devices became more understandable to the responding children.

Empirical Assertion 2: The Children are Able to Generate a Wide Range of Ideas for Turning the Pressure Pad on in Different Ways

When asked to think of different ways to turn on the pressure pad switch, the Finnish children came up with the following ideas:
- turn it over
- step on it
- lean on it
- knock on it

- put something on it
- throw at it
- somersault on it
- blow on it
- drop something on it
- drive over it
- put a can on it, when rains fills it to a certain extent - the switch goes on
- put it between the pages of a book.

The English children came up with the following ideas:

- step on it
- sit on it
- squeeze it
- pinch it
- head butt it
- put some weight on it
- belly flop on it
- elbow it
- punch it
- touch it with your tongue
- fart on it
- flick it
- kneel on it
- kick it
- throw the pressure pad against the wall
- blow on it
- stamp on it
- drop something on it
- put some water on it (meaning squirt water on it)
- slap it
- run over it
- tiptoe on it
- close the window on it
- lay on it.

These were added to the flipchart and acted out by the teacher researchers to repeat and reinforce the suggestions.

Commentary. The ideas did not rely on previous knowledge or experience since this was a new situation for the children. They were already being creative as they came up with plenty of ways to close the circuit with the pressure pad. Through this brainstorming session the children were establishing a basis for a wide variety of uses for pressure pads, including possibly novel and innovative ones. This was important as it gave a fertile basis for generating ideas for using the pressure pad later. Some unusual or less obvious ideas came up: for example the Finnish child's: *put a can on it, when rain fills it to a certain extent - the switch goes on. The English child's throw the actual pressure pad against the wall*, shows an interesting reversal of the normal pressing or throwing things on to the switch. This child seems to have stumbled across a recognised strategy for generating innovative ideas.

Empirical Assertion 3: The Children are Able to Find Possible Uses for the Pressure Pad Switch in their Own Environment

When asked to think of as many uses as possible for the pressure pad switch, the Finnish children came up with the following ideas:

- doorbell
- burglar alarm
- it could be used in a game – thrown at on a wall
- under bicycle tyre (e.g. to warn of theft)
- it could tell you it was raining, even if you were reading
- a kind of wind meter
- put by the side of the bed to tell when you have fallen out
- could tell you when something was full
- put pressure pad in door handle (to warn of sleepwalking)
- knocking doorbell
- used inside the mailbox to tell when newspaper has arrived - indicates inside the house
- put pressure pad on bird table to tell when birds come
- warns that a car is at your gate and you need to go and open it or in the road (in rural Finland, often small roads off the main one lead to houses).
- to control a torch.

The English children came up with the following ideas:

- control a remote control car
- under the door mat to turn on a tape recorder to scare people at Halloween
- stand a glass on the pressure pad to keep a night light on if you're scared in the dark. You could easily find your drink and you could use it as a light to help you read.
- an automatic door bell that no one would need to ring it and you'd know people were there… hide it under the mat
- put a weight on it and it would give you light to work in the garden at night …use the light as a signal, they used it in the war and out at sea
- a car goes over it and the bulb comes on instead of speed cameras
- use it to tell which model car has won as they roll down a slope
- a game for children... like a play mat
- when they stop a lorry (truck), they might want the light on. When the car goes quiet.
- if the driver was really tired there could be a buzzer to wake him when he drops off
- a different burglar alarm so that if he comes in the window and the window shuts the buzzer would go on
- when burglars put their hand in the letter box and try and push the door then when the letter box shut the thing would go off
- detecting when a dog gets out of its basket when it has been told to stay in.

Commentary. These 27 examples seem to show that some of the children were able to combine the concept and functioning of a pressure pad to produce innovative product ideas. Importantly, some of the children's ideas can be regarded as innovative and novel applications of the pressure pad concept, e.g. the mailbox that tells when the newspaper has arrived and the night light operated by the weight of the glass. This is in accordance with the definition of technology as 'human innovation in action' (ITEA, 2000). It is important to notice also that many of the children's ideas are feasible in practice and could be a basis for their actual projects in Design and Technology education. This was the purpose of the *spa* from the start. Infrequently, ideas came up that probably could not be made to work, at least at first sight. This was often because the essential nature of the pressure pad had been forgotten. Other children quickly reminded the group that pressure was needed and an idea based on sound, for example, would not be practical. This needed careful handling to preserve the idea giver's enthusiasm. Making an apparently impractical idea feasible was another chance for the adults and children to use their creativity. The children were encouraged not to dismiss ideas too readily. It must be realised that each idea listed could be the starting point for many different designs. The children went on to explore these through drawings, modelling and discussion and resulted in some being made into finished artefacts.

One of the Finnish pupils applies the concept of the pressure pad to the context of a mailbox and remote sensing. In Finland it is common for mailboxes to be at the boundary of a property. In this idea, a pressure pad in the bottom of the mailbox would activate an indicating light when the mail arrives, so alerting the householder. This is an example of combinational creativity. Michalko (2001) devotes a chapter in his book on idea generating strategies that involve making novel combinations. It seems that the 'mailbox child' did this naturally. The mailbox case also illustrates the importance of the 'audience': the Finnish teacher/researcher knew the context and was able to appreciate the usefulness of the idea. However, teachers may sometimes need to get children to explain the context for their ideas if they are to appreciate them.

Significantly, most of the above ideas seem to occur as a response to the children's own needs, interests and purposes, true to the nature of Design and Technology as it should be.

DISCUSSION

In this study an effort was made to add to the children's understanding of the made environment. The data indicated that some of the children were able to make meaningful connections with a pressure pad by identifying existing uses for it in the made world around them. This in itself has value, demystifying the technology by having the children build their own examples. When they were making the pressure pads, the children acquired skills and knowledge of basic issues in electricity (open/closed circuit, conductor, etc.).

It was evident from the data that some of the children were able to apply the pressure pads in a creative and innovative manner as a response to the problems they identified. Importantly, it was not known beforehand what applications of pressure pads would emerge from the children's creative minds. The technological process did not aim just at discovery (as in science), but rather and more essentially, at children's innovations. In this regard, many of the children who took part in the study acted in accordance with the idea put forward by Adams (1993):

> Successful inventors that I know are extremely problem-sensitive. They are tuned to the little inconveniences or hardships in life that can be addressed by the technology they know. (p. 87)

Importantly, this is in accordance with how the made environment has developed and still develops through human activity. Ingenuity, innovation, problem finding and problem solving are part of the basic essence of technology (Sparkes, 1993; Järvinen, 2001). This could also be crystallised in the definition: 'technology is human innovation in action' (ITEA, 2000). Consequently, teaching Technology should not be mere study of how technology works. Children need to be given opportunities for creative and innovative action. This is why the researchers wanted to focus the study on the innovative use of pressure pads in applications arising from the pupils' own ideas. This relates also to the concept of situated learning (Lave, 1988).

The *spa* seems to facilitate children's creativity in Technology education to a greater extent than an approach where the teacher specifies the purpose of the project for all. However, it is not so open that children have to search for a need or problem to solve without any support. Although the making stage of the *spa* is close to focused practical tasks in the English National Curriculum, these usually lead to projects with the same purpose within a class. The authors do not claim that using the *spa* is the only worthwhile approach to Technology teaching nor that it is the only way to foster children's innovativeness. Using the *spa* need not hinder the ability of children to have an open and sensitive mind to identify needs and problems without a starting point.

However it seems to the researchers that the *spa* offers a compromise between what the teacher and student can manage, what needs to be done and what the student would choose to do. By giving opportunities for children to identify their own problems and design their own solutions, the *spa* seems likely to increase their perception of Technology education as relevant. This approach is primarily aimed at maximizing creativity but it may also help motivation and behaviour. The children can be said to have greater ownership than when the purpose of the projects is imposed. The *spa* seems to offer a way of allowing individual children to identify their own design problems and for outcomes with different purposes to be designed and made within a class. Thanks to the shared starting point, this can be done while maintaining the sanity of the teacher.

REFERENCES

Adams, J. L. (1993). *Flying buttresses, entropy, and O-rings. The world of an engineer*. Cambridge, MA: Harvard University Press.

Benson, C. (2004). Professor John Eggleston Memorial Lecture 2004. Creativity caught or taught? *Journal of Design and Technology Education, 9*(3), 138–144.

Craft, A. (2002). *Creativity and early years education*. London: Continuum.

DATA. (2005, September). Survey of provision 2004/2005. *Datanews, 30*, 18–20.

Davies, D. L., & Howe, A. (2004). How do trainee primary teachers understand creativity? In E. W. L. Norman, D. Spendlove, P. Grover & A. Mitchell (Eds.), *Creativity and innovation: Proceedings of DATA international research conference* (pp. 41–54). Wellesbourne: DATA.
Erickson, F. (1986). Qualitative methods in research on teaching. In M. C. Wittrock (Ed.), *Handbook of research on teaching* (3rd ed.). (pp. 119–161). New York, NY: Macmillan Library Reference.
Good, K. (1988). *Starting CDT – Projects*. London: Heinemann.
Good, K. (1999a). *Amazing machines, Design challenge*. London: Evans Brothers.
Good, K. (1999b). *Exciting electrics, Design challenge*. London: Evans Brothers.
Good, K. (1999c). *Super structures, Design challenge*. London: Evans Brothers.
Good, K. (1999d). *Moulding materials, Design challenge*. London: Evans Brothers.
Good, K. (2000). *Teachers' book, Design challenge*. London: Evans Brothers.
National Advisory Committee on Creativity and Culture. (1999). *All our futures: Creativity, culture & education*. London: Department for Education and Employment.
ITEA (International Technology Education Association). (2000). *Standards for technological literacy: Content for the study of technology*. Reston, VA: ITEA.
Järvinen, E.-M., & Twyford, J. (2000). The influences of socio- cultural interaction upon children's thinking and actions in prescribed and open-ended problem solving situations (An investigation involving Design and Technology lessons in English and Finnish primary schools). *International Journal of Technology and Design Education, 10*, 21–41.
Järvinen, E.-M. (2001). *Education about and through technology. In search of more appropriate pedagogical approaches to technology education.* Acta Universitatis Ouluensis/Scientiae Rerum Socialium E50. Oulu: Oulu University Press. Retrieved from http://herkules.oulu.fi/isbn 9514264878/
Kimbell, R. (2001). Creativity, risk and the curriculum. *Journal of Design and Technology Education. 5*(1), 3–4. Wellesbourne: DATA.
Lave, J. (1988). *Cognition in practice. Mind, mathematics and culture in everyday life*. New York: Cambridge University Press.
Michalko, M. (2001). *Cracking creativity: The secrets of creative genius*. Berkeley, CA: Ten Speed Press.
Miles, M. B., & Huberman, A. M. (1994). *Qualitative data analysis*. Thousand Oaks, CA: SAGE Publications.
O'Sullivan, G. (2005). Creative problem solving in technology education, A juggling act. In C. Benson, S. Lawson & W. Till (Eds.), *The proceedings of the fifth international Design and Technology conference – Excellence through enjoyment*. Birmingham: CRIPT at UCE Birmingham.
QCA, Qualifications and Curriculum Authority. (1998). *Design and Technology – A scheme of work for key stages 1 and 2*. London: Qualifications and Curriculum Authority.
Ritchie, S. M., & Hampson, B. (1996). Learning in the making: A case study of science and technology projects in a Year Six classroom. *Research in Science Education, 26*, 391–407.
Sparkes, J. (1993). Some differences between science and technology. In R. McCormick, C. Newey & J. Sparkes (Eds.), *Technology for technology education* (p. 36). London: Addison-Wesley Publishing Company.
Spendlove, D. (2003). Gendered perceptions of creativity and Design and Technology. In D. Spendlove & E. W. L. Norman (Eds.), *The proceedings of the DATA international research conference* (p. 100). Wellesbourne: DATA.

Dr Keith Good
University of Greenwich
England

Dr Esa-Matti Järvinen
University of Oulu, Oulu Southern Institute
Finland

GILL HOPE

18. TAKING IDEAS ON A JOURNEY

Researching Designing in a Kent Primary School

RESEARCH CONTEXT

Designing and researching are both teleological.
Journey's end, if not in view, is somewhere over the horizon.

This couplet, along with other bits of idiosyncratic terminology, enabled me to place-mark my journey through the maze of learning to be a researcher: from being an angry middle-aged infant teacher (demanding to know on what research data were based the English Design and Technology National Curriculum's (1990) claims of what six year olds will do) to being a university lecturer with a Ph.D. behind her (2003), wanting to move into reflective "scholarly" work, developing "conclusions" rather than reporting "findings".

My research focused on young children using drawing to develop design ideas "Drawing as a Tool for Thought". This chapter reflects on the journey as well as reporting some of the findings along the way.

DESIGN AND RESEARCH

My research split into two phases, which I called the *Exploratory* and *Structured* phases. These seemed to parallel two ways of designing: hands-on "design-as-you-go" which typified children, novices and craft-workers, and planned ahead "design-before-you-start" style typical of industrial practice apparently underpinning National Curriculum expectations (Hope, 2000). The Exploratory Phase was firmly in the "design-as-you-go" camp.

Middleton's (2000) model (Figure 1) captured not only the children's design processes but also my research experience (especially those arrows going back to the "Search and Construction Space" after falsely assuming arrival at the "Satisficing Zone").

In the Exploratory Phase, I collected more than 500 design drawings from about 350 children aged 5–9 years and devised a classification system (Hope, 2001b, 2005). It became clear that although the younger children could not use the more sophisticated types of drawings, once children had gained an understanding of using drawing for designing, they were able to choose appropriately for their needs of the moment. For instance, if they saw many possible solutions they would make many quick sketches, whereas one good idea would be drawn more carefully and then re-drawn with changes.

C. Benson and J. Lunt (eds.), International Handbook of Primary Technology Education: Reviewing the Past Twenty Years. 221–234.

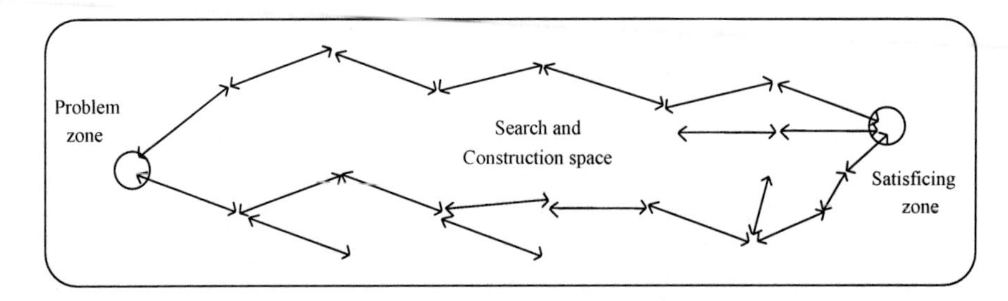

Figure 1. Middleton's (2000) model of problem solving.

As a researcher, this collection enabled me to say with relative confidence what children were likely to be able to do at different ages (Hope 2001b, 2005). If I had not also been a teacher, I might well have stopped there. But as a teacher, the inevitable next question was: can I get them to do it better? This demanded that I excavate the reasons *why* children were producing the different types of drawings, not only at different ages, but in response to different activities. Also: what was meant by "better"?

At the crux of the problem seemed to be the children's lack of understanding of the role of drawing within the process of designing (Egan, 1999), which led to the next question: if I were to teach them the purpose of design drawing, would they understand and be able to do it? The Structured Phase of the research, therefore, involved answering these questions.

CONTAINERS AND JOURNEYS

The journey metaphor seemed to come with me and coloured the way I viewed both my own research fumblings and the design activities of the children. Lakoff & Johnson (1980)'s book *Metaphors We Live By* contributed greatly to the development of my thought on many fronts but one of their examples (p. 90, see Hope 2001a) related quite specifically and led to the development of the model shown as Figure 2:

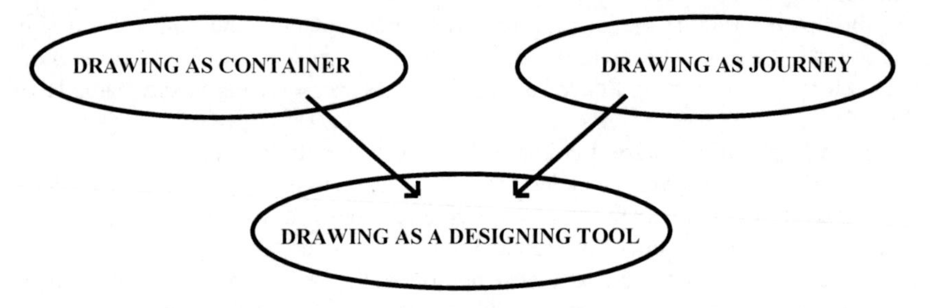

The container is the product of thought, the journey is the process.

Figure 2. The container/journey metaphor of designing.

The plan for the Structured Phase was to explain the purpose of design drawing to the children in terms of this metaphor. My knowledge of the mechanics of conducting research activities had moved along on its own journey across the Exploratory Phase and so comparing children in 2001 with children in 1996 was of questionable validity. Additionally, received wisdom and personal teaching experience suggested that Year 3 (aged 7–8 years) children could cope with linguistic metaphors (even poor ones) and that this kind of explanation would be beyond the grasp of Year 1 children (aged 5–6 years). Year 2 (aged 6–7 years) was, therefore, the choice. If the metaphor was a good one, it would work; if not, then no.

A programme was devised in which I taught Design & Technology to one class on a weekly basis throughout Year 2 and into Year 3 (my "Focus Class") whose developing design capability was compared at the end of each school term to the "Comparison Class" who learnt Design & Technology with their class teacher. She had worked with me in Year 1 and had been privy to my research activities throughout the Exploratory Stage and I had used children in her classes as research subjects during that phase. She was quite confident, therefore, that our teaching skills were not being compared and that I was comparing my new ideas with what I had done before and that her class represented the "before". The children had been my class in Year 1, which meant that they were quite saturated with "before". They had also been the class that Sue Hammond had used (as 4–5 years old) for her M.A. dissertation research, so between us we had a large amount of rich data on these children.

The programme taught to the Focus Class had to:

- parallel the making skills being taught to the Comparison Class otherwise they would go into Year 3 with very different levels of practical capability;
- cover the full range of Design & Technology knowledge, skills and understanding specified in the National Curriculum;

The programme had *not* to be:

- a design drawing programme and at no time were the children taught specific drawing techniques.

I was aiming to teach them the *purpose of the drawing* not *how to draw*. Any drawing that moved design thinking forward was a valid means of doing so.

The Comparison Class were also using drawing to plan what they wanted to make. The difference was that the Focus Class were taught the purpose of drawing for designing (using the Container / Journey metaphor) and the Comparison Class were not. This was introduced to the children part-way through the programme rather than at the beginning, to ensure that I had sufficient data from the first two Assessment Tasks to demonstrate that the two classes began from a level position.

Stan: the Metaphor

The context for introducing the Container / Journey metaphor to the children was puppet-making based on the story of Flat Stanley by Jeff Brown, who goes on a journey to America through the post in an envelope. This was not chosen because of the metaphorical link but because it was the design activity that I knew best. It had been my starting point, my first research activity (1996), conducted as a small-scale study for my M.A. dissertation. Conducting the activity across every year group in Jan.1998, experimenting with as many different ways of presenting it as I could, meant that I knew its possibilities and pit-falls well.

I showed the Focus Class children a selection of drawings from my collection. On one side were design drawings clearly showing a development of ideas and on the other were drawings done in other contexts. I asked if they could tell me the difference between the two sets. After a moment's silence, one girl tentatively asked "Are those ones planning drawings?" And the other set are? "Pictures" said the majority. Not only had they identified the difference but they had produced appropriate terminology.

I invoked my metaphor. I said that planning drawings were a bit like going on a picnic, you put the things you think you need into a carrier bag and off you go. You get to the first place, sit down and take things out of your bag, you might eat a sandwich and have a drink but then see some blackberries on a bush or pick some daisies and put them in your bag and set off again. At the next place you have a look in your bag to see what it contains now, lay it out on the ground, eat some, drink some, collect a few more things, put them in the bag, off you go, and so on, until by the time you get back home, your bag contains far more than when you started and you have lots of useful things to play with. These planning drawings were a bit like that. The children who drew them had started with some ideas and drawn them, but then moved on and used some of those ideas (along with new ones they just thought of) and that their ideas had gone on a sort of journey across the paper. Each drawing on the page contains their ideas at each point on their journey. The children have taken different journeys, some have had lots of completely different ideas and some had one good idea which they had changed several times. That was fine because the end of the journey was not on the paper but in the thing they made, which might or might not look like what they had drawn because although the journey began on the paper, it went off the paper into the making and continued throughout the making until they were happy with their finished product.

The children looked totally bemused, as if they had not one single idea what I was talking about but they drew planning drawings that contained design ideas that went on journeys across the paper (better than many Year 3s had produced for me before) and the puppets they made looked like a development of what they had drawn. I went back to my own classroom and said to Jackie, my teaching assistant: "I now know how to teach Design & Technology." She replied: "Well, you should do, you've been doing it long enough."

Place-Marking: the Container

I inherited the term "place-marking" from my colleague Sue Hammond's M.A. dissertation (Hammond, 1997). Her study focused on children's emergent writing skills and the way in which 4–5 year olds used a combination of drawing, pre-writing mark-making and single letters or parts of words to record the development of a narrative on the paper. These looked like proto-design-drawings: the easy combination of graphic and written symbols, the use of the paper space in non-conventional ways, the priority of developing ideas over other considerations. These were the skills that children brought to the early writing process but were left undeveloped because of schools' inevitable focus on swift movement towards conventional linear recording in readable script as the means of expressing ideas. These early intuitive combinations of graphics and text are foundational to design drawing. Given the freedom and encouragement to use these latent skills, my Focus Class did so.

How neatly or precisely the children communicated their ideas to other people was not, in my view, the purpose of them using drawing for designing. In my view, drawing was a staging post on the way towards a destination that existed in another medium (card, fabric, food, etc.). The Comparison Class was taught how to do properly labeled diagrams to show what they wanted to make. Although they produced convincing diagrams in the Assessment Task immediately after this teaching, they did not use them effectively for developing design ideas. It seemed as if they were more concerned about producing a good *drawing* than about using drawing as a means of developing a good *design*.

Encouraging the Focus Class children to talk as they drew was important for developing their use of the drawing as a discussion document. Instances such as the following were recorded on video:

"What you could do is…" (child prodding friend's paper with own pencil)
"What I'm going to do…." (child holding pencil off paper half way through drawing)
"Look…" (pointing to own paper in animated conversation with group)

In response to "I'm going to…" came either:

"Oh, yeah, and.." indicated the sparking of creative ideas from one child to another, or
"No, because…" indicated a reference back to the task criteria.

The Comparison Class, however, drew quickly, quietly, and came to show me (not their friends) the drawing, almost as if it were a permission ticket to be allowed to go on to the next stage, fetching the materials, at which point the real designing started to take place.

The final Assessment Task of the programme was to design and make a model of a maze that would help Theseus escape from the Minotaur. The children were in Year 3, average age 7.5 years at this stage. It became obvious that the Comparison Class were not using drawing to develop design ideas, not only by their haste, but by the sudden increase in noise level as, materials in hand, they began designing by discussion and prototyping. The Focus Class, by contrast, was so talkative whilst drawing that I was worrying whether there would be time to make any mazes at all. What emerged, however, was significant.

Not having sufficiently defined a possible solution before cutting into the materials, the Comparison Class began to develop their design ideas once engaged with the materials and made a product which related to, rather than answered, the problem as set. Many children used drawing simply to produce a "puzzle book" maze, far too complex to make as a three dimensional structure from card. One boy, Alex, said to me "Look, I've got to make all that!" as if just realizing the implications of the over-complexity of his design. Zara, drew a simple room with a single barrier and a river outside. She then made a cut-out crocodile and a boat for Theseus to escape from this creature after the escape from the Minotaur. Effectively, she was constructing Episode 2 (see Hope 2004), which really challenged my views on what creativity within Design and Technology should look like. How far from the plot should a child be allowed to stray? What if the product does not answer the design brief at all? Is that not how post-it note glue was invented – by accident? Many of the boys were engaged in glorious flights of fantasy, making staircases for Theseus to fall from and snake pits to fall into. What was not happening in this classroom was a sense of movement towards solving the problem that had been initially set to them: helping him to escape. They were, corporately, seriously off-task, regardless of the diversity (divergency?) of individual responses.

My evaluation notes on the Comparison Class session read:

> Some of them were in a sort of narrative mode - making bones, water etc. for the Minotaur. They did not define & solve the problem at drawing stage & then find they needed to make changes (as did Focus Class) but developed ideas in the making - and these then diverged from the "model for Theseus so he knew the way out" scenario. They were making a model of the maze as a personal play object. The different understandings and working methods of the two classes were immediately apparent. Some of the Comparison Class were still at the "drawing a picture to define the problem" stage. There were considerable numbers of the Comparison Class who ignored their plans completely, also quite a few drew a maze on the yellow base and then built a set of walls round the outside only.

The Comparison Class had left too many possibilities hanging and un-addressed before handling the construction materials. They were then still unfocussed on the problem to be solved and so used the materials to make what they fancied rather than solve the task. The Focus Class had a far higher level *of engagement* with the task, understood what was to be done and set themselves to satisfy the criteria of the problem. The Comparison Class were not grappling with the problem about making Theseus a model of a maze. They were doing something else, parallel to that: playing with the idea of a maze or making themselves something with maze-like characteristics and figures to move around within it.

In contrast, for the Focus Class the process of drawing enabled objectification of their ideas to themselves whilst also exposing ideas to public scrutiny and comment. By having these discussions across the drawings, the children were honing their ideas and developing a realistic idea in their heads of what they would make. The Comparison Class developed their inner images as they made the product, discussing with friends and incorporating new features into their design without reference to the task criteria. Focus Class children got up from their seats *as a group* to fetch materials once they had solved the conceptual problems. This did not mean that their products were alike. Each child had their own idea of final form but they looked at each other's and made comments and suggestions. The design energy, however, happened across the drawings. Not surprisingly, the Focus Class' final product resembled their drawings to a much greater extent than did those of the Comparison Class, yet the final products often showed far greater variations in form across the class as well as far more closely answering to the task criteria.

Multiple Routes to an Uncertain Destination: the Journey

There appeared to be a level of symbolic manipulation implicit within the children's design drawing capability, which once achieved could be exploited at will. Drawings that exhibited journeying displayed an understanding that drawing could be used to represent ideas that could be changed and developed, "seen as" an object imagined in the mind's eye, a place-marking from which a design journey could be continued. Prior to this realisation, the recording of design ideas is static: a possibility is drawn but does not represent or support the flow of ideas towards possibilities or solutions.

The child's earliest understanding of a drawing is as a *product*. They do not understand that drawing can be used to develop ideas about a product to be made in another medium. These drawings frequently have pictorial features such as trees and flowers, rainbows, or even a person holding the object that the child has been asked to design. Once children realise that it is only the object to be designed that should feature on the page, then this is what they will draw. It is abstracted from context (no rainbows etc.) but the child is using the drawing to *clarify for themselves* what it is that they are being asked to make. They have no grasp of the idea that drawing can be used as a tool for the development of a design. It takes another cognitive leap of imagination to realise that drawing is an *abstract system* that can be used to support the development of design ideas, that the paper can be used to record lots of alternatives or a good idea can be re-drawn with modifications and improvements. In essence, the child needs to realise that the purpose of using drawing for designing is to work out what will be made and how to make it. This is represented graphically in Figure 3.

Crossing the bridge between clarifying the task and designing solutions means that children can then choose the recording technique that they feel most appropriately fits their level of clarity about the task in hand. Arriving on the bridge indicates arriving at an understanding of the journeying aspect of the genre of design

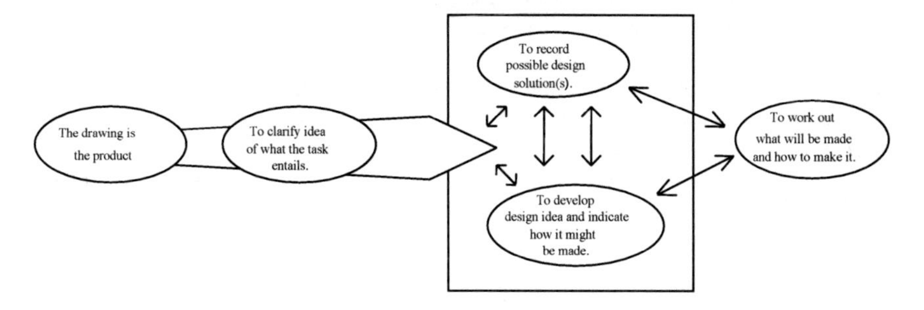

Figure 3. Developing understanding of design drawing.

drawing, which is the key factor in understanding the purpose of drawing for designing that allows children to cross the "bridge" into fluency with the genre of design drawing. Once children understand that drawing can be used to develop initial ideas and work out how to make a product that satisfies the design brief, then they are able to choose the kind of drawing that suits their response to the task.

Gaining such understanding enables children to choose to record multiple possibilities if swamped with ideas or to record and develop their instant "I know what to do" reaction towards a design resolution to see if it would work. Understanding the purpose of drawing for designing, as an aid to generating and developing design ideas, sets children free from having to produce any particular sort of drawing, even to the extent of knowing when not to draw at all (e.g. writing a list of materials). Several possibilities recorded as quick sketches represent multiple possible directions in which the design could go (like roundabout exits). The development of a single idea towards making is more like the unfolding of a route with few side-turnings.

THE RULES OF THE DESIGN GAME

Liddament's (1991) article Design Talk rested on Wittgenstein's (1969) assertion that every field of human endeavour has a set of underlying assumptions and metaphors with which the discipline is framed and through which it is subsequently seen by those who engage in it (which he called "language games"). This is close to Polanyi's (1958) position, whose "tacit knowledge" is the ill-defined, frequently unspoken, underpinnings of a subject or discipline that needs to be absorbed in order for a community of practice to function effectively. Parallels exist with paracosms: the ability to imaginatively create a whole alternative fantasy world for personal exploration. The emergence and burgeoning popularity of computerized simulations such as "Second Life" demonstrate that the ability to create and inhabit paracosms is common and enjoyable, not confined to fiction writers and/or eccentrics. The skill is used to greatest effect by authors, inventors and designers. What is required is the ability to construct and mentally inhabit a logically closed system. Strokes of genius and high level creativity occur when parallels are seen between features within two such systems and leaps of imagination are made between the two.

Winnicott (1971) viewed play as the ability to combine the outer reality of the world with the inner dream world of the imagination, and Liebermann (1977) noted parallels between children's playfulness and the adult trait through devising a series of research tasks, many of which are design based (e.g. How could you make this toy more appealing for other children to play with?).

At playtimes, Year 1 boys often run around in small groups, kicking a football wildly across a school field, throwing themselves dramatically on the floor, touching hands and shoulders, dancing sideways, and when the ball occasionally hits the fence between their designated posts, running around wildly with their hands in the air, leaping on each other with hugs of joy: role-playing being footballers as seen on TV. Perhaps it follows, therefore, that they could also learn to role-play being designers, if they knew what that was. However, to do so convincingly, their game-playing needs to develop into the kind that Year 3 and 4 children play: using rules to maintain systematic play within the confines of a delineated physical space.

There seem to be parallels with the human ability to play games with complex rules, to construct and maintain paracosms, understand and tell jokes, read between the lines and make inferences in story texts and work creatively within the given constraints of a design activity, all of which seem to emerge at about the same age or stage of cognitive development. Translated into designing, this means being able to solve the problem or see the opportunities within the situation as given. Donaldson (1992) called this the ability to solve "this problem and this problem only." One of my Assessment Tasks involved designing and making an Easter Egg Holder from a piece of card tubing that was far too wide to securely hold the sample egg. The task centred on solving the technical problem of creating an internal structure that would hold the egg securely. Several Comparison Class children simply discarded the card tube; which may have been common sense but it was not the problem they were asked to solve. These children had not come to terms with having to solve "this problem and this problem only."

For the Focus Class, to be a designer meant having ideas about what might be made, developing them on paper by drawing, talking with friends about their ideas, making changes, perhaps adopting or adapting someone else's better idea, in order to create a workable design solution to a specified problem or opportunity, that could be made with the materials provided. Design drawing had become one of the tactics of the game they were learning to play. This internalized tacit understanding of what being a designer was about had not, it would seem, occurred within the Comparison Class. They had not been able to second-guess how to play the game.

Standing between the inner image and the outer reality, drawing's clarity and ambiguity enabled the Focus Class children both to see clearly their ideas and allow for multiple re-interpretations of them. Using drawing effectively as the primary modelling tool for their ideas put the Focus Class one step ahead of their peers in the Comparison Class for the simple reason that drawing is such a temporary, ambiguous and adaptable medium. They could have as many attempts as they liked to find a solution or design a product. They could discuss with their friends, borrow ideas from each other, ask each other which of their ideas they liked best and act on the response: (e.g. "Craig said that one looked like Superman.")

This opening up of their ideas to public scrutiny made the Focus Class children more confident in their own abilities as designers. I would circulate with my notebook, sketching their products in progress and jotting down their comments, and they would tell me about it at the speed at which I could write it down. It would be couched in clear language, *written* rather than *oral* genre. So, I think that my interest in what they were doing also had a significant effect. Frequently, I would be sitting at one table talking to children about what they were doing, with a little group standing next to me waiting for help or advice, and a child would have waited (sometimes for several minutes) to say to me "I have done something really interesting I want you to look at and put in your notebook." In effect, the very act of researching their designing helped to make them better designers.

OFF THE END OF THE MAP: *WHERE NEXT?*

If the children in the Focus Class had only become better at drawing, there would have been little real point to the whole endeavour. In this article I have attempted to give a flavour of the research from the perspective of the fumbling, stumbling researcher, who began with a box of card puppets and their accompanying drawings and went off on "a terrific (even epic) struggle of understanding and improvement" (Kimbell, 2003).

Figure 4 shows the analysis model used to assess the children's design capability in the Structured Phase. It demonstrates how understanding the purpose of design drawing was seen as central to the ability to use drawing as a basis and support for design thinking. Children such as Zara, quoted above, often scored well on Exploring Possibilities but low on Addressing Constraints. Children who produced well-drawn labeled diagrams might score highly on Communicating Ideas, but lower on Evaluating Whilst Planning.

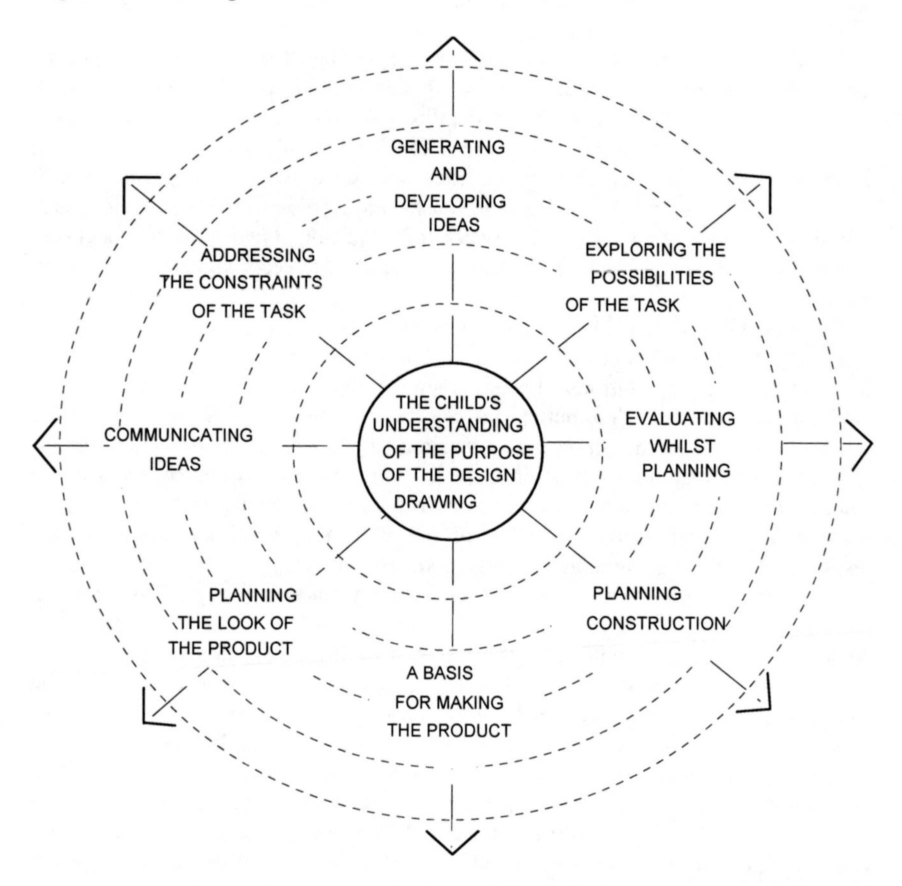

Figure 4. The dimensions of drawing for designing.

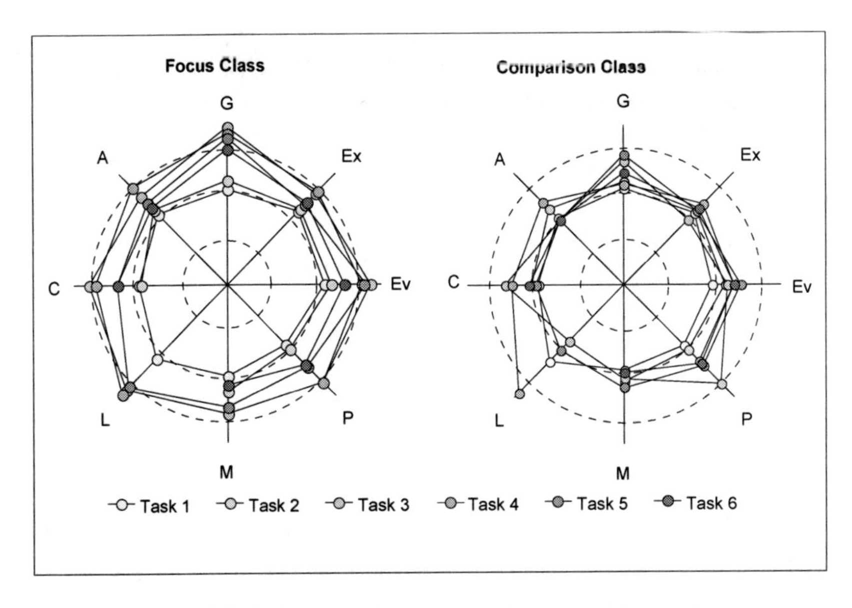

Figure 5. Radial plots to show progress in design understanding.

Figure 5 shows how, across the course of the programme, the Focus Class' understanding developed and progressed in a way in which that of the Comparison Class did not (see Hope 2003).
The following abbreviations are used for the dimensions:
G = Generating and Developing Ideas
Ex = Exploring the Possibilities of the Task
A = Addressing Task Constraints
L = Planning the Look of the Product
C = Communicating Ideas
P = Planning Construction
Ev = Evaluating Whilst Planning
M = Basis for Making the Product

The Tasks were:
Task 1 (Oct 2000) = design a pizza
Task 2 (Jan 2001) = design a way to help Frosty the snowman cross a thawing lake to reach a shop on the other side
Task 3 (April 2001) = design and make packaging for an Easter Egg
Task 4 (June 2001) = design a surprise card for a friend or relative (Kay Stables conducted this task from the suite of research activities developed at TERU, Goldsmiths College, London)
Task 5 (July 2001) = design and make a mock-up of a travel bag for a toy panda
Task 6 (Dec 2001) = design and make a 3-dimensional model of a maze (to help Theseus escape from the Minotaur)

Not all dimensions were considered for all tasks. Planning the Look of the Product was inapplicable to Tasks 2 & 6 (since these were problem-solving scenarios rather than product design tasks) and the children did not make the product for Task 4.

Figure 6 shows the assessment criteria on which these radial plots are based.

Continuum layer	1	2	3	4	5	6
Generating and Developing Ideas	Drawing a picture of an object, not designing a product	Single, simple drawing	Several attempts to improve drawing of single idea	Several unrelated drawings showing a range of ideas	Progression of ideas across drawings	Using drawing reflectively to generate new ideas
Exploring the Possibilities of the Task	Not exploring design problems	Minimal use of drawing to explore ideas	Aiming to improve the drawing, not explore solutions	Brainstorming onto paper without developing a solution	Using drawing to develop a design solution	Combines previous ideas with new ones to produce the best solution
Addressing Task Constraints/ Client Needs	Minimal understanding of the clients' needs or wants	Drawing conveys partial understandin g of addressing the task	Drawing shows understanding of addressing needs of a client	Several ideas for satisfying client needs are recorded but not developed	Clients' needs and wants are considered as the design proceeds	Client needs and wants are part of iterative process of designing
Planning the Look of the Product	Views the drawing as the product	Minimal consideratio n of the look of the product being designed	Only one overall finishing scheme considered	Experiments with several finishing schemes	Ideas about finishing schemes are added to the design at drawing stage	Ideas about finishes are added interactively within overall design development
Communicating Ideas	Use of narrative or other drawing genre	Minimal communicati on of design ideas	Several drawings of same idea	Quick sketches of a range of ideas	Conveys sense of object to be made e.g. by labeling	Clearly conveys ideas about object to be made e.g. multiple viewpoints
Planning Construction	Not planning to make the object drawn	No evidence of thinking about materials or construction	Minimal consideration of construction	Indicates which idea will be made but not how	Materials or constructio n features shown on drawing	Constructional issues shown on route to final design
Evaluating Whilst Planning	No evaluation	Minimal evaluation	Rejected earlier attempt(s) at drawing single idea	Considered and rejected range of ideas	Decisions made about the object whilst drawing	Changes made as a result of reflecting upon previous design drawings
Basis for Making the Product	Making the object seen as separate, new activity	Minimal relationship between drawing and making	Object made is the same as the object drawn	Object made is one of the objects drawn	Progression from drawing into making	Uses drawing as a resource during making

Figure 6. Assessment criteria.

The three rings shown in Figure. 5 relate to the first three levels of capability on each continuum. The continua were constructed on the basis of evidence from the drawings collected during the Exploratory Phase (i.e. across the 5–9 age range). The Comparison Class performed at the capability level expected for Year 2 children based on previous experience and assessment of children's use of design drawing during the Exploratory Phase.

The Focus Class' profile is remarkable, therefore, because:

- Their understanding has expanded beyond that which would be expected in such a short time frame.
- The expansion of understanding happens between Tasks 2 & 3 (i.e. directly after the introduction of the Container/Journey metaphor) and is sustained.
- Their capability is balanced, expanding equally across all dimensions. Close scrutiny of each task plot for the Comparison Class plot indicates that they do well on different dimensions in different tasks. For instance, labeled diagrams that did not relate to making are a feature of Task 3.
- There is a much stronger direct relationship between designing a product and making it amongst the Focus Class children (Tasks 2 & 5) and a much freer playing with ideas, supported by drawing, in solving a problem scenario (Task 6).

This last point is, to me, the crux of the matter. If drawing is an important component in the design tool box, then children need to learn how to use it. This is more than simply being able to do the drawing. Children can readily learn to do all sorts of styles and genres of drawing. Each new cartoon or video game craze brings new style graphic images that children emulate. The essential difference in design drawing, as Egan (1999) so rightly identified is "What is it *for*?" The role of the design drawing is not to bring closure to thought but to open it up, to objectify and play with possibilities, to develop ideas, evaluate and think again, and finally come to a place of clarity about what will be attempted to be made with the materials provided. This evolving open-endedness seemed to me to be equally true of the research process in which I was engaged:

Journey's end, though not quite yet in view, lies somewhere just over the horizon.

ACKNOWLEDGEMENTS

In looking back over the Ph.D. journey, I am indebted to Prof. Kay Stables and Prof. Richard Kimbell at Goldsmiths University, for turning this whole venture into something worth doing. I am also indebted to my colleagues at the First School (for children aged 4–9 years) in South East England where I worked, who allowed me to take their classes for activities that were not in their plans, and especially to Sue Hammond, Maggie Robinson and Therese Smith, whose insights into children's learning and their passion for discussing them at the photocopier made significant and important contributions to my developing thinking.

REFERENCES

Brown, J. (1964). *Flat Stanley*. London: HarperCollins.

Donaldson, M. (1992). *Human minds*. Harmondsworth, Middlesex: Penguin Books.

Egan, B. A. (1999). Children talking about designing: A comparison of year 1 & year 6 children's perceptions of the purposes/uses of drawing as part of the design process. In P. Roberts & E. Norman (Eds.), *International design and technology educational research conference IDATER 99*. Loughborough: Loughborough University of Technology, Department of Design & Technology.

Hammond, S. (1997). *Acknowledging the signposts in young children's writing development.* Canterbury, Christchurch University College; unpublished Postgraduate Diploma paper.

Hope, G. (2000). Why draw anyway? The role of drawing in the child's design tool box. In P. Roberts & E. Norman (Eds.), *International Design and Technology Educational Research conference IDATER 2000*. Loughborough: University of Technology, Department of Design & Technology.

Hope, G. (2001a).Taking ideas on a journey called designing. *The Journal of Design and Technology Education, 6*(3), Wellesbourne: DATA.

Hope, G. (2001b). The emergence of understanding of the relationship between planning and designing amongst young children. In C. Benson & W. Till (Eds.), *Second international Primary Design and Technology conference 2001*. Birmingham, UK: CRIPT at the University of Central England.

Hope, G. (2003). A holistic view of assessing young children's designing. In C. Benson, M. Martin & W. Till (Eds.), *Third International Primary Design and Technology conference 2003*. Birmingham, UK: CRIPT at the University of Central England.

Hope, G. (2004). Little "c" creativity and "Big I" innovation within the Context of Design & Technology. In E. Norman, D. Spendlove, P. Grover & A. Mitchell (Eds.), *International research conference DATA*. Wellesbourne: DATA.

Hope, G. (2005). The types of drawings that young children produce in response to design tasks. *Journal of Design and Technology Education, 10*(1), Wellesbourne: DATA.

Kimbell, R. (2003). *Private communication.*

Lakoff, G., & Johnson, M. (1980). *Metaphors we live by*. Chicago: University of Chicago Press.

Liddament, T. (1991). *Design talk; Design & Technology teaching* (Vol. 23, No. 2). Trentham Books.

Lieberman, N. J. (1977). *Playfulness.* New York: Academic Press.

Middleton, H. (2000). Design and Technology: What is the problem? In R. Kimbell (Ed.), *Design & Technology international millennium conference 2000.* Wellesbourne: Design and Technology Association.

Polanyi, M. (1958). *Personal knowledge*. London: RKP.

Winnicott, D. W. (1971). *Playing & reality*. London: The Tavistock Press.

Wittgenstein, L. (1969). *Philosophical investigations; the blue book; the brown book.* Oxford: Basil Blackford.

Gill Hope
Canterbury Christchurch University
Kent
England

STEVE KEIRL

19. PRIMARY DESIGN AND TECHNOLOGY EDUCATION AND ETHICAL TECHNOLOGICAL LITERACY

HUMANITY AND THE PLANET

We humans are a dominant species. We live on the planet, we live off it, we occupy and colonise it, we use it. It can be argued that we are taking much of value from it and giving little of value back. Such observations can fall under a heading of 'the environment' – a rather loose term that is often used without clarification. I value Franklin's distinction between

> ...the built and constructed environment, which is truly a product of technology; (and)... nature which is not. (She asks that)...we get away from the egocentric and technocentric mindset that regards nature as an infrastructure to be adjusted and used like all other infrastructures. (Franklin, 2004, p. 118)

It seems to me that there are four truisms about our planetary existence. In fact, they amount to four ways in which we *co-exist*. First, there is us and the planet. Second, there is us and other species. Third, there is us with each other. Fourth, there is us and our technologies. Thus, to say that we are human is to also acknowledge planet, other species, other humans, and technologies. These all contribute to our existence and we have a *co*-existence with them all. Equally, we could have no *existence* without them.

In all of these co-existential relationships there is human agency and, thus, ethical questions arise about behaviours and actions. The planet is not ours alone and how we live affects other people and other species. So far as technologies are concerned, they are our creations. We bring them into being. We give them existence and, in turn, they shape our existence.

In this chapter, I present a case for a particular kind of technological literacy – one that considers our ethical behaviours in relation to technological practice and the technologies we create. Such an ethical technological literacy, I argue, reaches across a spectrum from our personal lives to complex global issues. It is in primary Design and Technology education that the seeds of such a rich literacy can be sown.

ETHICS IN THE TECHNOLOGIES OF DEMOCRACY AND SCHOOLING

Ethics is a particularly human phenomenon. Whether it is just a few humans on the planet or over six billion, we need some form of ethical organisation. This is a matter of health, wellbeing, harmony and longevity - of self, of each other, of other species,

C. Benson and J. Lunt (eds.), International Handbook of Primary Technology Education: Reviewing the Past Twenty Years. 235–246.

and of the planet. In turn, our technologies demand an ethical framing. Questions of how we should live and how we should act abound in our technological existences albeit that we often just take technologies for granted.

While both technology and ethics are fundamentals of humanity, an irony emerges when scrutinising the two. We need both ethics and technologies for our existence and for our co-existence yet, curiously, the two have quite different attention in our history and in our education systems.

Technology has been a matter of practical action, tangible, indisputable in its reality, and given little academic interest. By contrast, ethics has enjoyed thousands of years of academic debate. While ethics has been a major field of philosophical enquiry, the notion of a philosophy of technology is a mere infant of around a century and still far from the public eye. One deduction is that ethics is often perceived as a matter of theory while technology is seen as a matter of practice. This is reflected in our education systems today. Fortunately, we are beginning to realise that ethics needs to be understood as practical action (Singer, 1995; Warnock, 1998) while technology must be understood ethically (Keirl, 2006).

There are also some notable commonalities between technology and ethics: both are contestable fields begging rational discourses; both are values-rich; both share interests with democratic theory; both have an interest in matters of determinism and free will; both beg sophisticated understandings about 'choice'; and, neither is an explicit or properly understood educational reality. Below, I hope to bring these points together in a particular understanding of technological literacy.

If we accept that living ethically is the best way to conduct ourselves then we should surely *apply* ethics to our technologies as much as to our society and our education systems. I would argue that we can view 'democracy' and 'schools' as technologies.

To do this I must say something of what I understand 'technology' to mean. Simple dictionary definitions are unhelpful and there are almost as many attempts to define technology as there are theoreticians. For now, I take technology to be *anything that our species has created.* Of course, immediately, problems arise. Many other species are technological too. Quite what I mean by 'created' is problematic but I do mean some sense of intentional act that attempts to bring about change. Here, of course, further questions emerge but they must wait.

Perhaps, then, we can see why a system of government or of education are technologies. They are of human design and creation and, like all technologies are far from perfect. And here the ethical emerges because, as with every technology, there are no perfect solutions – merely continuous compromises. How the compromises are resolved calls for, I would argue, ethical approaches.

The running of society is a matter of government, through politics. A society might be run by dictatorship – benevolent or otherwise. Whilst the former of these may be ethically preferable to the latter, it still involves a denial of certain basic human rights. Many societies claim to be democracies but almost all of these have their ethical fault-lines. The concept of a democracy is one of an ideal. It is something striven for and never wholly attained. Nevertheless we view democracy as the most ethically defensible form of government so far devised.

Key to healthy democracy is active, thinking, well-educated citizens. While all systems of government use schools to serve their own ends, within a democracy schools must serve the individual (and their community), the society, and the democracy itself. To fail to serve one of these is to fail them all and it is a major educational challenge. However, we can still view educational systems and schools as technologies – ethically determined and designed within and for democratic purposes.

I have made these points because, I argue, if we are able to see the *commonalities* of some of the attributes of education, democracy and technologies we will be better decision makers about each as well as about our planetary coexistence - with each other, with other species and with the planet itself. However, it is essential to look at technologies in greater detail.

TECHNOLOGY IN THE WORLD – THE IMPORTANCE OF META-PERSPECTIVE

When we start to look at technologies in the world we begin to open up what seems like a host of problems so vast as to be incomprehensible. This should not deter us. It is surely education's job to help make the incomprehensible comprehensible and good primary education leads the journey in this. First, though, it is necessary to look across a range of technologies to understand the world our children perceive and grow with.

I stress that I am not saying that we identify a range of technologies to teach about – this is a matter for the professional judgement of the teacher and their pedagogical decision-making. What I am presenting is a brief *meta-perspective* to contribute to teacher professional awareness and judgement. It is a perspective that is problematic because of technology's complexity and because there is no evidence that formal education has taken this approach before.

Such a perspective seeks to remind us of intimate human-technology relationships, of the ways in which technologies shape our existence and, of the richness of the technological phenomenon. This is not to show how daunting and incomprehensible the revealed technological world is. Rather, it is to give us – society, teachers *and* students – heightened awareness of the commonalities of technologies by holding multiple examples (realisations) of them up to scrutiny. If this is possible, it may then be easier to develop an educational position of benefit to all.

Whenever and wherever each of us was born and spent our early years, we were profoundly influenced by the technologies we encountered. They shaped our thinking and perception; they enabled us; they defined self and place. Take, for example, an easy chair in a room in a house. For most of the time it is a chair – just that – and we sit in it. For a child who is crawling and learning to walk this is something else altogether. It has texture, colour and form. It is a tool to enable walking – first to become upright and then to move laterally. It may be a special place where a grandparent always sits and, better, sits and reads or sings to the child. Alongside other chairs it is a reference for much language development and perceptual activity.

Taking a different slant, one can trace continua across the development of particular technologies and then 'place' ourselves and others along such a continuum. One example is sound recording which might be represented a century ago by a shellac and ground slate record for a wind-up gramophone, then subsequently a vinyl album for an electric record player, reel-to reel tapes, cassette tapes, compact discs, personal digital sound systems (of various types, emerging simultaneously). Depending on our age or living circumstances we look back to a point at which we had direct personal experience of one of these technologies in our formative years. Beyond that point the technology was 'old - of another generation'. 'Our generation', whoever's it is, can be dated by its technologies. The current youth generation has been tagged 'digital natives' (Prensky, 2005/6).

In similar ways we can see the influence of more integrated technological *systems* in our lives – the home, the house (they are different), the street or farm, the school, or, most powerfully, a growth-driven economy. All such technologies are *formative* as well as informative and enabling. This is important to this discussion because too often we view things as just that – things, objects – when in fact they shape our being, our existence, and our behaviours – positively and negatively.

It is helpful to look at a range of technologies and to critique them. That is, to get behind their face value(s) and to interrogate their other values and the uses to which they are put, and how they shape our collective, as well as individual, lives. Here are some examples:

There are three that seem to be 'must have' technologies – ones that most children in minority world societies are born to – they are seen as just part of 'normal' life. They are the car, the computer and the mobile phone. Each can be quite radically interrogated (Keirl, 2006) and, were a debate of their merits to be held, I believe that the rational case could go either way to justify their use. Because these things empower us we also make sacrifices and alter our behaviours to accommodate them. To suggest to an individual that their lives could be richly lived *without* car, computer, television or mobile phone is to invite derision, charges of Luddism or naïvety. Yet, little more than two generations ago most people lived well without such technologies.

In two generations time (when our current primary school children are perhaps halfway through their expected lifespan) it will be a very different world again. Our current explorations of nanotechnologies, xenotransplantation, genetic engineering and continued pursuit of artificial intelligence (including consciousness) mean that the post-human or trans-human world is now a prospect (see e.g. Kurzweil, 1999; Somerville, 2000; Broderick, 2001; Scientific American, 2001). Here, matters of our history (which has always been technological) and our future (which, as 'humans', is potentially limited) emerge and both warrant more education.

Within these developments we find *design* at work. We design babies, lifestyles and life-lengths, and death (whether euthanasia, execution or at war). We design crops and we design animals. We design robots that are furry and furry toys that are robotic. Lines blur and perceptions blur. It is a rolling process by which we unquestioningly accept many technologies and their consequences into our lives. Thus, as with the sound recording example, a continuum will run from teddy bears to trainable anibots. Such innovations, taken alongside spare-part replacement technologies, acculturate us and our children to our transition from humanity to post-humanity in a way that seems 'natural'.

In our pursuit of these designs-inventions-innovations-creations new ideas and new knowledge emerges and, in a capitalist system, that knowledge is harnessed for profit first and any common good later. We have patent systems to protect 'ownership of knowledge' but patenting is now applied to the DNA of people(s) (Penenberg, 1996). Patents and knowledge are also frequently suppressed – to eliminate market competition, for 'security' reasons or to maintain power and its distribution (Eisen, 1999). The organised suppression of new knowledge has been named 'elite Luddism' by Sclove (1995).

We are also creating societies of mass surveillance and this is not just surveillance by those in power. We also monitor each other. The extent of technological surveillance is now arguably near total in some societies (ABC, 2001; NI, 2005) with digital trails (through payments, billing, booking, registering, swipe-cards, emails, phones calls, etc.), satellites, outdoor and indoor cameras, media snooping and interpersonal (mobile phone, answering systems) and workplace computer monitoring all contributing to our vulnerability as citizens and as targets of cyber-crime and identity theft.

Perhaps the most transparent impact in the world of primary school children is the phenomenon of technological abundance as overblown consumerism. The sheer volume, (low) quality and cultural pervasiveness of the mass-produced world are key to shaping the majority of children's (and their families') existences (Suzuki, 1997; Klein, 2001; Quart, 2003). It is nowhere more apparent than in the consumer world that we really are 'spoiled for choice'.

I collect brushes, mops and scourers used for utensil cleaning in the kitchen sink. I have over 900 – all different, and just one case of consumer choice gone mad – but only for those who actually have sinks, plates, taps, and even water. Today, most products are created for profit – not to fulfil a genuine need. No matter the desirability, ethically or otherwise, the product must be generated and sold. In 2001, the Kraft global 'food' company budgeted $800m in two years for the marketing of 100 new products (Schiller, 2001). In 2004, Australian company Dairy Farmers had a range of 800 dairy product variations (Sinclair, 2004).

The excesses of production are not the only aspect at play. When designed obsolescence, poor quality, high energy use, non-recyclability, and harm to the wellbeing of people are also taken into account, we have a negative situation indeed. Schumaker, a clinical psychologist, notes:

> ...high degrees of materialism have a toxic effect on psychological and social wellbeing. A strong materialist orientation has been associated with diminished life satisfaction, impaired self-esteem, dissatisfaction with friendships and leisure activities, and a pre-disposition to depression...(a) worrying rash of 'consumption disorders' such as compulsive shopping, consumer vertigo and kleptomania...
>
> Hyper-materialism also features predominantly in the emerging plague of 'existential disorders' such as chronic boredom, ennui, jadedness, purposelessness, meaninglessness and alienation... (Schumaker, 2001, p. 35)

ATTRIBUTES OF TECHNOLOGIES

This *very brief* meta-perspective of technologies illustrates the world into which we bring children. For many of us such a brief overview – especially when all consequences are considered – is mind-boggling but, to eschew a term used earlier, it is not incomprehensible. One may still be charged with saying this is beyond the remit and capacities of the primary educator. Thus, to help frame matters, it is important to identify commonalities of all technologies. While no educator, at any level, can address thousands of technologies in all their nuances and permutations, perhaps it is possible to approach the challenge though a more grounded approach. Curriculum and pedagogy can be informed by, for example, knowing that:

- technologies are *integral* to our lives and cultures. We can hardly define our existence without reference to them yet they remain *outside* of common critical discourse;
- all technologies have contested values. No technology is neutral or universally good;
- all technologies are created by a manufacturing or enabling process resulting from human intention and design;
- a technology cannot 'be' in any functional sense without a relational human engagement. This may well be less the case in the future;
- technologies often undergo 'function creep' – uses other than those originally intended;
- technologies converge and gain greater technological efficacy than the sum of the parts;
- the post-human condition or era is emerging, where the balance between our human identity as we have known it and the engineered human is shifting;
- technologies almost always emerge faster than the necessary associated ethical and legal considerations;
- personal and collective identities are shaped by the technologies with which we interact; and,
- as the raison d'être of technology, power and empowerment are subject to attribution, distribution and ownership – in equitable or inequitable ways.

One succinct position might be that... technologies are bundles of competing values.

For teaching purposes, I have found it is useful to consider five phases through which technologies move in their presencing with us (for more detail see Keirl, 2009). The first is the one perhaps most often neglected by enquiry into technology – it is that of intention. At the very outset, someone has an idea or a goal or a purpose in mind for a technological development. Invariably today, it is profit-driven, but there may be altruistic reasons too. There are many issues to address at the stage of intention but, so far as the public is concerned, engagement does not usually come until much later – often by enacting laws in reaction to the effects of this or that technology.

The second phase is that of design. Once a decision has been made (the intention) to create a technology a pre-production process is engaged – one perhaps of problem-solving, research, ideation, creativity, or invention. Designing is the resolution of competing variables into a whole. It is also the articulation of competing (and contestable) values into a reality. It is in the design phase that the values of the designer, the culture, the economy, and the society become embedded in the technology.

The third phase is that of realisation – the bringing-into-being of a technology from the design phase. (From an educational perspective, this is the phase we mimic mostly - it's about 'making things' whether puppet, game, sign or website.).

Fourthly, there is the use or life of the technology. Many issues arise here as every technology has consequences seen and unforeseen. As has been shown, power relations, access, behavioural change, adaptation, identity, personal and planetary environmental impacts, all come into play when technologies are put into use.

Finally, the fifth phase is that of consequences – which may be intentional or unintentional. Consequences may be patently obvious, subtle or invisible and, of course, they may have differing values attributed to them.

ETHICAL TECHNOLOGICAL LITERACY

It is apparent that technology is not just 'things' or 'hi-tech' or computers. Nor is it 'applied science' or 'making products'. Furthermore, technology is not 'inevitable' – beyond our capacity to do something about – in any of its phases. And technology is not incomprehensible – beyond our understanding. It is rich, complex, ever-present yet seemingly invisible. It is just there, usually unquestioned and yet has mind- and life-shaping consequences. I argue that we can dismantle these orthodoxies but a quality education is needed.

Thus, if we want a democracy holding rich understandings of technology then we are challenged to formulate a correspondingly rich technological literacy for children of and for the twenty-first century. This, I contend (Keirl, 2006), calls for both *ethical critique of technologies* (and how they affect our lives) as well as a need for an education to enhance the *practice of critiquing in general in society*.

For a rich technological literacy to occur there must be both opportunities and questions in regard to technology ethics. These can occur by engaging with each of the five phases of technologies and, pedagogically, the engagement may be through exercises in critiquing or designing or creating. This is not to say that all are engaged at any one time. This would be both impractical and confusing. It is through such pedagogy that ethics is practised and the practical gains ethical context. The interplay of these feeds students' democratic thinking and being. It is the subtle curriculum that enhances the kind of citizenry that schools can produce.

As with naïve or restricted views of literacy – that it amounts to technical abilities in reading, writing, and spelling – so it is the case with technological literacy. If we constrain our views to the technical we position technology education as making and skilling. These are necessary, of course, but they are only a part of the whole. Similarly, those who would say that students who can use computers effectively are technologically literate are doing both the students and education a serious dis-service. This is *one* technology, often taught uncritically and often limited by the software in use.

The challenge can be approached in different ways but they can all converge in quality Design and Technology education. Democratic society needs an educated populace – one that uses debate and critical thinking, one that is sceptical (Postman, 2000) and does not take everything fed to it at face value. Nowhere better can these traits be exercised than over technologies and how they are used in our lives.

Designed technologies warrant ethical interrogation and as children explore the values embodied in their worlds and in the technologies around them they learn that things are only the way they are because someone designed them that way. Things could be different and design is a way to achieve a difference. It can be said that design is about changing one set of circumstances into another. This is the case whether designing a birthday party or a political party. As a result of designerly behaviours change occurs. So children can readily learn about personal efficacy, can-do, and change.

In the process of designing, children also build their understanding of values and how these are designed into technologies. The values within a design are as much variables as are materials, costs, aesthetics, function and so on. Children learn that particular colours, shapes, materials and images are all merely one person's (or a team's) choice and are used in particular ways for particular purposes. Children's senses of fairness, care, access and reasonableness can all be applied to their own work (when *designing*) and to others' or already existing designs (when *critiquing*). Part of the development of technological literacy is the enhancement of a range of critical designerly attitudes towards all technologies whether they be a democracy, a duvet or a dinner.

As with all good democratic education (remembering it is both for and about democracy as well as being for both the individual and society) quality Design and Technology education potentially develops both *design intelligence* (Cross, 1995; Keirl, 2002) and a *design culture*. Design intelligence is something of value to the individual but, well taught, it has both a social dimension and a social purpose. In a world where the technological shapes our existences it makes sense to develop a design culture that keeps critical design awareness as part of everyday discourse and not apart from it. Such is the case for democratic discourse too.

As children learn that there is no such thing as a perfect design or technology and that all designs and technologies have downsides, so the values of questioning, scepticism and critique are valorised. We know that such activities can be discomforting but good pedagogy manages discomfort constructively. Whilst norms and preconceptions (even prior learning) may be challenged so opportunities arise for creativity, imagination and empathy.

Creativity, too, can be discomforting. Here, mental risk may be needed and the teacher's guidance and facilitation is paramount but this is potentially no different from many other teaching challenges that are intended to bring about fulfilling human activity. As Csikszentmihalyi (1997) argues in his valuable text, we are programmed for creativity and when we are being creatively productive we are actually happy too. Perhaps such fulfilment could be antidotal to Schumaker's (2001) 'plague of existential disorders' and we could use more of it in the classroom and in life.

Imagination is something we value and nurture in the primary years but we have to admit that, in the secondary school phase, it is less valued and celebrated. Design and Technology offers great opportunities for the explorations of the 'what if' and '...to look at things as if they truly could be otherwise' (Green, 1988, p. 55). It is an indictment of our times that it is tests, rather than imagination, which are celebrated by politicians of education. As Patten puts it at the end of his eloquent poem:

> Well, let me set a test.
>
> Q1. How large is a child's imagination?
>
> Q2. How shallow is the soul of the Minister for Exams? (Patten, 1993, p. 12)

Interestingly, links between ethics and imagination emerge in the literature and these can readily be woven into rich technological literacy pedagogy. Raphael (1984, p. 66) on moral philosophy talks of 'imaginative sympathy'. Mackay (2004, pp. 239–240) suggests that 'Morality is the work of the imagination: making moral choices is a creative act that, like all creative acts, requires courage and involves risk.'. Warnock (1998, p. 120) sees teachers helping students

> ...to discover that there is such a thing as private morality, the ethics of conscience and of possible ideals.

Addressing moral education she comments:

> Important though...explicit ethical teaching is, I would give a high priority to the development of a child's imagination; indeed without this a child will have no safeguard against the deadening cynicism which is the enemy of morality.

When these authors write they have in mind the wellbeing of society – its members and its future. It can be useful to consider futures education through Design and Technology by using the device of Future (big F) and futures (little f) (Keirl, 2003). There is a Future – big, unknown, collective, and seemingly beyond our personal power to shape. This might be contrasted with our personal future – that which is our own. Thus, while the Future comprises all our personal futures it is possible to conceive of shaping the Future by collective action and opinion. Here we return to notions of intention, design, personal and collective efficacy, and choice-making. It is the last of these that I believe can be key to an ethical technological literacy that serves both Design and Technology activity and our personal and collective being (the four truisms of humanity and the planet, above).

THE SIGNIFICANCE OF CHOICE

As individuals, communities, societies, and as a species, we have choices and there are many ways we may exercise them and in which they are inhibited or denied us. We have a human capacity to choose – it is something of our essence. We can choose to choose and we can create choices. We can also choose not to choose. It would seem that we have plenty of choice.

The existence (and defence) of democracy and democratic processes is what provides the arena for our freedom to choose. Thus, the practice of choosing (and the teaching of it) is itself a practice of democracy. It may be that low voter participation is a manifestation of Warnock's (1998) 'deadening cynicism' or that there is even a 'who cares?' generation but the democratic challenge for education lies in exploring and promoting choice-making as human enterprise. A most fundamental educational achievement is to teach children and students that they have choices - from the seemingly insignificant to the highly significant - and that there are different ways they can exercise choice.

Design and Technology has a particular educational role to play. Choice-making and choice-creating are both key aspects of designing. Design is intentional and pro-active. It involves defending decisions and imagining alternatives. By exercising and understanding *choice* in its richest sense, students enhance their designing capability and efficacy. They are also engaging in democratic practices – as must the teacher. Democratic pedagogy is needed to facilitate design and critique, to engage in the ethical and to question technologies and designs rather than to accept blindly and reinforce a status quo.

There are 'whole person' and general education benefits to be gained in equal measure from choice education through Design and Technology. For the former, students develop their 'intentional intelligence' (Gregory in Cross, 1995, p. 106), their efficacy and their identity. For the latter, their choice education works synergistically with ethical education, civics and citizenship education and their political education. Too often textbook and hypothetical approaches remain removed from the student – too often these approaches are teacher-centred.

IN CONCLUSION...

Primary school students of all cultures live and engage with their built environments and can use their sense of fairness and justice to explain when things are not 'right'. They can learn through quality Design and Technology education that technologies do not *have* to be adopted, that all designs and technologies can be critiqued and, that alternatives are possible. Students can learn that will can be expressed and that ethical action is a matter of choice. They can learn that there are consequences of both action and inaction. Their design decisions and all that those decisions entail provide an ideal grounding for learning that futures, and the Future, can be influenced by their choices.

Design and Technology today has a highly defensible role in both future-orientated curriculum policy and in future-orientated classrooms but our innovative curriculum area continues to need deep, and more interactive, theorising and practice. In a world presenting massive technological and existential challenges to humanity it would seem that there is a place for a rich ethical technological literacy for all the planet's children. The ingredients of such a technological literacy include critique, design, imagination, efficacy, creativity, ethics, choice and more – all of which the primary classroom offers.

If the demands on curriculum continue to try to embrace futures education, sustainability, environmental education, civics and citizenship and whatever else may be deemed of value to a society then these cannot be addressed without taking account of technologies, their designs, the intentions behind them and the consequences which follow. Like literacy itself, ethical technological literacy is not a subject, nor a discipline, nor is it static as some kind of body of knowledge. It is naturally integrative and cross-curricular and, like the democracy it can serve, it is fluid and constantly evolving.

Times change and the current curriculum pressures created by calls for more accountability, more testing of children and greater educational reductionism must change. Society, democracy and planetary concerns for humanity are presenting complex challenges. By helping students develop critical design intelligence, and society a critical design culture, education maintains its democratic purposes. The primary school is where this educational journey begins and Design and Technology articulating ethical technological literacy is a key to this.

REFERENCES

ABC (*Australian Broadcasting Corporation*). (2001). *'Aftershock'*. URL: http://abc.net.au/aftershock/guide.htm

Broderick, D. (2001). *The spike: How our lives are being transformed by rapidly advancing technologies*. New York: Forge.

Cross, N. (1995). 'Discovering design ability'. In R. Buchanan & V. Margolin (Eds.), *Discovering design: Explorations in design studies*. Chicago: University of Chicago Press.

Csikszentmihalyi, M. (1997). *Creativity: Flow and the psychology of discovery and invention*. London: HarperPerennial.

Eisen, J. (1999). *Suppressed inventions and other discoveries*. New York: Avery.

Franklin, U. M. (2004). *The real world of technology*. Toronto: Anansi.

Green, M. (1988). What happened to imagination? In K. Egan & D. Nadener (Eds.), *Imagination and education*. New York: Teachers' College Press.

Keirl, S. (2002). Hedgehogs, foxes, a passing crow, and other 'intelligent' beings: Explorations of the relationship between multiple intelligence theory and design & technology. In H. Middleton, M. Pavlova & D. Roebuck (Eds.), *Learning in technology education: Challenges for the 21st century*. Proceedings of the 2nd Biennial International Conference on Technology Education Research. Queensland: Centre for Technology Education Research, Griffith University.

Keirl, S. (2003). Looking behind while designing the future: Can practice meet theory in Primary Design and Technology curriculum? In C. Benson, M. Martin & W. Till (Eds.), *Fourth International Primary Design and Technology conference 2003*. Birmingham, UK: CRIPT at the University of Central England.

Keirl, S. (2006). 'Ethical technological literacy as democratic curriculum keystone'. In J. R. Dakers (Ed.), *Defining technological literacy: Towards an epistemological framework*. Basingstoke: Palgrave Macmillan.

Keirl, S. (2009). Seeing technology through five phases: A theoretical framing to articulate holism, ethics and critique in, and for, technological literacy. *Design and Technology Education: An International Journal*, *14*(3).

Klein, N. (2001). *No logo: Taking aim at the brand bullie*. New York: Picador.

Kurzweil, R. (1999). *The age of spiritual machines: When computers exceed human intelligence*. St Leonards, N.S.W: Allen & Unwin.

Mackay, H. (2004). *Right and wrong: How to decide for yourself*. Sydney: Hodder.

New Internationalist (NI). (2005). *'They are watching YOU'*, (376), 12–13.

Patten, B. (1993, October). The taste of the moon. *New Internationalist*, p. 12.

Penenberg, A. (1996). Gene piracy. *21.C Scanning the Future*, #2, 44–50.

Postman, N. (2000). *Building a bridge to the eighteenth century: How the past can improve our future.* Carlton North, Vic: Scribe Publications.
Prensky, M. (2005/6). 'Listen to the natives' *Educational leadership, 63*(4), 8–13.
Quart, A. (2003). *Branded: The buying and selling of teenagers.* London: Arrow Books.
Raphael, D. D. (1984). *Moral philosophy.* Oxford: Oxford University Press.
Schiller, D. (2001, May). 'Globe with a logo – The big sell: The world takeover'. In *Le monde diplomatique*. (p. 15).
Schumaker, J. F. (2001, July). Dead zone. *New Internationalist,* (336), pp. 34–35.
Scientific American. (2001, September). Special Edition: *'Nanotechnology', 285*(3).
Sclove, R. E. (1995). *Democracy and technology.* New York: The Guilford Press.
Sinclair, L. (2004, November 25). Shake-up for milk giant. In *The Australian.*
Singer, P. (1995). *How are we to live? Ethics in an age of self-interest.* Port Melbourne: Mandarin.
Somerville, M. A. (2000). *The ethical canary: Science, society and the human spirit.* Harmondsworth: Viking/Penguin.
Suzuki, D. (1997). *The sacred balance: Rediscovering our place in nature.* St. Leonards, NSW: Allen & Unwin.
Warnock, M. (1998). *An intelligent person's guide to ethics.* London: Duckworth.

Steve Keirl
Centre for Research in Education
University of South Australia

JULIE LUNT

20. RESEARCH INTO PRIMARY-AGED CHILDREN'S DESIGNING

A Review of the CRIPT Conference Papers 1997–2009

INTRODUCTION

Learning to design through the activity of designing is a fundamental element of children's experience in Design and Technology education in many countries across the world. In England, for example, the significance of designing has been reinforced through successive versions of the programmes of study outlined in the National Curriculum: children should be taught designing skills; they should be involved in designing and making assignments; and assessments of children's performance should be made based on their capability in designing and making (DES/WO, 1990; DFE, 1995; DfEE, 1999). However, despite its centrality, designing has frequently been identified as a weakness and has consistently been highlighted as a critical area for development (Ive, 1997; Benson, 2007). So what do we actually know about primary-aged children's designing? In her critical review of ten years, Benson (2007) poses three fundamental and very practical questions to be addressed: How do children design? Is this different at different stages of development? What strategies are useful?

The biennial CRIPT conferences 1997–2009 have provided a unique focal point for researchers in the field of primary Design and Technology from around the world to come together to share their work and to discuss issues affecting the development of Design and Technology education for young children. This chapter provides a review of research into primary-aged children's designing based on the CRIPT conference papers which focus specifically on this topic. Although many of the CRIPT papers touch on designing or discuss it in theoretical terms, the papers selected for this review have been limited to those which either involve close observation of children designing or investigate the perceptions of children and their teachers of designing and learning to design.

OVERVIEW OF THE CRIPT CONFERENCE PAPERS 1997–2009

In Design and Technology, designing can be viewed as an activity that involves children in the process of generating, developing and communicating ideas for functional products which serve a need or purpose and have an intended user. Papers focused on designing have featured in each of the seven CRIPT conferences

C. Benson and J. Lunt (eds.), International Handbook of Primary Technology Education: Reviewing the Past Twenty Years. 247–259.

1997–2009. They all report on small-scale research studies and relate to the following four themes:

- the nature of children's designing;
- children representing and developing design ideas;
- children's perceptions of designing;
- teaching and learning strategies.

THE NATURE OF CHILDREN'S DESIGNING

There are many alternative definitions of the term 'designing' and various theoretical models of designing feature in the literature for Design and Technology education (Kimbell et al., 1991; Johnsey, 1995). Curriculum policy documentation is often based upon such models which can then become enshrined in classroom practice. Baynes & Johnsey (1997), in their paper reporting on a special interest group for research into pre-school and primary Design and Technology, highlight this issue, and, from their analysis of research papers from the IDATER conferences[1], identify two major mismatches between the model presented in the National Curriculum of that time (DFE, 1995) and the classroom reality of children's designing. The first is that in real practice when children are designing they follow a reiterative process rather than a linear one; the second is that they use making as a way of designing. This is further developed by Johnsey (1998) whose close observation of primary children engaged in designing and making activities suggests that a single model with an implied linear design procedure is misleading to both teachers and pupils. He also notes the centrality of working with materials in children's processes of designing. He provides an alternative model based on the metaphor of a toolbox in which design process skills such as investigating, making and evaluating are seen as compartments that can be filled with appropriate strategies. It is suggested that these can be drawn upon when a child or teacher feels it is appropriate, similar to the way in which tools might be selected and used for particular purposes (Johnsey, 1998).

The papers from the CRIPT conferences 1997–2009 which report on in-depth observation or analysis of children's activity further substantiate the claim that children do not adopt a uniform, consistent procedural approach to designing. It is apparent that children's designing is far more messy and iterative than a simplistic linear model would suggest (Welch, 1997; Welch & Lim, 1999). This is confirmed by Anning & Hill (1998) who, in a comparative study of designing in primary/ elementary schools in England and Canada, found that children did not design in similar ways, within or across age groups. They argue that designing skills are also often linked to specific contexts so that generic design skills and processes are therefore not always appropriate.

Baynes & Johnsey (1997) suggest that people of any age use a variety of methods and approaches when designing. The research described below poses serious questions about established practices of teaching designing to primary-aged children. For example, 'What is the role of drawing in the design process?' 'Should children be asked to make a design drawing before they begin making?' 'What are effective ways for children of different ages to generate and develop their design ideas?'

'Should children always enter a designing and making assignment at the same starting point?'

CHILDREN REPRESENTING AND DEVELOPING DESIGN IDEAS

The process of representing and developing ideas is a key element of designing. A variety of modes of representation can be used such as talking, drawing, gestures, working with materials and writing. These are often used in conjunction with one another, for example, talking with a design partner while working out ideas with materials, or adding labels to a design drawing. Drawing has traditionally been ascribed a particular importance in this area and has been seen by experienced designers and design educators as a valuable method of modelling ideas (Egan, 2001).

> Images are our prime instrument of technological expression. The things we can draw are in effect the things we can think. Models are the terms of our thinking as well as the terms in which we present our thoughts, because they present the objects of thought to the thinker himself. Before a drawing communicates ideas, it gives them form, makes them clear and in fact makes them what they are. (Kimbell, Stables & Green, 1996, p. 23)

It is not surprising therefore that drawing features prominently in curricula for Design and Technology and associated teaching and learning materials. However, research findings from studies which have observed children designing suggest that drawing is not always highly valued by children as a modelling tool. The CRIPT papers which focus on this area provide us with valuable insights about the role of drawing in children's designing and raise important questions about school-situated design activity and ways of enhancing children's ability to design.

Welch (1997) and Welch & Lim (1999) report findings from two studies in Canada in which pairs of Year 7 students, inexperienced in designing, were set a design and make task in which they had autonomy over how they designed. Each design and make session was recorded and analysed to describe the nature of the activities they were engaged in over time – understand, generate, model, build and evaluate. In the earlier study Welch found that these novice designers did not use sketching as a way to explore and communicate ideas, but moved immediately to three-dimensional modelling. Neither did they generate multiple solutions in order to develop the one with the most promise, a drawing task often required of children in school-situated designing. In the later follow-up study, half the students were given instruction in freehand isometric sketching before the design and make task, whilst the other half acted as a control group. Analysis of the activity of both groups showed that whether they were taught drawing skills or not, these novice designers did not use sketching in order to develop a proposal, choosing instead to model with materials and simultaneously discuss their ideas. This apparent rejection of sketching as a modelling tool might have been caused by a number of factors, for example, the nature of the sketching skills taught or the novelty value of working with materials. However, the preference for three-dimensional modelling and discussion supports the earlier findings of Johnsey (1998).

Two researchers from England have focused more specifically on drawing in extended studies of children's designing (Hope, 2001/2003; Egan, 2001). Both researchers emphasise the importance of children understanding the purpose of drawing in designing. Hope (2001) found that children will not naturally use drawing for modelling ideas ahead of engaging with the materials of construction unless they understand the purpose for doing so. In this paper, she reports on her analysis of children's design drawings over several years. She takes a developmental approach, suggesting that children progress through various stages in the development of their ability to use drawing as a method of modelling in designing. She categorises the children's drawings according to genres of drawing which relate to these various stages. The genres progress from 'the picture' in which the child makes a drawing related to the subject but not used to support making or the development of their design, through four additional stages to the 'interactive' genre associated with the final stage. This is where the child sees the drawing as a means to work out what will be made and how to make it. It suggests a dynamic relationship between internal and external representation as in a conversation between designer and paper. More than one idea is recorded, and these are then thoughtfully evaluated and discarded or developed through more drawings, combining and discarding ideas from several drawings. The genres demonstrate children's increasing ability to consider the client's needs, design problems and constructional issues through their drawing. It is not until the penultimate stage 'progressive' that children appear to realise they can use their drawing to progress their ideas about the design solution and work out how the object will be made or fit together. It is at this stage that children start to include verbal annotations or expanded drawings to show small or separate details or diagrams to show different viewpoints. Hope (2001) comments that once children begin to annotate their drawings they start to seriously consider them as plans for making. Unlike Welch (1997), she did find evidence that children were able to use drawing to produce multiple ideas. However, their selection of a final idea might be made on 'best drawing' criteria rather than the most effective design solution.

In this study children were given free rein to explore their ideas on paper prior to making and used blank sheets of paper rather than structured design sheets. Questions to explore further might be, 'Are these stages children pass through in more structured situations as is often the case in primary classrooms?' and 'Could these linked genres and phases be used to inform the development of more appropriate design planning sheets for children?'

Hope has also contributed to our understanding of teaching children how to use drawing as a tool in designing as opposed to teaching particular techniques or a process order (Hope, 2003). She has developed a strategy for talking to children about drawing in designing using metaphor that distinguishes between drawing as a container (product) and drawing as a journey (intellectual process). This use of metaphor is discussed more fully in Hope (2001) and is shown to be an effective strategy in an evaluative study (Hope, 2003). She observes that:

> It is when children begin to realise that drawing is not just a product but can also be a process and that they can go on an intellectual journey with it and

> through it, that they begin to use it for developing ideas and, therefore, as a genuine design tool. (Hope, 2001, p. 199)

In her paper on constructing ideas through drawing, Egan (2001) shows how drawing can be used constructively in interactions between pupils and teacher. She worked with groups of children aged 7–8 as a participant observer. The children were given a design task and as they came up with ideas they were encouraged to make a drawing to show her what they meant. In this way she framed drawing as a communication tool with a genuine purpose in this context. These tentative initial drawings were used as a focus for discussion during which she was able to suggest improvements and identify areas in which children lacked the necessary skill or knowledge to carry out their ideas. She planned appropriate focused practical tasks to support the children and they returned to the original sketches in order to review them. The majority of children chose to rework their sketches at this stage as they had clearer ideas of what they wanted to make as a result of their focused practical tasks. Whilst acknowledging that this is not a typical teaching situation, Egan suggests that there are aspects of this work that offer a key to the successful use of drawing as a way of expressing and refining the initial mental models of children. These are summarised below:

- there was no expectation on children to come up with a 'finished' idea in the early stages;
- the focus of the discussion was on function and construction rather than appearance;
- the drawings had a clear and valid purpose – to communicate with the teacher;
- by using the initial drawings as a focus for discussion, children were enabled to clarify their mental models for themselves as well as for the teacher;
- the drawings and associated discussion helped the teacher to plan appropriately to support the children to develop successful outcomes.

Egan and Hope's work reminds us how having a model of an idea in the external world can be useful as an object of thought and discussion. This notion is also supported by Welch (1997) and Welch & Lim (1999) in their finding about the significance of three-dimensional models in enabling discussion about ideas. This concurs with earlier findings of the APU project with older students. Kimbell et al. (1991) found that the opportunity for learner-designers to discuss ideas, with each other and with teachers, at specific points in the designing process was critical for the development of high-quality outcomes. This raises issues about how we organise designing activity to ensure that children have sufficient opportunity to discuss ideas (their own and other people's) - with one another and with the teacher.

Writing might be considered to be a less significant mode of representation in designing when compared to drawing or modelling with materials. However, Hope (2001) found that the use of annotations with drawings was a significant step forward in the use of drawing as a planning tool. Lunt (2005, 2007) reporting on research into 9–11 year old children's perceptions of writing tasks in Design and Technology found that the majority of children perceived writing as helpful in conjunction with other modes of representation in recording and developing their design ideas.

CHILDREN'S PERCEPTIONS OF DESIGNING

There has been a growing body of work in educational research and school improvement in the last decade which seeks to listen to children's views of their educational experiences. Children are increasingly seen as key stakeholders in education and central actors in teaching and learning situations with very particular insights to offer our understanding. In terms of developing our understanding of children's designing, there are four papers in the CRIPT conferences 1997–2009 which adopt this approach: Lunt (2005/2007/2009) focuses on children's perceptions of writing tasks, while Barlex et al. (2005) focus on children's perceptions of design portfolios.

Lunt (2005) reports on an investigation into children's writing tasks in Design and Technology with children aged 9–11. Writing tasks are defined as any activity initiated by the teacher requiring the children to write – often in conjunction with other forms of representation such as speaking, working with materials and drawing. The majority of children regarded writing tasks as helpful. They recognised that writing tasks could help them as designers and makers in three important ways:

- to create a record to refer to when making, such as an annotated drawing or plan;
- to help them to plan and make their product by giving them the opportunity to think through their ideas and avoid making mistakes;
- to help them to learn and understand so that they would be better equipped to design and make successfully.

The children also identified that writing tasks helped them to prepare for the future and improve their writing, but these were far less frequently mentioned than those directly related to designing. 'Unhelpful' writing was described as that which they themselves did not use again, unnecessary or inauthentic writing (i.e. without a genuine purpose or audience) and writing which was in an inappropriate form, e.g. paragraphs when bullet points would be more functional.

In a later report of the same study, Lunt (2007) asked the children to grade particular writing tasks in terms of helpfulness. Design drawings were ranked as the most helpful form of writing task except in one case where the teacher asked the children to make a final version of their design drawing which was in effect a neat copy. In contrast to this, in a different case, the final design drawing was ranked by the children as the most helpful writing task as this was preceded by an activity in which the children made a three-dimensional mock up and so the final drawing was an elaboration of their design using new information gained through three-dimensional modelling. This highlights the importance of children seeing a relevant purpose for their activity within designing, an issue raised by Hope (2001) and Egan (2001) in relation to drawing. Lunt suggests that children's views of writing tasks (which could be extended to any activity related to designing) are influenced by their construct of Design and Technology as a goal-directed activity built around the creative act of designing and making a product. She identified three issues of concern to children in their designing: time, relevance and control. Some of the children in her study felt a pressure on time which threatened their ability to complete their products successfully. This led to the sense of time being at a premium and therefore any designing activity which was not perceived as being directly relevant to their goal of designing and making a successful product was an unwelcome distraction. Some children also expressed a desire for greater autonomy over their designing processes, for example, being able to set out their own work, or make more decisions themselves. A sense of autonomy has been found by many researchers to be closely associated with a greater level of learner engagement (Rudduck & Flutter, 2004).

The issue of autonomy also featured in a small-scale case study by Barlex et al. (2005) in which twelve pupils aged 10–11 were interviewed in groups to investigate their views of design portfolios (referred to as design booklets by the school). The children enjoyed the opportunity to develop a portfolio and to design and make their own products but they were critical of the limitations placed on these by the teachers. For example the teacher mainly prescribed the work to be included and the children were not permitted to take their portfolios home unless homework had to be completed. The children regarded the purposes of a portfolio as:

a) a record of their ideas;
b) a reference to help them make a product;
c) a historical record, and
d) as a source of ideas for future work. (Barlex et al., 2005, p. 12)

While some pupils used their portfolio as a 'job bag' (a collection of everything to do with a specific project), others used their portfolio as a showcase in which only 'final ideas' or 'best drawings' were stored (see Welch & Barlex, 2004). It would appear that in this case the children had not yet had the opportunity to learn how to use their portfolios in a designerly way as a product development tool. Perhaps Hope's metaphors of 'container' and 'journey' could also be useful in relation to the portfolio (Hope, 2001/2003).

Although this is a small-scale study carried out in one school, it raises some important questions to be addressed by teachers and researchers in primary Design and Technology: How can primary teachers reconcile the three purposes of the portfolio: as a learning tool, a teaching tool and an assessment tool? What instruction needs to take place to encourage children to regard the portfolio not only as a repository of their work, but as a tool that can help them develop design ideas?

TEACHING AND LEARNING STRATEGIES

Investigating, disassembling and evaluating products is often used as a starting point for children's designing and making, particularly in England as it is part of the pattern of activities used as a structure in the exemplar schemes of work for Design and Technology (QCA, 1998). Working with existing products not only helps children to develop an understanding of the designed and made world, an important outcome for its own sake, but enables them to develop designing and making skills and knowledge and understanding that assists them in their designing and making. The Designerly Thinking Project (Benson, 2003/2005) aimed to develop opportunities for very young children to actively engage with the designed and made world. Although not specifically focused on the activity of designing, this project is referred to here as it is a major curriculum development project that has implications for practice throughout the primary age group in relation to designing. The project encouraged teachers to use a collection of products and the school and local environment as a focus for children's attention in order to develop designerly thinking. In-service training enabled teachers to develop their understanding of Design and Technology concepts and increase significantly the range and depth of questions they used in interactions with children. This in turn enabled children to develop their understanding of concepts such as user and purpose, structures and mechanisms. Teachers also reported that as a result of focusing on developing their own questions when working with children (open/closed; lower/higher order), the children began to ask questions for themselves (Benson, 2005).

Two teachers report on insights gained from their involvement in the Designerly Thinking project. Taylor (2005) highlights the central importance of time, space, materials and relationships in creating the necessary conditions for developing designerly thinking. He argues that children need time to build, time to return to their designs, time to change and modify, time to develop their creativity and ideas, time to problem solve and time with adults readily available to children to use as a 'tool'. An appropriate environment is also required within which children can engage with designerly concepts and develop designerly thinking through well planned interactions with practitioners. Treleven (2007), in an action research study with nursery children, found that through activities focusing on products, children developed their understanding of design issues such as user, product and purpose. These concepts are essential for effective designing and the implications of this project should be considered throughout the primary age range when considering ways of improving teaching and learning in designing (See Chapter 12).

A common strategy for organising children's learning in Design and Technology is to plan a range of short tasks intended to help children to gain knowledge and skills which will be useful to them in a related designing and making task. Barlex & Welch (2009) carried out an empirical investigation with twelve Grade 8 students to investigate the use they made of learning gained from support tasks and the nature of the design decisions they made during a 2 day design and make task. In this small study they found that the students were able to make use of the learning to support their designing and making. They made a range of design decisions related to construction, aesthetic and technical matters even though the designing and making task itself was relatively closed.

Teachers often ask children to draw or write as a teaching and learning strategy within designing. In her investigation into writing tasks, Mantell[2] (2003) identifies a range of purposes which teachers articulated for using writing tasks in Design and Technology. Four of these have particular relevance to designing: 'remember this', 'think about this', 'take it step by step' and 'put your heads together'. 'Remember this' and 'think about this' are categories which confirm the capacity of writing to:

- create a useful record, e.g. a list of what you need or a plan of what to do;
- aid the processes of memory, i.e. the cognitive and physical effort of writing something down can help us to remember it;
- think things through, i.e. to see our thoughts and therefore be able to reflect upon them.

'Put your heads together' suggests joint writing tasks or writing tasks which engage children in discussion with one another. Activities involving shared planning sheets or tasks in which pairs of children discuss before writing a single response are useful tools in promoting discussion, a key activity in designing (Kimbell et al., 1996; Welch, 1997; Egan, 2001). Mantell comments that in her research there was actually little evidence of children discussing as most writing tasks were carried out individually despite children's declared preference for working together and the acknowledged educational and designerly value of doing so. 'Step by step' refers to tasks which encourage children not to rush into making their product but rather to think through their designing more carefully. Although this can be a useful strategy for teachers managing designing in the classroom, a potential danger can be that the writing tasks become a series of hoops for children to jump through in a linear process. This can limit the potential of the designing and making assignment and adversely affect children's motivation. Mantell also suggests that there are particular genres of writing or writing and drawing which can be seen as significant forms of expression and communication in designing practice which children should be introduced to as part of becoming a member of a community of practice in designing, e.g. annotated sketches, specifications, comparison charts and mind maps.

In a search for authentic experience in designing and learning to design, Lunt (2009) continues this work by drawing on an analysis of children's perceptions of writing tasks in Design and Technology to describe the types of writing tasks which children are more likely to regard as authentic elements of their experience. She identifies three dimensions of authenticity: personal, cultural and pedagogical. Children are more likely to regard writing tasks as authentic if they meet their criteria for personal authenticity, i.e. they are perceived as meaningful, engaging and appropriate. They also need to be well integrated into the overall designing and making task and for the writing itself to be in an appropriate genre. Finally, writing needs to be regarded as 'the best tool for the job' by the children. Writing has a number of characteristics which distinguish it from other forms of representation such as speaking, drawing and three-dimensional modelling. It is suggested that teachers need to be aware of these characteristics and the affordances they offer in order to plan writing tasks in Design and Technology which enable children to use writing as an authentic tool in designing.

DISCUSSION

The work represented by these CRIPT papers provides us with some important insights into children's designing and contributes to some extent answers to Benson's three questions cited at the opening of this chapter. We learn that children's designing is complex and varied, and does not follow a neat, linear model of design. Designing is iterative and does not conform to a consistent pattern for each designing context, or indeed for each designer. One major implication of this is that teachers need to be given sufficient opportunity to develop their own understanding of designing processes, rather than follow over-simplified prescriptive models, if they are to be effective in supporting children in developing their designing capability.

Simplistic models of designing can lead to an impoverished experience for children and can have a demotivating effect (Stables et al., 2000).

A number of the researchers stress how important it is for children to understand the purposes of various activities they encounter as part of their designing (Hope 2001, 2003; Egan, 2001; Mantell, 2001; Lunt, 2005/2009). This is particularly significant in many primary classrooms where teachers are following a set of pre-planned, prescribed activities such as the national exemplar scheme of work (QCA, 1998). It is not enough for children to be led passively through a set of activities if they are to become active designers in their own right. One of the responsibilities of an effective teacher of designing is to make explicit the designing processes and the purposes for them. This might serve to meet the desire for greater autonomy of some children, although this has to be balanced out against the issue of time. Although making is acknowledged to be the most popular element of Design and Technology for the majority, most children have been found to value designing activity when they can see that it can help them in the successful making of their product (Lunt, 2007; Benson & Lunt, 2007).

The research papers here are conclusive in their acknowledgment that children need to have access to a range of methods of representing and developing ideas. Welch (1997) and Welch & Lim (1999) argue that there is often an over-reliance on drawing as a way of developing ideas before making in curricula documentation. Teachers might like to consider increasing opportunities for children to discuss ideas and model using three-dimensional materials before drawing designs for making. However, Hope (2001/2003) has shown that children can develop their skills in drawing before making and can develop their understanding of how to use drawing as a tool for designing. Drawing offers a powerful mode for representing and clarifying one's own thinking and for communicating ideas to others and therefore is an important skill for children to develop within designing. It is also a skill central to many professional designing practices. There are many purposes for drawing within designing, but in order to use it effectively as a modelling tool children need to develop fluency and understanding of genres of drawing that can help them to develop designerly thinking behaviours, e.g. sketch pads, notebooks, annotated drawings, story-boards, architectural or engineering drawings from the world of work (Anning, 1997).

There are a number of possible avenues of enquiry which are not represented in the CRIPT papers 1997–2009. Mantell (1999) discussed the possibility of teaching designing techniques to junior-aged children in order to help them to develop a repertoire of designing skills they could use in their designing and making. This has not been followed up with research although the notion has been developed in more recent curriculum materials and curriculum development projects. Elements of designing such as investigating, evaluating and generating ideas are noticeably absent from the body of research discussed here, as is the use of ICT in designing, an increasingly popular area of practice. This is an aspect particularly worthy of research attention as, like writing, schools have been urged to incorporate ICT into all subjects and therefore there is a need for teachers to have an understanding of what is appropriate to support children's learning in designing, rather than using ICT merely to fulfil an external requirement.

At a time when primary schools are charged with implementing a wide range of educational initiatives, research which takes account of teachers' and children's classroom experience is greatly needed if we are to ground our practice in reality. Designing is still a relatively new area of learning in the primary curriculum and therefore our understanding of its complexities is still in its infancy. The work of the researchers presented here from the CRIPT conferences 1997–2009 makes a valuable contribution which it is hoped will support primary teachers and researchers to build upon these foundations and ultimately enhance children's experience of this dynamic and exciting element of Design and Technology education.

NOTES

[1] The IDATER conferences (International Conference on Design and Technology Educational Research and Curriculum Development) were held annually at Loughborough University 1988–2001. Papers can be retrieved electronically from www.dater.org.uk.

[2] Julie Mantell was the author's pre-married name.

REFERENCES

Anning, A. (1997). Drawing out ideas: Graphicacy and young children. *International Journal of Technology and Design Education*, *7*(3), 219–239.

Anning, A., & Hill, A. M. (1998). Designing in elementary/primary classrooms. In J. S. Smith & E. W. L. Norman (Eds.), *International Design and Technology Educational Research 1998* (pp. 5–10). Loughborough, UK: Loughborough University.

Barlex, D., Welch, M., & Taylor, K. (2005). 'Never keep your ideas in your head': Elementary pupils' views of portfolios in technology education. In C. Benson, S. Lawson & W. Till (Eds.), *Fifth International Primary Design and Technology conference 2005* (pp. 11–14). Birmingham, UK: CRIPT at University of Central England.

Barlex, D., & Welch, M. (2009). Revelations of designerly activity during an immersion experience. In C. Benson, P. Bailey, S. Lawson, J. Lunt & W. Till (Eds.), *Seventh International Primary Design and Technology conference 2009* (pp. 14–24). Birmingham, UK: CRIPT at Birmingham City University.

Baynes, K., & Johnsey, R. (1997). Research in pre-school and primary design and technology - The formation of a new IDATER special interest group (SIG). In R. Ager & C. Benson (Eds.), *International Primary Design and Technology conference 1997* (pp. 43–45). Birmingham, UK: CRIPT at University of Central England.

Benson, C. (2003). Developing 'designerly' thinking in the Foundation Stage. In C. Benson, M. Martin & W. Till (Eds.), *Fourth International Primary Design and Technology conference 2003* (pp. 5–7). Birmingham, UK: CRIPT at University of Central England.

Benson, C. (2005). Developing designerly thinking in the Foundation Stage - Perceived impact upon teachers' practice and children's learning. In C. Benson, S. Lawson & W. Till (Eds.), *Fifth International Primary Design and Technology conference 2005* (pp. 15–18). Birmingham, UK: CRIPT at University of Central England.

Benson, C. (2007). 10 years on: Critical reflections and future aspirations. In C. Benson, S. Lawson, J. Lunt & W. Till (Eds.), *Sixth International Primary Design and Technology conference 2007* (pp. 12–17). Birmingham, UK: CRIPT at University of Central England.

Benson, C., & Lunt, J. (2007). 'It puts a smile on your face!' What do children actually think of design and technology? Investigating the attitudes and perceptions of children aged 9–11. In J. R. Dakers,

W. J. Dow & M. J. de Vries (Eds.), *PATT 18 international conference on Design and Technology 2007* (pp. 297–305). Glasgow: University of Glasgow.

Department for Education (DFE). (1995). *Design & Technology in the national curriculum*. London: HMSO.

Department for Education and Employment (DfEE). (1999). *The national curriculum: Handbook for primary teachers in England*. London: Author.

Department of Education and Science/Welsh Office (DES/WO). (1990). *National curriculum order for technology*. London: HMSO.

Egan, B. A. (2001). Constructing ideas through drawing. In C. Benson, M. Martin & W. Till (Eds.), *Third International Primary Design and Technology conference 2001* (pp. 61–66). Birmingham, UK: CRIPT at University of Central England.

Hope, G. (2001). The emergence of understanding of the relationship between planning and designing amongst young children. In C. Benson, M. Martin & W. Till (Eds.), *Third international primary design and technology conference 2001* (pp. 80–86). Birmingham, UK: CRIPT at University of Central England.

Hope, G. (2003). A holistic view of assessing young children's designing. In C. Benson, M. Martin & W. Till (Eds.), *Fourth International Primary Design and Technology conference 2003* (pp. 61–66). Birmingham, UK: CRIPT at University of Central England.

Ive, M. (1997). Primary Design and Technology in England - Inspection evidence and 'good practice'. In R. Ager & C. Benson (Eds.), *International Primary Design and Technology conference 1997* (pp. 1–5). Birmingham, UK: CRIPT at University of Central England.

Johnsey, R. (1995). The design process – Does it exist? – A critical review of published models for the design process in England and Wales. *International Journal of Technology and Design, 2*.

Johnsey, R. (1998). *Exploring primary design and technology*. London: Cassell.

Kimbell, R., Stables, K., Wheeler, T., Wosniak, A., & Kelly, V. (1991). *The assessment of performance in design and technology*. London: SEAC/HMSO.

Kimbell, R., Stables, K., & Green, R. (1996). *Understanding practice in design and technology*. Buckingham: Open University Press.

Lunt, J. (2005). Pupils' views of writing tasks in design and technology: Purposeful activity or just more paperwork? In C. Benson, S. Lawson & W. Till (Eds.), *Fifth International Primary Design and Technology conference 2005* (pp. 65–68). Birmingham, UK: CRIPT at University of Central England.

Lunt, J. (2007). Investigating pupils' perceptions of writing tasks in design and technology. In C. Benson, S. Lawson, J. Lunt & W. Till (Eds.), *Sixth International Primary Design and Technology conference 2007* (pp. 77–82). Birmingham, UK: CRIPT at University of Central England.

Lunt, J. (2009). Towards more authentic writing tasks in design and technology: Investigating the perspectives of children aged 9–11. In C. Benson, P. Bailey, S. Lawson, J. Lunt & W. Till (Eds.), *Seventh International Primary Design and Technology conference 2009* (pp. 75–79). Birmingham, UK: CRIPT at Birmingham City University.

Mantell, J. (1999). Teaching designing skills at key stage 2: Is there a role for techniques? In C. Benson & W. Till (Eds.), *Second International Primary Design and Technology conference 1999* (pp. 90–93). Birmingham, UK: CRIPT at University of Central England.

Mantell, J. (2003). Teachers' perceptions of purposes for writing tasks in design and technology at key stage 2. In C. Benson, M. Martin & W. Till (Eds.), *Fourth International Primary Design and Technology conference 2003* (pp. 112–115). Birmingham, UK: CRIPT at University of Central England.

Qualifications and Curriculum Authority. (1998). *Design and technology: A scheme of work for key stages 1 and 2*. London: QCA.

Rudduck, J., & Flutter, J. (2004). *How to improve your school: Giving pupils a voice*. London: Continuum.

Stables, K., Rogers, M., Kelly, C., & Fokias, F. (2000). *Enriching literacy through design and technology evaluation project: Final report*. London: Goldsmiths College.

Taylor, R. (2005). The 'designerly thinking project': The beginning of a pedagogical journey. In C. Benson, S. Lawson & W. Till (Eds.), *Fifth International Primary Design and Technology conference 2005* (pp. 180–183). Birmingham, UK: CRIPT at University of Central England.

Treleven, T. (2007). Developing designerly thinking in the Foundation Stage - A case study. In C. Benson, S. Lawson, J. Lunt & W. Till (Eds.), *Sixth International Primary Design and Technology conference 2007* (pp. 97–102). Birmingham, UK: CRIPT at University of Central England.

Welch, M. (1997). Thinking with the hands: Students' use of three-dimensional modelling while designing and making. In R. Ager & C. Benson (Eds.), *International Primary Design and Technology conference 1997* (pp. 13–17). Birmingham, UK: CRIPT at University of Central England.

Welch, M., & Barlex, D. (2004). Portfolios in design and technology education: Investigating different views, In E. W. L. Norman, D. Spendlove, P. Grover & A. Mitchell (Eds.), *The Design and Technology Association International Research conference 2004* (pp. 193–198). Wellesbourne: DATA.

Welch, M., & Lim, H. S. (1999). From stick figure to design proposal: Teaching novice designers to 'think on paper'. In C. Benson & W. Till (Eds.), *Second International Primary Design and Technology conference 1999* (pp. 136–141). Birmingham, UK: CRIPT at University of Central England.

Julie Lunt
CRIPT
Birmingham City University
England

GARY O'SULLIVAN

21. TECHNOLOGY EDUCATION AND EDUCATION FOR ENTERPRISE (E4E)

INTRODUCTION

This chapter will report on the findings of a Professional Development and Research Contract funded by the Ministry of Education (MOE) in New Zealand which ran between June 2005 and June 2007. The project was mainly organised through running workshops and providing schools with access to facilitators over a two year period. There were four workshops in each of the three regions Auckland, Taranaki and Manawatu. Data was collected from the participants at all the workshops. The results presented here are based on the fourth set of workshops which were undertaken close to the culmination of the project. These workshops were the most significant in terms of reporting because all the participants had attempted units incorporating technology education and E4E.

RESEARCH SIGNIFICANCE

In New Zealand technology education is undergoing a significant period of change as we move from the old to the new curriculum (see Chapter 8). Since its introduction as a compulsory subject in 1999 there has been little research carried out to identify the impact technology education or school community partnerships have had on developing an enterprising culture in the classroom. It is timely and important for academic research to be conducted in to this area. An interesting article by Clark (2004) highlights the differing views about the word 'enterprise' when used in association with education. This research will clarify and exemplify these discussions and offer some insight into what is actually taking place in the classroom under the technology education umbrella.

RESEARCH METHOD

Educational research can be carried out in a variety of ways; the selection of methods often depends on a philosophical perspective held by the researcher. There are essentially three main viewpoints taken. Although different, each is valid; they are positivist, interpretive and critical theory.

The positive viewpoint is based on scientific and experimental research where the gaining of knowledge is the key. Any research carried out by a positivist will focus on the observable and measurable and deal with factual matters. According to

C. Benson and J. Lunt (eds.), International Handbook of Primary Technology Education: Reviewing the Past Twenty Years. 261–273.

Clark, (1997) positivism is supported by educational researchers who believe that natural scientific methods can be equally applied to social research. One of the main differences of this philosophical position is that the researcher acts as an observer whose values are kept discrete from the research. Positivists maintain a strict view of science and place no emphasis on subjective or interpretive meaning.

The interpretive researcher places more emphasis on values. Social research invariably involves working with people, and includes some study of human behaviour. To gain an insight the interpretive researcher will incorporate meanings including their own into the study. These meanings are paramount in this research paradigm and are seen as a social reality based on common sense. The interpretation of these actions along with recognition of values forms the basis of any findings.

Critical theory as a philosophical framework combines the strengths of the other two positions and tries to go beyond what they can offer individually (Clark, 1997). Thus critical theorists must engage in research which is not just for the researcher but should be for education. It usually involves some critique of ideology as well as the observable.

This research will use critical theory as its background philosophy and a mainly qualitative, constructivist methodology to explore what is actually taking place in the sixteen schools associated with the professional development programme. The number of schools involved in the programme was determined by the Ministry of Education. The eleven primaries and five secondary schools were selected by location and recommendation from the programme facilitators. For the purposes of this chapter I will concentrate on the eleven primary school findings.

The research was carried out alongside the professional development programme; as such it is deemed to be an evaluative study. Various definitions of evaluation have emerged. In particular Worthen, Sanders, & Fitzpatrick (1997) have provided a useful summary. Evaluation has been described by Fort, Martinez, & Mukhopadhyay (2001) as the periodic assessment of the relevance, performance, efficiency, and impact (both expected and unexpected) of the project in relation to stated objectives. To facilitate the critical theory philosophy a particular type of evaluation will be used. This evaluation method is what Guba & Lincoln (1989) describe as fourth generation evaluation.

> Fourth generation evaluation is a marriage of responsive focusing using the claims, concerns, and issues of stakeholders as the organising elements and constructivist methodology - aiming to develop judgement; consensus among stakeholders. (Guba & Lincoln 1989, p. 184)

For evaluation research to be effective it must not only be collaborative, it must also involve critically examined action by individuals involved in the process. Wadsworth (1997) describes this process as being one where action is intentionally researched and modified, leading to the next stage of action which is then again intentionally examined for further change, and so on.

There have been a number of researchers who have been influential in the development of pluralistic approaches to methodology, i.e. using the constructivist paradigm along with qualitative methods when conducting this type of evaluation, e.g. Guba & Lincoln (1989); Patton (1997); House & Howe (1999); and Stake (2004). According to Stake & Schwandt (2006), evaluation studies are fundamentally a search for, and a claim about, quality.

Responsive evaluation as described by Guba & Lincoln (1989) is organised through claims, concerns and issues. The research described here used four basic methods for generating information and making decisions with regard to these organisers. These four methods have been widely used and tested by applied social scientists. They are facilitated group meetings and exercises; participant observation; individual interviewing; and focus group interviews. Dialogue played a key role in the process via the hermeneutic dialectic (Guba & Lincoln, 1989). The research was conducted by:

– using an iterative process which alternates between action and informed critical reflection;
– continuously refining methods, data and interpretation in the light of the understanding developed by all the stakeholders during the research;
– facilitating an emergent process which takes shape as understanding increases.

The research report was developed by interactive negotiations moving towards a better understanding of what was happening - both cause and effect.

The main research foci were:

– Focus 1: What are the teacher practices that support or undermine the development of enterprising attributes?
– Focus 2: What are the school wide practices that support or undermine the development of enterprising attributes?
– Focus 3: What is the influence and impact of school community partnerships on teaching and learning?

All these foci were researched in technology type programmes in the participant schools.

ETHICAL ISSUES

According to Anderson (1990) the practice of research is subject to ethical principles, rules and conventions. Clark (1997) also identifies that educational researchers conduct their research within a framework of ethical deliberation. These ethical deliberations are relevant both in the collection and dissemination phases of the research. The researcher has to decide how much information to give the parties involved in the research and how much about the parties to reveal in the findings. Obviously to receive informed consent the parties involved must have sufficient information on which to base their decisions.

In addition to informed consent, confidentiality of both the data and the individuals providing it must be ensured. Minimising of harm to, for example, participants, researchers, and schools is paramount. Truthfulness, the avoidance of unnecessary deception, and social sensitivity to the age, gender, culture, religion and social class of the subjects must also be strictly adhered to. This was achieved by supplying all parties with accurate information about the research and giving regular updates on its direction and findings.

RESEARCH INDICATIONS

This chapter includes data collected from the final set of workshops carried out in May 2007, the main focus of which was a professional development package built around a detailed seven page questionnaire. The questionnaire was completed by twenty teacher participants. Additional evidence is taken from four transcribed focus group interviews.

Question 1: (School Wide Practice)

> The participants were given a copy of the current Ministry of Education (MOE) definition of Education for Enterprise and asked to edit it in light of their experience on this project.

Many of the responses were similar in content; a sample is shown here:

- learning directed towards developing in young people those skills competencies, understandings and attributes which equip them to be innovative and prepared for the challenges of life;
- learning directed towards developing in young people those skills competencies, understandings and attributes which equip them to be innovative;
- learning directed towards developing and enhancing in young people those skills competencies, understandings and attributes which equip them to be innovative, creative, motivated and inspired. Encouraging students to successfully manage personal opportunities which will automatically extend in all facets of their lives for themselves and therefore for others.

These responses indicate that participants have developed a positive understanding of the value of education for enterprise.

Question 2: (School Wide Practice)

> How important do you think the following aspects are if more teachers and students are to be involved in E4E as a way of learning in your school?

Twenty statements were constructed and the participants were asked to respond using a Likert scale ranging from very important to not important.

The statements graded as very important by the highest number of participants were:

- the selection of learning activities across the curriculum that challenge students to connect enterprising skills to practical situations;
- the promotion of authentic contexts across all curriculum areas to help students see relevance and purpose in what they do and the link to the wider world;
- showing teachers that E4E is an effective vehicle for delivering the key competencies;
- getting a clear understanding of the underlying principles of E4E and how this can and should positively impact on teaching practice, engagement and relevancy;
- celebrating all successes and achievements.

The statements graded as very important by the lowest number of participants were:

- focusing on the 'end product', i.e. the development of students with the skills and attributes of enterprising people, before considering the programmes, activities and structures within the school;
- building working relationships with the local community;
- building working relationships with local businesses;
- led by teachers from subjects where E4E can be more easily applied - technology, business economics.

The participants indicated a belief in the value of authentic contexts for learning. They saw a strong connection between E4E and the key competencies. They felt that E4E was a good medium for cross curricular activities.

They indicated working with the community and business and having lead teachers from technology or business as less important. They focused on the benefits of E4E in general teaching terms rather than the economic imperative of the policy makers.

Question 3A: (Teaching and Learning Practice)

What priorities should be given to each of these practices?

Nineteen statements were constructed and the participants were asked to respond using a Likert scale ranging from very high to very low.

The statements receiving the highest number who indicated the very high or high key priorities were:

- encouraging students to gather information from a wide range of sources;
- putting less emphasis on curriculum content 'coverage', to allow time for deeper understanding or more relevant learning to occur;
- students having the opportunity to try out new and innovative ideas and take risk;
- encouraging students to see 'mistakes' as learning opportunities.

The statements receiving the lowest number who indicated the very low or low key priorities were:

- involving student in assessment decisions - what should be assessed and how;
- students having the opportunities to use experts from the community as mentors;
- students presenting the results of their learning/activities to an audience other than their teachers or classmates;
- supporting the development of students' business knowledge and skills.

The participants indicated a belief in the value of collecting information from a wide variety of sources. They saw a strong positive connection between E4E and risk taking. They felt E4E was a good opportunity for students to see 'mistakes' as learning opportunities. E4E allowed the students to learn more through trial and error. They indicated a lower priority rating for students developing business skills and knowledge, i.e. seeing this as less important.

Question 3B: (Teaching and Learning Practice)

How often did these practices occur before and after E4E project involvement?

Nineteen statements were constructed and the participants were asked to respond using a Likert scale ranging from most of the time to hardly ever.

The statements showing important and significant, i.e. number ten or higher, shifts were:

- students learning the curriculum through 'real-life' projects;
- putting less emphasis on curriculum content coverage to allow time for deeper understanding or more relevant learning to occur;
- students having the opportunity to try out new and innovative ideas and take risks;
- encouraging students to gather information from a wide range of sources;
- students taking leadership in planning and organising learning activities in the classroom/outside the classroom and gathering and managing the resources they need.

These results indicate a marked and significant shift in emphasis in regard to teaching and learning as a result of being involved in the E4E project. Teachers felt more comfortable with shared ownership and acknowledged the importance of unexpected learning outcomes.

Question 4A: (Teaching Practice)

> Show how strongly you agree or disagree with the following statements about your teaching practice with regards to E4E.

Fourteen statements were constructed and the participants were asked to respond using a Likert scale ranging from strongly agree to strongly disagree.

The statements receiving the highest number of strongly agree or agree responses were:

- I have a good understanding of the principles of E4E;
- I am enthusiastic about making E4E happen at this school;
- E4E has had a positive impact on student learning;
- E4E has increased decision making opportunities for students at this school.

The statements receiving the lowest number who indicated the strongly agree or agree key statements were:

- E4E has helped students see the relevance of the curriculum;
- E4E has helped build positive relationships with local businesses;
- E4E has heightened students' interest in the community;
- E4E is harder to plan for than conventional planning approaches to curriculum teaching and learning and it takes more time.

These results indicate that the professional development programme has given the teachers a confidence and enthusiasm for E4E. Less successful have been the connections made with the community. Successful meaningful connections can be difficult to establish and hard to maintain.

Question 4B: (Teaching and School Wide Practice)

> In your view does E4E have any special relevance or relationship to the values, philosophy or culture of your school?

Responses:

Yes	No	Not sure	Nil response
12	0	6	2

A qualitative expansion was sought and the following were received:

Indicative sample positive statements:

- it should be underpinning all enquiry learning;
- we have a very strong community and E4E involves them in the learning experiences a lot more;
- in our mission statement we refer to 'providing students skills for life' - E4E fits soundly into that concept;
- we want out kids to have real life meaningful contexts. We want our kids to be problem solvers, have ownership, increased responsibility. I see that E4E is an excellent 'vehicle' for this;
- E4E has great relevance to the philosophy of our school. Our school aims to encompass each child as a 'whole' person and develop them across the curriculum. E4E focuses largely on encompassing and including the various learning styles and thus creating motivated engaged learners.

Indicative sample negative statements:

- it could do if Principal, Boards of Trustees and other teachers took it on board. At the moment, the answer is probably more a no;
- it could have if staff and management wanted it to.

These results showed there has been real progress integrating E4E into the school culture. The staff felt that E4E allowed them to address the philosophical aims of their schools in a practical and meaningful way. Those that have made progress indicated it was a positive measure.

Question 5: (School Wide Practice)

Show how strongly you agree or disagree with the following statements about your school's teaching practice with regard to E4E.

Thirteen statements were constructed and the participants were asked to respond using a Likert scale ranging from strongly agree to strongly disagree.

The statements receiving the highest number who indicated they agree or strongly agree were:

- E4E is likely to change the way the curriculum is planned at a whole school level;
- E4E is likely to change the way the curriculum is planned at a department/ syndicate level;
- E4E has helped more students see the relevance of the curriculum;
- E4E has increased decision making opportunities for more students at this school.

The statements receiving the highest number who indicated they disagree or strongly disagree were:

- all teachers at my school have a good understanding of the principles of E4E;
- all teachers at my school are enthusiastic about making E4E happen;

- E4E is a central organising concept for curriculum and teaching at this school;
- teachers' workloads will increase if they get involved with E4E as a way of learning.

The positive responses show E4E to be somewhat of a change agent in both teacher planning and pupil ownership. The negative responses indicate there is still some way to go in convincing teachers outside of the project of its worth. However teachers involved did not see it as significantly increasing their workload.

Question 6: (School Wide Practice)

How would you rate the following challenges for E4E if it is to be sustained and extended upon as a way of learning in your school?

Thirteen statements were constructed and the participants were asked to respond using a Likert scale ranging from 'so challenging won't proceed' to 'little challenge easily solved'.

Most responses were in the 'challenging but likely to be overcome' column. There were three responses for 'so challenging won't proceed' worthy of a mention:

- the time demands of compulsory programmes and initiatives such as Keeping Ourselves Safe or Decision making, Assertiveness, Responsibility and Esteem (DARE). These programmes are designed to challenge children, parents and their communities to develop skills in Decision making, Assertiveness, Responsibility and Esteem;
- the need for some teachers to change their teaching style and give students more ownership of the learning process;
- a perceived 'side-dish' idea from the Ministry rather than an official emphasis, push and alignment with the key competencies.

There were three responses for 'challenging but likely to be overcome' worthy of a mention:

- the demands of assessment;
- organising people and groups outside the school to act, for example, as mentors, helpers;
- the need for teachers to be very enterprising themselves when planning units of work.

These results indicate a real need for approbation of E4E from the MOE; workload issues in other areas will take priority unless this happens.

TRANSCRIBED INTERVIEWS AND FEEDBACK

The following section of the report will be grouped under the research foci with an additional section looking at the both the positive and negative impacts of the professional development programme.

Focus 1:

Teacher practice that supports the development of enterprising attributes, capabilities and competencies of students.

Positive:

– Well that is the basis of it and why the unit is so successful. It's because it is developed within an authentic context. And the unit is to work on what the students have gained with their knowledge building - it goes on from there but the students expect to see a need - "What are we doing this unit for?"
– All decisions were made by the class but at this time they were totally focused. They were regularly brought back to the mat during the day to discuss options and decide on the next step.
– All students have goals for the day which are written out at the beginning of the session. This helps them stay on track and meet their deadline. The day is broken into blocks and each block has goals. They all have different jobs which are listed, with goals set.
– Making the learning intention and purpose transparent to the students right from the start seems to be very important. In this case the students know they are not just learning animation - another skill. They are learning it to use for a purpose.

Negatives:

– I did have a brief but I must admit I did more talking and calling their attention to it. That wasn't wonderful. The other thing was that it was a rush at the time of the year.
– We will finish this unit but the stop start nature of it probably doesn't maximise the potential for building enterprising learners. Unfortunately it is very difficult for the teacher to lessen the amount that needs to be done in a school day.
– It's the other factors that come into the process that can upset the time management of it. And also what I found hard was making the kids try not to go so broad - too far. I found sometimes they were way out there and having to actually find out for themselves took a lot of time.
– This teacher wasn't prepared to relinquish ownership to anyone, children, parents or a community member.

Focus 2:

What are the school wide practices that support or undermine the development of enterprising attributes?

Positive:

– Probably the best professional conversations I've heard with staff is when we are planning stuff we are thinking about how things are going to work. What's our purpose? Have we got a purpose and how do we make the learning more purposeful? – Realistic.
– So that's what I'm finding that the staff feel quite excited about it and we are planning things now. We are planning to build creative kids and to create the drive.
– The kids are taking pride in the fact we are pushing them.
– One of the most interesting things out of it though is that the behaviour management has come down to next to nothing.

Negative:

- One or two staff members looked at the barriers to the students' ideas rather than seeing them as clever ideas that could be changed slightly to make them safe.
- The whole literacy thing – we've done that contract and now we are in the embedding stage. There are changes the whole time. If you try and do things on the side (as in this case) you can see why it doesn't happen. The demands on a classroom teacher are huge these days, but something like that (E4E) can make it easier because the students can take over but they need the guidance.
- One of the problems with E4E is that not all teachers are that creative or enterprising in themselves. Perhaps principals should set up situations where they must become enterprising to get a task done. Then we hope they apply the same approaches to their kids.
- Teachers find this too, so this becomes a barrier for successful E4E. Some adults find it incredibly difficult to stand back.
- We need to move parents and teachers away from doing things when kids can do it.

Focus 3:

What is the influence and impact of school community partnerships on teaching and learning?

Positive:

- Getting the mentors into the school is really important. We use them on a regular basis and the mentors add that extra quality to a lot of units because they are the experts. The students have developed some good models of questioning and it's good to see these experts coming in and being responsive to the kinds of questions the students are asking.
- Having those experts around does make the unit more interesting because they can bring a real life perspective to the theme and what the students are learning.
- She was wonderful and the kids had so many questions they wanted to ask. But they ran out of time they were so enthusiastic and at the end of the time there were 20 or 30 hands up and the bell rang and they just wouldn't go.
- Having the builder on the panel worked well. He was able to give real life feedback to the students. For example, many had under priced their materials and Errol knew exactly the cost of timber etc. He was also able to explain to the students why some of their processes e.g. concreting wood to a post wouldn't work.

Negative:

- Some of the pitfalls of course are that the experts may not turn up – that's happened - or they direct the answers above the heads of the students and that is the danger.
- They need to be children friendly in terms of being able to relate information in a way children can understand and some aren't. The teacher actually does a better job.

- It is difficult to get mentors in work time.
- The mentors weren't always delivering what teachers wanted because we would be inside and they would have a group of kids outside.

Evidence of a positive shift in practice:
- E4E has made a difference in this school since the first term. We always knew kids could run with things if they got the chance and they have. They are working cooperatively doing their own research and working out their own ways to share it with others. It's how I do everything in the classroom now.
- The biggest difference has been in that everything has become that much more authentic.
- E4E gives us the structure to allow us to develop some higher order skills that adds value to our work in literacy and numeracy. It's helped us engrain what we are trying to do in our school and developing the enterprising attributes will be a strategic goal in our school.
- Teachers are taking the big topic and working it down to something relevant. The staff are planning for the kids and with the kids a lot more.
- There is a more conscious effort for staff to utilize the community people around them and although this has always been an important part of the school the shift has been on enhancing their potential and enhancing their experience in the school.
- Probably the best professional conversations I've heard with staff is when we are planning stuff we are thinking about how things are going to work. What's our purpose? Have we got a purpose and how do we make the learning more purposeful? – Realistic. All the links with planning anything in our class now we think about that as an aspect. I don't think I've noticed that kind of thinking before and that's quite cool.
- The parents noticed that the students' attitude to school and learning had changed there was greater enthusiasm.

CONCLUSIONS

This chapter has reported on a Professional Development and Research Contract funded by the Ministry of Education (MOE) in New Zealand. Through this contract a programme was developed to try and facilitate the development of E4E within technology education. Some success criteria have emerged and these have been grouped under the following headings:

Teacher Practice:

There is evidence to support E4E as part of technology education when:
- the context for the activity is shared, authentic and real;
- the activity is linked to practical undertakings and includes tangible outcomes;
- students are given a controlling function within the project, i.e. ownership of individual learning;
- the student contribution is encouraged, mentored and acknowledged;

- students are provided with flexible frameworks to facilitate project management;
- the approbation of E4E is evident from the teacher;
- the teacher reflects on their delivery and involvement from the beginning to ensure an enterprising approach is taken and modelled;
- the teacher encourages and values reflection from the students and this is incorporated in to progression and assessment;
- time management is paramount; therefore the teacher must operate both as a facilitator and goals chaser.

School Wide Practice:

There is evidence to support E4E as part of technology education when:
- support for participation and monitoring comes from the senior management team of the school;
- participation is shared to prevent burn out and remains consistent in the event of staff changes;
- the understandings of E4E are shared amongst all staff not just those involved in particular projects;
- E4E is not seen as another extra but is interwoven with key learning intentions from numeracy, literacy and the technology curriculum area;
- time allocations are flexible enough to allow for appropriate research and enough time to see the projects through;
- time allocations are concentrated, not too disjointed or disrupted by other activities within the school (suspended timetable alternatives);
- consideration is given to how to place E4E within the whole school;
- recognition and utilisation of quality facilitation and advisory programmes.

School Community Partnerships:

There is evidence to support E4E as part of technology education when:
- boards of trustees and parents are involved in the planning stages;
- experts and mentors are sought as soon as possible and time commitment established;
- co-operation and co-ordination of assistance occurs between mentors and teachers;
- due consideration of student participation and decision making is kept central to the project;
- there is community pride established in the activities undertaken;
- reports and updates are provided to all parties regularly using a variety of media.

There is an early indication that a quality E4E approach as part of a technology education programme can help to improve:
- behaviour management and motivation;
- participation of boys;
- stronger ties between school and the community, i.e. a connected curriculum;
- meaning and therefore a better learning experience to aspects such as numeracy, literacy and developing specific curriculum knowledge.

The findings support notions of authentic situations, shared ownership, and the integration of E4E within existing curriculum areas such as technology. Teachers with support from senior managers and facilitators or advisors can make useful connections with the wider community to enhance their teaching and ultimately the learning that occurs has more meaning.

REFERENCES

Anderson, G. (1990). *Fundamentals of educational research*. Lewes: Falmer Press.

Clark, J. (1997). *Educational research: Philosophy, politics, ethics*. Palmerston North: ERDC Press.

Clark, J. (2004). Enterprise education or indoctrination? *New Zealand Journal of Educational Studies, 39*(2), 321–332.

Fort, L., Martinez, B. L., & Mukhopadhyay, M. (2001). *Integrating a gender dimension into monitoring and evaluation of rural development projects*. Washington, DC: World Bank.

Guba, E. G., & Lincoln, Y. S. (1989). *Fourth generation evaluation*. Thousand Oaks, CA: Sage.

House, E. R., & Howe, K. R. (1999). *Values in education and social research*. Thousand Oaks, CA: Sage.

Ministry of Education (MOE). Retrieved from http://www.tki.org.nz/r/education_for_enterprise/definition_e.php

Patton, M. Q. (1997). *Utilization-focused evaluation* (3rd ed.). Thousand Oaks, CA: Sage.

Stake, R. (2004). *Standards-based & responsive evaluation*. Thousand Oaks, CA: Sage.

Stake, R., & Schwandt, T. (2006). On discerning quality in evaluation. In I. F. Shaw, J. C. Greene & M. Mark (Eds.), *Handbook of evaluation*. London: Sage. Retrieved from http://www.ed.uiuc.edu/circe/Publications/Discerning_Qual_w_Schwandt.doc

Wadsworth, Y. (1997). *Everyday evaluation on the run* (2nd ed.). Sydney: Allen & Unwin.

Worthen, B. R., Sanders, J. R., & Fitzpatrick, J. L. (1997). *Program evaluation: Alternative approaches and practical guidelines*. White Plains, NY: Longman. Retrieved from http://www.tki.org.nz/r/education_for_enterprise/definition_e.php

Gary O'Sullivan
School of Curriculum & Pedagogy
Massey University
New Zealand

MAGGIE ROGERS

22. EMBEDDING EDUCATION FOR SUSTAINABLE DEVELOPMENT IN PRIMARY DESIGN AND TECHNOLOGY EDUCATION

Reflections on a Journey

INTRODUCTION

This chapter charts not only the journey undertaken but also reflects on the experience of embedding Education for Sustainable Development (ESD) in the teaching of Design and Technology (D&T). The context is teacher education in England and the changes that have taken place in recent years. As a series of research projects the journey has been supported by Oxfam Education initially and World Wide Fund for Nature (WWF-UK) for the main part with funding, advice and support. There has also been support from colleagues in two other institutions of Higher Education: London South Bank University and Greenwich University. As a personal journey of leaps in understanding and pitfalls in carrying out what is now a longitudinal study within this framework of support, it is hopefully a fitting tribute to student teachers, past and present who have studied to be early years and primary teachers at Goldsmiths, University of London taking the opportunities to examine and develop their awareness and understanding of ESD through Design and Technology.

BACKGROUND TO THE RESEARCH

From 1989 until 1999 the Department of Educational Studies at Goldsmiths offered a four-year honours degree in primary education jointly with other departments in the university. As part of this programme student teachers were able to study Design and Technology alongside a qualification to be early years and primary teachers. The degree programme was innovative in that the student teachers had access to Design Department facilities and tutors during the first and second years of their degree, which gave them the opportunity to develop a deep understanding of the subject. They also studied the Design and Technology knowledge and understanding elements of the programme jointly with their secondary colleagues and engaged in joint projects. One of the elements in the second year of their degree programme was a study of alternative energy where the students were given the brief to research and design and make a system to use alternative energy sources. Through the author's participation in a WWF-UK initiative (Rogers, 1996), all of the student teachers

C. Benson and J. Lunt (eds.), International Handbook of Primary Technology Education: Reviewing the Past Twenty Years. 275–283.

were introduced to aspects of ESD during their curriculum Design and Technology application course. By 1997 this course was well established and a presentation entitled 'Preparation for teaching Design and Technology in primary schools: Changing models' was made by the author at the first Centre for Research in Primary Technology (CRIPT) Conference in Birmingham, UK (Rogers, 1997a).

In response to changes to requirements for teacher training in England, a new three year degree was introduced by the department at Goldsmiths, University of London in 1996 which still offered student teachers the opportunity to specialise in Design and Technology but on a much reduced timescale. Instead of two years of full time study in the subject and its application in school, student teachers experienced courses which were progressively more specialist, but shorter, over the three year programme. This new degree programme, however, retained aspects of the four-year programme in that student teachers in their third year still had access to facilities and workshop tutors in the Design Department and were able to develop their design and make skills through a study of the nature of the subject. ESD became part of the second year option course initially with support from colleagues at Oxfam Education (Rogers, 1997b).

In 2000 the author joined the WWF-UK Partners in Change project, working alongside colleagues at London South Bank University (LSBU) under the directorship of Professor Sally Inman, Head of Education at LSBU and Director of the Centre for Cross Curricular Initiatives (CCCI). This was the beginning of a seven-year partnership that has included two publications to which the author has contributed, 'Teaching for a sustainable future: Embedding sustainable development education in the initial teacher training curriculum' (Rogers, 2002) and 'Building a sustainable future: Challenges for initial teacher training' (Inman & Rogers, 2006). As further changes in primary curriculum teacher training have taken place, there has been increasingly less time for foundation subjects such as Design and Technology on the BA (Ed.) degree at Goldsmiths. One positive aspect of changes in the then Teacher Training Agency (TTA) Standards for Initial Teacher Training (ITT) (TTA, 2002) unexpectedly resulted in a more comprehensive embedding of ESD in D&T across primary initial teacher training programmes. Although given much less time on specialist courses the pre-service generic courses in Design and Technology give all participants the opportunity to look at issues of ESD (Rogers, 2003).

THE IMPERATIVE FOR THE RESEARCH

The imperative for this research was brought about mainly by a subject response to environmental issues and social injustice which are inextricably linked to Design and Technology activity. Awareness of these issues has grown since the publication of Agenda 21 after the Rio Earth Summit in 1992. Much has been written about values education within D&T by members of the Values in Design and Technology Education (VALIDATE) special interest group which was formed as a result of common concerns shared at the International Design and Technology Education Research (IDATER) Conferences held at Loughborough University between1989 and 2001 (Martin, 1996; Conway, 1999).

A key feature of the research into embedding ESD in D&T courses carried out at Goldsmiths, has been that of investigating the values and dispositions necessary to develop awareness of ESD. The framework developed by the Partners in Change (Inman, 2006) team was invaluable in identifying aspects of the subject that could most effectively support the development of these.

- valuing the physical environment
- valuing and celebrating diversity
- commitment to justice and equality
- empathy with others
- respect and caring for ourselves and others
- openness and commitment to individual and collective change
- commitment to lifelong learning
- commitment to a lifestyle consistent with sustainable development

Figure 1. Partners in change framework – values (Inman, 2006, p.39).

As Layton stated, value judgements '...reflecting people's beliefs, concerns and preferences are ubiquitous in Design and Technology activity' (Layton, 1992). Kimbell et al. also stress the teaching and learning about values which is a critical dimension of the subject (Kimbell et al., 1996). In 1996 Martin added to the values debate by suggesting that an important part of technological literacy was looking at values within products and 'reflecting on the effect of products on people and society' (Martin, 1996).

Huckle reported in 2006 that ESD has been largely developed and carried out by non-government organisations and 'interested university tutors'. He suggested that ESD could be shifted from the margins to the mainstream due to key initiatives such as the Decade of ESD (Huckle, 2006). At that time Huckle also drew attention to the divisions between the focus of Environmental Education (EE) on sustainability and the focus of Development Education (DE) on 'the global dimension' despite both groups of educators sharing common language and interest in ESD. In 2006 the DfES introduced the Sustainable Schools initiative, which provides a tool for sustainable school self-evaluation in addition to 'doorways' through which to address issues of sustainable development (DfES, 2006).

THE JOURNEY

The research described and reflected on in this chapter is part of a journey that involved awareness raising, implementation and measuring impact of these interventions. Data from each aspect has been collected using a methodology that is appropriate for that particular stage of the journey. Initial work carried out as product analysis was used to raise issues of sustainability through D&T. This was conducted through case studies, while on subsequent projects, questionnaires were used alongside interviews and round table discussions. The methodology also reflects the experience of an experienced researcher new to the field of ESD developing awareness of the issues involved and learning alongside the students.

The first phase of embedding ESD in Design and Technology education drew on the work of Stables, Rogers, Kelly & Fokias (2001) in that product analysis was used as part of the assessment of children who had taken part in a project researching the development of literacy through D&T. Issues of sustainability were introduced to student teachers on the second year option course through the use of a 'Bingo' activity of statements regarding energy sources, life style choices and the effect on the environment. Approaches to product analysis, developed through the Oxfam Education project (Rogers, 1997b), were used to support them to consider issues of ESD. By identifying questions that could be asked about various food and drink items, the student teachers highlighted issues without having to have any prior knowledge of the items. The student teachers were then asked to identify a product of their own choice to analyse. The main issues the students identified were to do with efficiency, use of materials and durability. Other issues considered were safety and cost (the latter in terms of value for money), aesthetics, consumer demand and life style choices.

At the end of the autumn term the students were given a questionnaire to fill in about the sustainability elements of the course. They were asked to comment on aspects they considered important before the course started and to indicate which aspects of the development education session they found effective and might use in the classroom. Data was collected in the following year and a comparison made between the responses of the two cohorts.

The second phase of embedding ESD in Design and Technology education occurred naturally when a new course was developed for the second year of the BA(Ed) at Goldsmiths. Artefacts were central to this new course, using and developing previous work with 'handling collections' through the choice of those with a global dimension. Working with Barbara Lowe from the Reading International Solidarity Centre, the author was able to introduce products which were not necessarily everyday items which would almost certainly raise issues of sustainability and social justice while fitting into projects the student teachers were familiar with. These included those from the Qualifications and Curriculum Authority exemplar scheme of work (QCA/DFEE, 1998). The products or artefacts offered opportunities to challenge stereotypical or negative views of their country of origin. The artefacts were used as a starting point for the student teachers' own designing and making. Subsequent work in schools was supported by the production of a resource pack for classroom use.

The development of this new course was monitored and the impact of the approach assessed using questionnaires and case studies to illustrate the impact on the student teachers' practice in school.

THE ARTEFACTS

The artefacts, all from countries of the South, were loaned from the collections at Reading International Solidarity Centre (RISC), and were chosen because they would challenge stereotypical or negative views about those who had made them, the countries these people live in as well as supporting the development of Design and

Technology activities. Two lengths of Adinkra cloth together with several calabash printing-blocks were supported by a poster explaining each symbol printed on the cloth. On another poster were photographs of a young boy from Ghana showing the process of making a tin car also provided within the collection of artefacts. A 'push along' bird, made by a child in Kenya offered the student teachers a toy, interesting technologically as well as for its story. The third pair of artefacts comprised of a hand and a shadow puppet, both from India. Within the collection on loan from RISC were books on toy and puppet making, and the children's storybook, 'Aani and the Tree Huggers' (Atkins, 2000).

USING QUESTIONS

The student teachers were asked to select one of the artefacts and identify questions they could ask of it. These questions were organised and developed through using the Development Compass Rose, originally developed by Teachers in Development Education (TIDE) based at the Birmingham Development Education Centre (TIDE, 1995). The students were then given the information and asked to research using the books available and the Internet. The chosen artefacts were used to base designs for the first design and make task. As the course progressed, the students compiled resource packs to support this activity. During their school experience placement the student teachers were required to teach Design and Technology. This was analysed and discussed in their written assignment.

At each stage (i.e. after the first session, after the completion of the design and make task, and on return from school experience), the student teachers were given a questionnaire to elicit their response to the project. As with any intervention, the questions asked also supported them in analysing the issues as they presented themselves. Feedback sessions were also used to monitor responses as well as supporting the student teachers' understanding.

RESPONSES TO THE ARTEFACTS

The student teachers were initially intrigued and not necessarily aware of the significance of the artefacts. One student thought that they were "...examples of children's work that Goldsmiths' students had brought back from teaching practice", while another thought they were great but thought that, "...they had been made by the children for the children and, on reflection, my opinions were possibly slightly patronising". This student teacher went on to explain that they had compared them "...with my knowledge of how children like making things - and thought that they would enjoy making them (more than playing with them!)". Some of the student teachers found the artefacts interesting immediately while the interest of others grew as they began to identify questions "as by finding out more explained that it had more meaning than we first thought". Using questioning as a strategy to explore and analyse the artefacts was also effective in developing their interest as evidenced by one response "I wanted to know more about them, as we were asked to think of questions, which was then tempting to find out about the answers."

As the stories behind the artefacts were revealed the students started to examine their initial responses with some significant insights gained.

> I felt quite humble given that we are so lucky to have at our disposal materials and money to buy new resources as teachers, but could fail to achieve the standards of craftsmanship seen in the toys. Most upsetting was the link between the design and manufacture of the toys and the need for them to be commercially viable to satisfy tourists. (Student)

The students who had chosen the toys to explore were particularly moved by the push along bird, which it had been assumed "...they were making it in schools for themselves (to take home)". As it emerged that the toys had been made for very different purposes. Children made both but one was made to play with, the other to sell to tourists to fund their schooling. Feelings ranged from "I suppose I was slightly disappointed to find that they had been made as part of a business and that the end-users would be tourists", to admiration for the producers: "...made you think about the work that went into make them". One student teacher expressed shock "...because I would never have thought that children/adults made these toys for tourists", while another found that the information "...prompted lots of thoughts about my own opinions".

The process of exploring before being given information was seen as a positive experience by the student teachers with one writing that they found it "fascinating, good to be able to compare what we thought about the puppets to where they were really from and made from etc.", while another wrote "because it was given bit by bit, it made me more curious and allowed me to take in more information than I would have done if I had all of the information all at once". The information also seemed to have supported the students in asking more questions as well as giving further insights.

> The information answered the questions that I had about how the puppets had been made and why they had been made. Knowing more was helpful, I appreciated having greater insight. (Student)

The students who had chosen the Adinkra cloth to explore had no idea of the purpose of the cloth and its design until they had been provided with the additional information.

IMPACT ON PRACTICE

The artefacts continued to offer a positive context for the students design and make projects with some very thoughtful and sensitive interpretations of the brief. Without doubt the artefacts had an impact on the students' approach to designing and making on the course. Further strategies were used to see how this experience would impact on the work in school, particularly the use of handling collections of artefacts supported by research in the form of stories wherever possible. To appreciate how difficult it is to take this experience directly into the classroom, it must be noted that with the National Curriculum supported by schemes of work, student teachers do not have the choice over topics for Design and Technology. Although the introduction of the Primary Strategy (DfES, 2004) opened up the possibility for more 'joined up' teaching and project based teaching, and QCA is encouraging schools to explore alternatives to the units of work, many teachers tend to use what feels familiar.

REFLECTIONS

In reflecting on the experience of raising awareness of issues of ESD through D&T at Goldsmiths during the second year course several key factors emerge. The first of these is about developing a critical awareness and understanding through the analysis of everyday products. From the experience of the research this can only take place when the student teachers know enough about the subject to be able to make connections between Design and Technology activities and ESD. Using questions without necessarily having all the answers at hand can support them in this process as they explore issues much larger than at first apparent. It is also essential that underlying principles must be made clear through the way the course is taught as well as the activities so that there are no contradictions between, for example, the use of materials and resources in design and make projects and the issues raised.

In terms of developing their practice student teachers must have the opportunity to develop their understanding of ESD through D&T during their school placements. This will allow them to take ownership of their understanding of issues and reflect this back into college-based discussions. Recognition of the student teachers' prior understanding needs to be explicit to develop a personal awareness of ESD issues and a respect for materials and resources.

Reflections on the development of a new course based on artefacts were presented at the fifth CRIPT conference (Rogers, 2005) and are also well documented by Rogers & Lowe in 'Building a sustainable future: Challenges for initial teacher training' (Inman & Rogers, 2006). What has not been documented, however, is how this initial research supported the development of a mature and well-established course.

Using quantitative and qualitative research methods to measure impact, record the 'student teacher voice' and archive the experience has been of immense value in developing the course further. Working with Non-Government Organisations (NGOs) at RISC and at WWF-UK has ensured a quality experience for tutor and student teachers alike. 'Handling collections' take on a different meaning when steeped in cultural traditions as well as socio-economic issues. In searching for an artefact that would not only raise issues of ESD but also stimulate designing and making using food, the author discovered a calabash dining set from Nigeria. Not only made from a sustainable source, the set, comprising of a large serving bowl, a smaller bowl and ladle, has insulating properties which suggest slow cooking and hay box cooking technology. The calabash bowl also has traditional and cultural meanings within regions of Nigeria. Added to the collection of Adinkra cloth are contemporary pieces of manufactured lengths, modern hand crafted designs and the incorporation of Adinkra symbols into a batik art work by a female Ghanaian artist. Puppets from traditions of Eastern Europe have also been used alongside the Sanchar puppet from India.

The impact of this work on future student teachers is uncertain as the three-year degree has finally been closed, replaced by a three year education degree without Qualified Teacher Status (QTS). What has been achieved however is ESD embedded in the Post Graduate Certificate in Education (PGCE) programme as every student in each cohort is introduced to using artefacts in their Design and Technology practice.

Perhaps the most important research carried out over the course of this journey has been accessing the 'voices' of the student teachers as they moved into their first posts as newly qualified teachers (Blunden et al., 2006). From final interviews it emerged that the artefact session had made a huge impact on the student teachers and was an experience that had stayed with them because it was very different from anything they had done before or since in their training. The artefacts and their 'stories' had inspired a variety of ideas for the classroom and had been used as a challenge to produce the equivalent local product, using and preserving 'local skills', resources and materials.

Suggestions for future development of support for student teachers to integrate ESD into their classrooms included the introduction of ESD-related activities, in the form of directed activities. Web-based resources were another suggestion. One of the student teachers suggested that a fresh perspective on D&T through ESD would support the status of the subject, which may be more of a possibility within the framework of Sustainable Schools (DfES, 2006).

The final message from these student teachers stressed that:

> ...the teacher's enthusiasm for and understanding of ESD was a key factor in children's capacity and willingness to engage in the issues. (Blunden et al., 2006)

thereby making the strongest case for initiatives of this kind.

ACKNOWLEDGEMENTS

Many thanks to the student teachers at Goldsmiths who have generously shared the journeys they undertook during this period of research into ESD through D&T. Gratitude is also extended to Barbara Lowe from Reading International Solidarity Centre for her valuable input to the Artefacts Project. Final thanks to WWF-UK and the members of the Partners in Change team based at the Centre for Cross Curricular Initiatives, London South Bank University and the Education departments at Greenwich University and Goldsmiths, University of London.

REFERENCES

Atkins, J. (2000). *Aani and the tree huggers*. New York: Lee and Low Books.

Blunden, J., Inman, S., Meadows, J., Norman, A., & Rogers, M. (2006). Reflective voices. In S. Inman & M. Rogers (Eds.), *Building a sustainable future: Challenges for initial teacher training*. Godalming: CCCI/WWF.

Conway, R. (1999). Lessons for technology education for social, ethical and environmental audits. In P. H. Roberts & E. W. L. Norman (Eds.), *International conference on Design and Technology Educational Research 1999* (pp. 72–78). Loughborough, UK: Loughborough University.
DfES. (2003 and 2006). *Sustainable development action plan*. London: Department for Education and Skills.
DfES. (2004). *Excellence and enjoyment*. London: Department for Education and Skills.
Huckle, J. (2006). *Education for sustainable development in initial teacher training: A briefing paper for the Teacher Training Agency*. London: TTA.
Inman, S. (2006). Meeting the challenges. In S. Inman, & M. Rogers (Eds.), *Building a sustainable future: Challenges for initial teacher training*. Godalming: CCCI/WWF
Inman, S., & Rogers, M. (2006). *Building a sustainable future: Challenges for initial teacher training*. Godalming: CCCI/WWF.
Kimbell, R., Stables. K., & Green, R. (1996). *Understanding practice in design and technology*. Milton Keynes: Open University Press.
Layton, D. (1992). *Values and Design and Technology, design curriculum matters: 2*. Loughborough: Loughborough: University of Technology.
Martin, M. (1996). Valuing products and applications. In J. S. Smith (Ed.), *International conference on Design and Technology Educational Research and Curriculum Development 1996*. Loughborough: Loughborough University.
QCA/DFEE. (1998). *A scheme of work for key stage 1 and 2 Design and Technology*. London: HMSO.
Rogers, M. (1996). Environmental education, technology and initial teacher training. In S. Inman & P. Champain (Eds.), *Thinking futures: Environmental education in initial teacher training*. Godalming: WWF.
Rogers, M. (1997a). Preparation for teaching Design and Technology in primary schools: Changing models. In R. Ager & C. Benson (Eds.), *International Primary Design and Technology conference 1997* (pp. 56–58). Birmingham, UK: CRIPT at University of Central England.
Rogers, M. (1997b). Development education through Design and Technology: A BA (Ed.), primary case study. In S. Inman & R. Wade (Eds.), *Development education within initial teacher training: Shaping a better future*. Oxford: Oxfam.
Rogers, M. (2002). Developing values and dispositions to education for a sustainable future through Design and Technology education. In *Teaching for a sustainable future; Embedding sustainable development education in the initial teacher training curriculum*. A joint WWF/South Bank University/Centre for Cross Curricular Initiatives publication.
Rogers, M. (2003). Exploring education for sustainable development through Design and Technology education. In C. Benson, M. Martin, & W. Till (Eds.), *Fourth International Primary Design and Technology conference 2003* (pp. 128–131). Birmingham, UK: CRIPT at University of Central England.
Rogers, M., & Lowe, B. (2005). Using artefacts and their stories to develop Design and Technology capability. In C. Benson, S. Lawson, & W. Till (Eds.), *Fifth International Primary Design and Technology conference 2005* (pp. 115–118). Birmingham, UK: CRIPT at University of Central England.
Stables, K., Rogers, M., Kelly, C., & Fokias, F. (2001). *Enriching literacy through Design and Technology, Evaluation project final report*. London: Goldsmiths College.
Teachers in Development Education (TIDE). (1995). *Development compass rose consultation pack*. Birmingham: Development Education Centre (DEC).
Teacher Training Agency. (2002). *Qualifying to teach: Professional standards for newly qualified teachers and requirements for ITT*. London: TTA.

Maggie Rogers
Centre for Cross Curricular Initiatives
London South Bank University

MARION RUTLAND, SALLY ASTON, DEBBIE HAFFENDEN, GILL HOPE, DOT JACKSON, BHAV PRAJAPAT, MAGGIE ROGERS AND MARTIN SEIDEL

23. PERCEPTIONS OF PRIMARY DESIGN AND TECHNOLOGY

Initial Teacher Education Students' Experiences

INTRODUCTION

In 2002 a joint seminar 'Developing and celebrating good practice in primary Design and Technology' (Nuffield Foundation, Design and Technology Association and the Centre for Research in Primary Technology (CRIPT)) formed the impetus for this research. One of the recommendations from the seminar was that a small working party of key players should develop a research framework and plan co-operative research activity utilising school and university links across participating universities. The main aim was to enable primary initial teacher education (ITE) trainers, together with teachers in schools, to use their normal working activities to generate data that can be used as the basis for academic papers (Barlex, 2003).

The first meeting of the National Research Group was held at the Nuffield Foundation, London (February 2004, Nuffield Foundation/Design and Technology Association). As a result, a group of ITE providers in South East England from the University of Brighton; Canterbury Christ Church University; Goldsmiths, University of London; Roehampton University and St Mary's University College met on a number of occasions. The aim of their research was to develop a clearer understanding of the position and character of Design and Technology (D&T) in the ITE providers' partner schools as experienced by the students. The data from the research project focused on concerns regarding the position and status of D&T in English primary schools since the introduction of D&T as a compulsory subject of the National Curriculum in 1990.

Art, Design, Craft, Technology and Science have a long history in the curriculum of primary schools in England, but it was not until 1990, with the introduction of the National Curriculum, that there was a legal obligation to teach D&T. The National Curriculum D&T Orders (DES, 1990; DfE, 1995; DfEE/QCA, 1999) programmes of study for Key Stages 1 and 2 (5–11 years) lay down the content that had to be covered including designing and making in a range of materials including food, textiles, electrical and mechanical components, stiff and flexible sheet materials and

C. Benson and J. Lunt (eds.), International Handbook of Primary Technology Education: Reviewing the Past Twenty Years. 285–303.

mouldable materials. It is acknowledged that children's achievement and progress has improved slowly but steadily since the beginning of the first Ofsted inspections in 1994:

> Pupils' making skills are better than their designing skills and both are better than their knowledge and understanding. (Ive, 1999, p. 16)

The planning and teaching of D&T in primary schools has developed considerably and as one teacher commented in the early stages of the National Curriculum:

> Design and Technology was our weakest curriculum area. This was due to a lack of confidence, expertise and understanding amongst staff. (Vaughan, 1997, p. 32)

D&T is a new curriculum area for primary teachers who qualified before 1990 when it was first included in ITE courses. In-service courses to accommodate these changes were varied (Benson, 1997) and it was not until 1993 that money became available to set up courses to enhance teachers' subject knowledge through government funded Grants for Education Support and Training (GEST) in-service courses.

The introduction of national literacy and numeracy strategies had a considerable impact on the classroom time allocated to D&T in school (Rogers & Davies, 1999), as did the Primary National Strategy (DfES, 2003) that advocated the linking of subjects. However, in-service courses funded by the Teacher Training Agency (TDA) and managed by the Design and Technology Association (DATA) have been very successful in helping primary teachers to develop the skills and knowledge to teach D&T in the classroom, and plan and co-ordinate D&T within their schools (Perry, 2003). Increasingly resources to support the teaching of D&T in primary schools have become available such as the Qualifications and Curriculum Agency (QCA) Exemplar Scheme of Work (QCA, 1998) (e.g. see Martin, 2001), the Design and Technology Association Help Sheets (DATA, 2000) and Lesson Plans (DATA, 2002), and the Nuffield Primary Project (see Mitra, 1999 Chapter 11).

It was apparent from the early days of the introduction of the National Curriculum that ITE institutions had very different time allocations for D&T, ranging from five to forty hours for each student (Ager & Benson, 1997) and this remains an issue. It is suggested that newly qualified teachers should be able to teach with advice from an experienced teacher where necessary (TTA, 2002), but in many cases the class teacher may not have the expertise or opportunity to provide them with such a rich experience (Davies et al., 2000). College courses, though limited, are frequently overridden in practice by the classroom with reduced flexibility of the curriculum and other constraints such as lack of resources, accommodation and limited time (Davies & Rogers, 2000). This can result in teachers with a 'lack of confidence and/or 'hands on' experience of designing and making that might settle for 'safe' and perhaps more prescriptive activities' (Rogers, 2004, p. 23). Finally, an additional complication is presented in the latest standards and requirements to prepare student primary teachers on ITE courses to teach a range of work across subjects including 'Art and Design or Design and Technology' (TTA, 2002, p. 11).

METHODOLOGY

It was against the background of these issues that the South East England group of ITE providers decided to gather data relating to the nature of D&T in their local primary schools. The decision to focus on the perceptions or impressions of their students, as carried out in Scotland (Dow, 2003), was taken because this would provide a picture of the position of D&T in the schools. It would help tutors identify issues that needed to be further developed or reinforced during taught university sessions.

In the summer of 2004 the university tutors developed pilot questionnaires based on their perceptions of good practice in schools. The content was discussed and agreed at two meetings. The main headings were:

- course details for the student:
- organisation of D&T
- accommodation
- materials available
- displays
- policies.

Details of the online questionnaire can be seen in Figure 1. This was made available to individual institutions through their university website. The results were held centrally at Goldsmiths' College before being passed to each institution. Though students were encouraged to complete the questionnaire it was done on a voluntary basis in the late Autumn and early Spring of 2004/2005.

SOUTH EAST PRIMARY D&T RESEARCH QUESTIONNAIRE
FOCUS: - IMPRESSIONS OF D&T IN A STUDENT'S PLACEMENT SCHOOL

1. COURSE

PGCE Stage 1 ☐ BA YEAR 1 ☐ BA YEAR 3 ☐ KS2/3 ☐

PGCE Stage 2 ☐ BA YEAR 2 ☐ BA YEAR 4 ☐

What previous design and technology experience do you have?

Degree ☐ FE ☐ School ☐ Work ☐ Other

What year group did you teach during your most recent experience?

Nursery ☐ Reception ☐ Year 1 ☐ Year 2 ☐ Year 3 ☐

Year 4 ☐ Year 5 ☐ Year 6 ☐ Year 7 ☐

Is D&T your subject specialism? Yes ☐ No ☐

2. ORGANISATION OF D&T IN SCHOOL

Was D&T taught in your class, while you were in school?

Yes ☐ No ☐

If yes, what D&T was taught?

How was the teaching of D&T organised in your school?

Weekly lessons ☐ 2-3 day blocks ☐ D&T weeks ☐
A combination ☐
Others

Did the school use the D&T QCA Scheme of work?

Yes ☐ No ☐

If yes, what units did you use?

Did the school use the Design and Technology Association's lesson and help sheets?

Yes ☐ No ☐

If yes, which ones?

Did the school use the Nuffield Primary Solutions units?

Yes ☐ No ☐

If yes, which ones?

Was there a D&T coordinator/subject leader in the school?

Yes ☐ No ☐ Don't know ☐

Did they coordinate/subject lead other curriculum areas?

Yes ☐ No ☐ Don't know ☐

If yes, which areas?

3. ACCOMMODATION

How was D&T organised and taught in the school?

In the classroom ☐ In specialist areas ☐ Shared areas ☐
Others

4. MATERIALS

Where were D&T materials stored in the school?

In a store cupboard ☐ Individual classrooms ☐
Boxed storage per project ☐

Other storage provision

Was your school resourced to teach with?

Food ☐ Textiles ☐ Wood ☐

Plastics ☐ Mechanical control ☐ Electrical control ☐

Construction kits ☐

Were ICT resources available to support D&T?

Yes ☐ No ☐ Don't know ☐

5. DISPLAYS

Did you see evidence of D&T other than in your classroom?

Yes ☐ No ☐ Don't know ☐

If yes, which?

6. POLICIES

Did you read the D&T policy?

Yes ☐ No ☐

Did you see links between D&T and other curriculum areas?

Yes ☐ No ☐

If yes, which ones?

Are there any areas of D&T that you would have liked to have taught or observe being taught?

Would you willing to be interviewed following the analysis of the questionnaire?

Yes ☐ No ☐

If yes, please give the following details so that we can contact you.

Name:

Telephone (mobile if possible):

Email:

Figure 1. Online questionnaire.

Overview of Data

Brighton University: total 37 students. Data was collected from 37 students on the primary undergraduate programme BA Hons with Qualified Teachers Status (QTS). There were 27 Year 3 and 10 Year 4 students. The majority of the students had experience of the subject at school level, only 1 student had studied D&T at further education level and 3 had no experience at all. The majority of the students were teaching at KS1 with even spread between Foundation Stage and KS2. In terms of the total number of D&T specialists (5 out of 8), the response was good.

Canterbury Christ Church University: total 35 students. Two very different groups completed the questionnaire immediately after their second school experience placement. The first group included 19 part-time Postgraduate Certificate of Education (PGCE) mature students, many of whom had worked as teaching assistants prior to joining the course, but who had received only 3 hours input on D&T within their course. The second group of 26 Year 2 BA (QTS) included 3 mature students (students not straight from school with previous experiences) part way through their D&T course, who had received 15 hours input in Year 1 & 30 hours in Year 2 when they completed the questionnaire.

Goldsmiths, University of London: total 34 students. Data was collected from 11 PGCE students on their return from their second placement in February 2005 at the beginning of their D&T course. In addition data was collected from 23 BA: Education (Ed) students in their second placement in March 2005. The majority had opted for a second D&T course, although only 7 were D&T specialists who would continue with D&T.

Both cohorts, taught in mixed phase groups, spent their first year/stage placement in their target age phase, with Early Years and Key Stage 2 specialists placed for their second placement in Key Stage 1 classes. A small percentage of the students did not teach D&T, or see it taught, though those following the second year option were required to teach D&T for their assignment.

Roehampton University: total 44. Data was collected from 44 PGCE students immediately following their first school placement and half way through their compulsory D&T foundation course. 36 of the students taught Key Stage 2 and 8 Key Stage 1. 42 students relied on their school experience of D&T. Their teaching experience in their placement schools was well spread across the year groups with the highest number (10) in Year 6. None of the students were PGCE D&T specialists. This had been common for Roehampton over a number of years, despite a strong tradition of a specialist D&T group on the BA Primary Education with Design and Technology programme.

St Mary's University College: total 35 students. All the students were in Year 2 of a BA (QTS) programme and had chosen D&T as their specialist subject. They had all attended a D&T non-specialist course in Year 1 and a specialist food technology module. The student group had all completed first and second year school experiences in Reception (children aged 4–5 years), Key Stage 1 (children aged 5–7 years) or Key Stage 2 (children aged 7–11 years) before filling in the questionnaire.

FINDINGS OF INDIVIDUAL PROVIDERS

Brighton University

Of the respondents, 71% saw D&T being taught in schools addressing a wide range of activities. Fifteen schools were identified as using QCA schemes of work (QCA, 1998) and eleven units were specifically named, although two were recognised as Art and Design units (QCA, 1999). Ten schools acknowledged using a combination of delivery modes but the majority of schools organised D&T teaching as weekly lessons. Only one school made use of the DATA helpsheets (DATA, 2000), and the Nuffield units (Nuffield Curriculum Project Centre,/DATA, 2001) were not utilised by any schools. Only 65% schools had a D&T coordinator, of which 37% had other curriculum responsibilities and this was often Art and Design.

The majority of teaching of D&T was taught in the classroom as opposed to a shared or specialist area, indicating a need for mobility in regard to practical resources. This raises questions about organisation and management of resources, tools and materials during the teaching of the subject. The majority of the equipment was stored in cupboards and a small number used racks and trolleys which may have addressed the issue of mobility. There were no indications that the resources seen were directly divided into areas or grouped together. There was some use of ICT linked with D&T and our questionnaire design did not provide an opportunity for a more specific answer. Twelve responses indicated that there was evidence of 'other' D&T displays outside the classroom. Finally, 60% of respondents took the trouble to read the policy though they did not see links between D&T and other curriculum areas.

Canterbury Christ Church University

The Year 2 BA group displayed their commitment to their subject with only 10% of the PGCE students reading the schools' D&T policy, compared to 50% of the Year 2 students. This demonstrated a greater awareness of links between subjects in their own classroom, evidence of D&T across the school and of the range of D&T resources available. This is what would be expected of students who envisage themselves as future D&T Co-ordinators.

The Year 2 group also seemed to be more certain about whether or not their school used the QCA scheme (QCA, 1998). In conversation after completing the questionnaire, several students said that schools had devised their own schemes and two reported schools moving to cross-curricular topic work. In general, the PGCE students did not appear to view D&T as an important subject. This may be due to their much lower personal experience of D&T as well as the low profile of all foundation subjects within their course.

The Year 2 students appeared to have been far more pro-active in creating space for D&T within their teaching. In answer to 'What D&T was taught?' one of the Year 2s wrote 'Only by me!' One Year 2 student reported her indebtedness to one of our Year 3 D&T specialists on placement in the same school, who stressed to her the importance of children making genuine design choices and they jointly planned their own scheme for bread-making. No such comments came from the PGCE group.

Goldsmiths, University of London

A wide range of activities were taught, with 'Fruit Salads' the most popular, followed by 'Vehicles' at Foundation and Key Stage 1. Construction was high on the activity scale with Celebrations used as a focus for Christmas cards and Chinese New Year cards.

More than a third of the students (39%) reported that D&T was taught in weekly sessions with only 20% reporting a combination of weekly and blocked time. The Year 2 students had more investment in whether the school used the QCA scheme of work (QCA,1998) because of assignment requirements; however there was little evidence of the DATA Guidance Materials (1995), DATA Helpsheets (2000) and Nuffield units of work (2001) being used, despite local support through an extended D&T co-ordinators' course funded by the TDA.

More BA Year 2 students than PGCE appeared to be aware that there was a D&T co-ordinator and a high percentage did not co-ordinate other areas. Storage of resources was most common in a cupboard, followed by a classroom. Food and textiles resources were the most common with mechanisms the least. ICT resources to support D&T were available in 50% of the schools. A high percentage of the students had not read the D&T policy with more evidence of links with other curriculum areas for example Literacy and Art.

Roehampton University

30 of the students taught D&T on their school placement. Food was frequently covered through healthy sandwiches and one student taught Christmas cookies. Electrical and mechanical components were taught through making a 'buggy' using a motor, clocks, wheeled vehicles, levers in pop-up Santa cards and 'moving jungle animal heads' and structures through making a shelter used during air raids, a money box and wire models. Puppets, purses and slippers covered the textiles element. This was encouraging for tutors as many of the examples are integrated into the compulsory D&T course for all PGCE and BA students.

Weekly lessons of D&T were the most common (20), followed by D&T weeks (6) and 2–3 days (2). 5 schools used the QCA Scheme of work (QCA, 1998), although 30 students did not know if they did or not, as was the situation with the Lesson plans (DATA, 2002), helpsheets (DATA, 2000) and the Nuffield Primary solutions (Nuffield Curriculum Project Centre,/DATA, 2001). 24 of the schools had a coordinator and the most common links were with Art and Design followed by Physical Education.

D&T was taught in the classroom in 33 of the schools and 2 schools had a specialist D&T centre. The store room was used to store D&T materials in 36 of the schools, 7 stored resources in individual classrooms and 3 had boxed storage.

An encouraging range of materials were used with textiles taught in 31 schools, food in 27, wood in 17, construction in 13, electrical control in 12 and plastics in 10 schools. ICT was available in 14 schools but 22 students said that they did not know if it was or not.

D&T was seen outside the classroom in 31 schools, for example cars, food, sewing, Art week display and making moving objects. A disappointing 41 students had not read the school D&T policy but links were seen with Literacy (4) Geography (4), Art (3), Science (2) and a range of other curriculum areas.

St Mary's University College

On their last school experience, 6 students were in classes for children aged 5 to 6 years, 3 were in children aged 5 to 7 years and the remaining 28 students were working in classes for children aged 7 to 11 years. 83% of the students were able to teach D&T. D&T was taught weekly in classrooms by 40% of the students. The areas taught most were food technology, mechanical control and textiles which is interesting as these topics featured in their university specialist courses. The results showed that 31% of students taught food technology, a topic that they had just completed as a specialist module and an area of national interest. It is possible that these students were able to select an area of focus in D&T and, having just completed a food technology course, felt confident to teach it.

43% of the students said their schools used the QCA Scheme of Work (QCA, 1998) for D&T and 26% said they did not. A significant number of students (30%) did not know if it was being used at all as many of college based lectures were linked to the QCA Scheme of Work (QCA, 1998) in Years 1 and 2 D&T modules. Only one student saw the Nuffield Primary Solutions (Nuffield Curriculum Project Centre,/DATA, 2001) being used, 15 students did not know if it was used and 9 were sure that it was not used, 10 did not answer this question. This is a concern as some of the Nuffield units would have been useful to support topics the students had seen in school. Art and Design was the most common subject that was shared with D&T as an area of curriculum leadership.

DISCUSSION

The total number of questionnaires completed across the ITE providers was 185. This was disappointing in some ways, but it was a pilot and a voluntary student task. The use of online facilities was appreciated and positive but there are issues for higher education institutions (HEIs) needing to retain hard copies to protect against computer problems. If the exercise was repeated individual HEIs may prefer to take responsibility of their own data collection and handling. However, the range of data collected was rich and highlighted some interesting issues across the ITE providers. 7 key issues are discussed below:

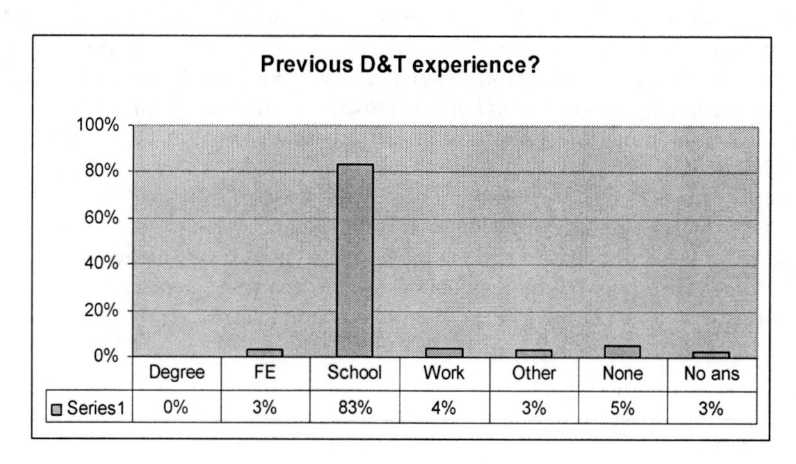

Figure 2. What previous D&T experience do you have?

A high percentage (83%) of the students had D&T experience in their schooling (Figure 2). None of the PGCE students had a D&T related degree. This does require further attention as there are very few specific D&T degrees and some students may not have been aware that their degree was related to the subject area.

It was encouraging to see that the students had taught D&T across the year groups mainly with a Year 1 (20%) and Year 2 (17%) the highest classes. It is important to note that only 56% of the students were D&T specialists (Figure 3), which has implications for future studies.

Again, it was very encouraging to find that 74% of the students had seen or taught D&T in their placement class (Figure 4). The impact of the content of the D&T courses taught by the ITE provider can be seen in the results from individual institutions and there is evidence of a relationship between the sessions, areas of D&T taught and the teaching taking place in the classes.

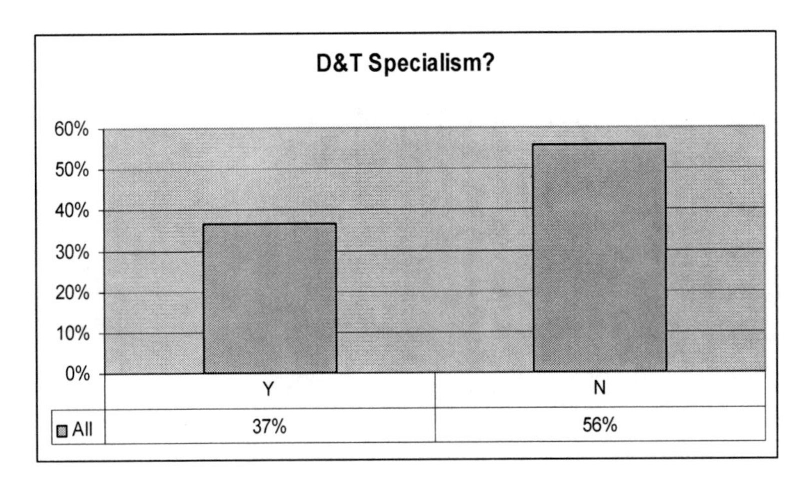

Figure 3. Is D&T your subject specialism?

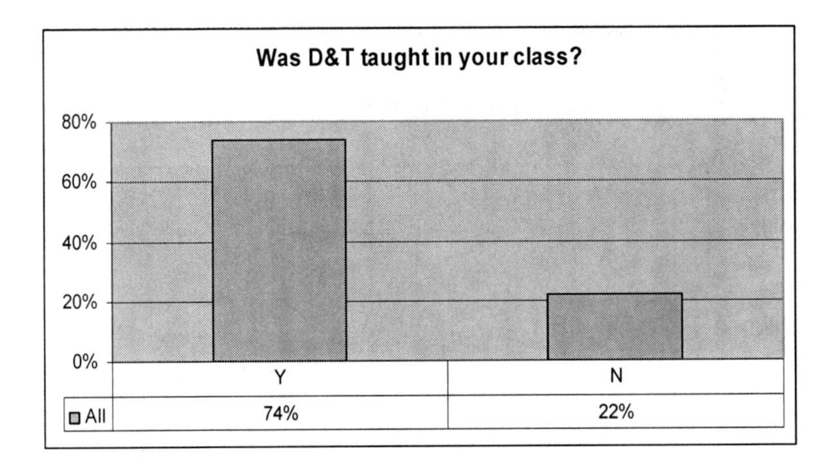

Figure 4. Was D&T taught in your class while you were in school?

The most common organisational approach used across the providers was weekly lessons (46%), (Figure 5), with 8% in D&T weeks and 7% in 2–3 day blocks. 13% of the respondents stated that lessons were combinational, but it was not clear exactly what this meant in practice. This wide variation indicates that schools are adopting different approaches taking into account their circumstances and needs. This seems to imply that the Primary Strategy (DfES, 2003), which promotes some integration of subjects, was not at that time being widely implemented. There were some worrying comments indicating that D&T was being taught 'when time allowed' outside the classroom and with the supervision of a parent.

62% of the schools had a D&T co-ordinator (Figure 6) and 21% of them were co-ordinating other curriculum areas, for example, most commonly Art and Design. This may be due to the commonality of the subject areas, similar resources needed

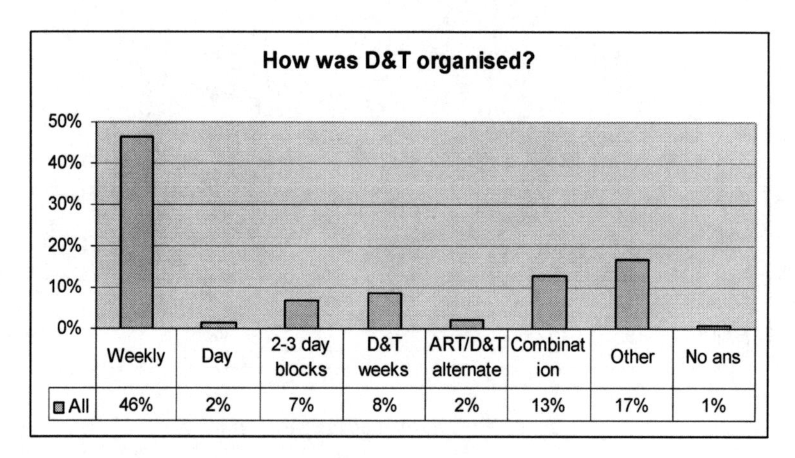

Figure 5. How was the teaching of D&T organised in your school?

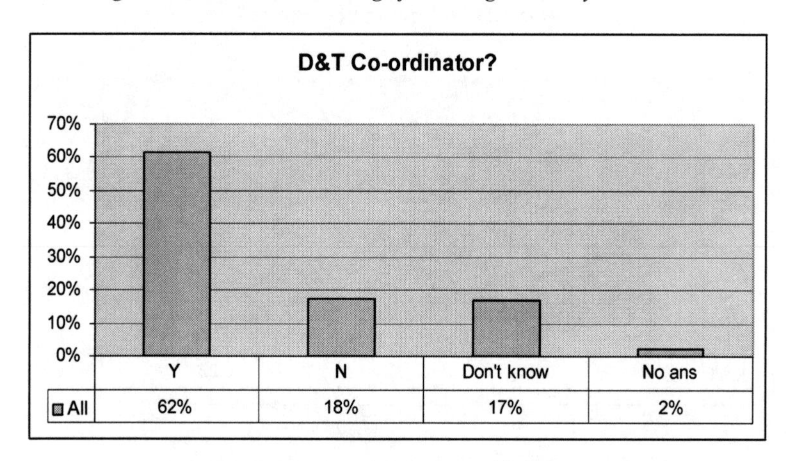

Figure 6. Was there a D&T coordinator/subject leader in the school?

to teach many aspects of both subjects or personal interests and skills. It would be interesting to find out if there was a clear distinction made between these two subject areas in schools. The concept of specialist D&T co-ordinators concentrating on the curriculum area in larger schools is a goal for the future and one to be encouraged.

The majority (85%) of the accommodation for teaching D&T was in the classroom with 5% in specialist areas (Figure 7), though this was probably due to the provision in middle schools which cross the primary (ages 5–11 years) and secondary (ages 11–16 years) in some areas of the country. This has implications for the management and pedagogy taught within D&T ITE courses. It is important that ITE providers ensure students are aware of issues related to managing D&T in the classroom, including pedagogy, resources, suitable activities and health and safety. As the storage of resources was most commonly in a store cupboard (65%) followed by the classroom (25%), students should be aware that a co-ordinator needs to be well organised.

The range of resources in schools to teach D&T varied across the ITE providers, though food and textiles materials were generally available and mechanical control was poorly represented. The range of topics seen and taught by the students appeared to be directly related to the D&T courses covered by the providers. This is encouraging in that students were applying what they had been taught, but it may imply that schools look to, and depend on, the subject background of the students. The concept of newly qualified teachers (NQTs) working 'with advice from experienced teacher where necessary' (TTA, 2002) is therefore of concern.

The issue of the use of teaching resources, such as QCA/DfEE Scheme of Work for D&T (QCA, 1998), Lesson plans (DATA, 2002) and Helpsheets (DATA, 2000), and Nuffield Primary Solutions (Nuffield Curriculum Project Centre,/DATA, 2001) by the schools is not fully addressed in the survey, as students may not have been aware of these resources or they may not have been used by their teachers for planning. It does highlight that ITE providers need to ensure that the students are

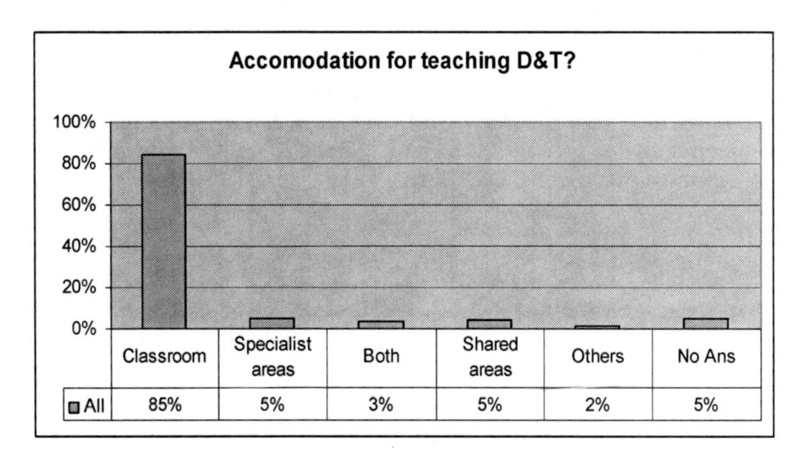

Figure 7. How was D&T organised and taught in the school?

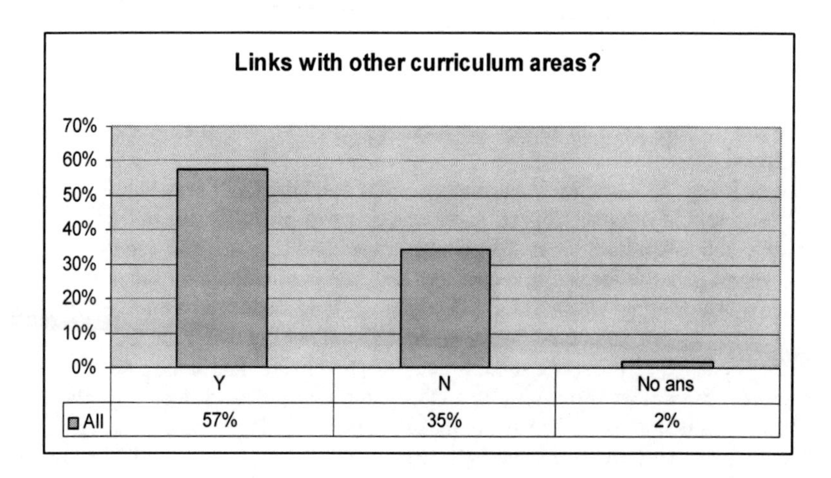

Figure 8. Did you see links between D&T and other curriculum areas?

fully aware of such resources and implement their use where appropriate on their school placements. It may be that some schools are unaware of what help is available, lack local education authority support or that in current practice there is not access to subject specific advisers locally. However, it could be that some teachers do not use the documents or only access these documents in the earlier planning stage so that the students were not aware of their use.

It is encouraging that the students saw a number of links between D&T and other subject areas in 57% of the schools (Figure 8). It would be interesting to see if secondary students, where traditionally subjects are very discrete, would see such links in their school placement. However, it was disappointing that only 59% of the students had read the school D&T policy in their placement school. This indicates a key issue for ITE tutors to raise with students. 36% of the schools had access to ICT to support D&T, but 35% of the students did not know if ICT resources were available. Finally, only 36% of the students commented that they had seen evidence of D&T other than in their classroom. There is a message here for students to utilise opportunities to highlight D&T in a variety of ways in school for example creating school displays with a D&T focus.

CONCLUSIONS AND CURRENT DEVELOPMENTS

This chapter presents the findings of a small scale research project in the South East of England in late Autumn and early Spring 2004/2005 using a voluntary pilot questionnaire in five ITE institutions. It highlights some common issues across the providers. The focus was the perceptions of primary ITE students of D&T in their placement schools. Future studies would require modifications to the questionnaire to reflect the findings of the pilot study. The findings were exploratory and they were used by the providers to refine their primary D&T courses and consider ways of working more closely with teachers in their partner schools. The pilot study showed that the students' perceptions of D&T practice in schools were varied, which has implications for courses provided by ITE institutions.

The student response to the initial questionnaire was limited, indicating that if the study was to be repeated, decisions need to be taken on the processes to be followed and the revisions to the questionnaire, such as including the students' background and level of D&T expertise. As indicated, students who have only experienced introductory courses in D&T are likely to have different impressions from those who have completed specialist courses. There is the potential for further research to look at these differences and assess the impact of specialist D&T students on practice in schools. Essentially, the questionnaire has proved to be a useful tool to draw together and build on an evolving picture of students' impressions of the nature of D&T in primary schools in five ITE institutions and could be used, following revisions, to address additional aims across a wider audience and track future developments. The desirability and potential for future in-service professional development in D&T for primary teachers has funding and resource implications.

Developments following the research project by the providers include the University of Brighton offering a 4 Year BA (Hons) Primary Education with QTS (3–7 years or 5–11 years). This 4 year course offers three school based placements and one complementary placement. Currently there is no D&T subject specialism and students learn about D&T in the foundation subjects' modules. The primary provision of D&T for undergraduates offers 10 sessions (1.5 hours each) in Year 1. These sessions introduce students to the underlying principles of good practice in D&T. The programme addresses all areas of D&T (food, textiles, mechanisms and structures) and the students are required to undertake a piece of practical work linking mechanisms to Citizenship, History or Music. This culminates in the form of interactive resource boards to support both learning and teaching. In Year 2, students are assessed on an evaluation of their practice of D&T in their placement and participate in a seminar where they explore their knowledge and developing understanding of the subject. In Year 3 there are a range of optional modules, many of which have cross-curricular themes where many students choose to re-visit an aspect of D&T. There is an expectation that all students will teach D&T in their final school placement. Student evaluations indicate that their experience of learning and teaching D&T makes an essential contribution to their overall profile as creative and innovative teachers.

From data gathered at Canterbury Christ Church University during the year following the pilot study a marked difference in attitude towards the subject was observed between students on the undergraduate and PGCE courses. It was clear that students on the second year option course had much more commitment towards D&T than those on the post graduate course. Since this research project was undertaken, the B.A. (Hons) with QTS course at Canterbury Christ Church University has moved towards a more explicitly cross-curricular approach. D&T is now, along with Science, Mathematics and History, in a module entitled 'The Child and Self as Enquirer', which whilst highlighting the importance of children's creative problem-solving, has meant that students are less aware of the distinctiveness of the subject.

In Year 3 there is now a compulsory cross-curricular module which includes single subject sessions (including D&T) as 'top up' for those who did not choose them in Years 2/3. This will hopefully enhance students' understanding and capability in the classroom. Since the introduction of the Primary National Strategy (DfES, 2003) this cross-curricular approach has become a more common practice in ITE. Although, this is an issue which is not exclusive to D&T it is particularly worrying as some teachers seem less keen on engaging in what might seem 'messy' practical activities. Further examination of the data gathered from part-time students at Canterbury Christ Church University suggests a correlation between those also working as teaching assistants who are mostly involved in supporting literacy and numeracy work in school and the lack of importance awarded to the subject.

Following the research study at St Mary's University College an enhancement of practice project was introduced which encourages Year 3 D&T specialists to spend a month in a school working with the subject leader to improve an aspect of Design and Technology in their schools Since the pilot study St Mary's University College has newly validated modules with both discrete D&T sessions for all Year 1 and PGCE students and also cross curricular modules for Year 2 undergraduates and PGCE students. These encourage making realistic and meaningful links between D&T and other curriculum areas (Aston & Jackson, 2009). It will be interesting to see if results from future studies reflect a greater awareness of D&T following the introduction of these courses. By way of contrast, the BA (Ed) course at Goldsmiths, University of London was finally phased out in 2007 with the phasing in of an Education degree course without Qualified Teacher Status (QTS) three years earlier. There are now no specialist students in D&T and a PGCE course reduced initially to four two-hour D&T sessions and then to three three-hour experiences. These changes indicate an increased importance and emphasis on the students' school experience, rather than on their university D&T sessions.

Roehampton University is at the time of writing undergoing a validation exercise for a new BA Ed degree that starts in September 2010. The modules within the programme are changing in that the credits awarded for each one will now be consistent across the university. The implication of this has not been too dramatic, with a slight reduction in credits awarded to the D&T subject specialists. All subject areas will have an equal award, and there will be an increase in the number of subject specialisms offered and students may now opt to specialise in Modern Foreign Languages (MFL) or Special Education Needs (SEN). The implications for D&T are that the subject specialism is to be continued, which is to be applauded in the current climate. In anticipation of the Primary Curriculum Review (Rose, 2009) being implemented it was intended that the non specialists will complete a module in conjunction with Science. This has resulted in both increased time and credits being awarded with the D&T and Science subject knowledge content continuing, but with clear cross curricular links made across both areas. The PGCE was recently revalidated with the inclusion of masters (M) level credits awarded for one module. There continue to be subject specialisms within this and the Masters level D&T specialists are required to research the teaching and learning of D&T from a social constructivist approach.

FINAL NOTE

Recent development at the individual teacher education providers has been varied, indicating an even wider range of the amount and quality of D&T taught to the students. This has implications for schools and could cause problems as their newly qualified teachers will have varied experiences of D&T during their training. It could result in schools reassessing their D&T teaching and reducing the quantity and quality of their children's D&T experience.

At the end of the research study the team envisaged the questionnaire being used as a basis for research in other institutions to address common issues across a number of schools and ITE courses. To this end, the team disseminated their findings at the Design and Technology Association Annual Conference in 2006 (Rutland et al., 2006) and a Joint Nuffield Curriculum Centre and Design and Technology Association seminar 'Enabling Research into Primary Design and Technology' (January, 2008)[1].

The authors of this chapter also recommended that the questionnaire should be used over a period of years by ITE institutions to track changes to students' impressions of D&T practice in schools. Initially, it was thought that extending the survey would be particularly interesting following the possible introduction of the recommendations of the Rose Report (Rose, 2009), for the teaching of discrete National Curriculum subjects as well as developing a more integrated approach to learning. The Report placed D&T in the area of learning 'Scientific and technological understanding'. However, following a change of government in May 2010 the Rose Report will not be implemented and there is an understanding that there will be no further changes for two years. A similar approach to the Rose Report is to be found in the Alexander Report (2009)[2], where one of the domains of the primary curriculum is cited as 'Science and Technology'. However, as with the Rose Report, future developments are unclear. Despite this, the development work begun in ITE institutions exploring linking between D&T and Science may continue in the context of the STEM (Science, (Design and) Technology, Engineering and Mathematics) agenda in schools, where subjects remain discrete but are encouraged to work together. Future developments in the structure of the primary curriculum, including the position of D&T, is currently unclear in England but it is interesting to reflect on the possible impact of these subject connections for primary curriculum developments in the future.

NOTES

1 www.primarydandt.org/data//files/itereport-15.pdf
2 www.primaryreview.org.uk

REFERENCES

Ager, R., & Benson, C. (1997). Primary ITE courses in the UK – Developing D&T education. In R. Ager & C. Benson (Eds.), *International Primary Design and Technology conference 1997* (pp. 6–11). Birmingham, UK: CRIPT at University of Central England.

Aston, S., & Jackson, D. (2009). Blurring the boundaries or muddying the waters? *Design and Technology Education: An International Journal, 14*(1).

Alexander, R. (Ed.), (2009). *Children, their world, their education: Final report and recommendations of the Cambridge primary review.* Abingdon, Oxon, UK: Routledge; Taylor and Francis Group.

Barlex, D. (2003). Developing and celebrating good practice in primary Design and Technology – A seminar, the recommendations and results. In C. Benson, M. Martin & W. Till (Eds.), *Fourth International Primary Design and Technology conference 2003* (pp. 2–4). Birmingham, UK: CRIPT at University of Central England.

Benson, C. (1997). In-service provision for Design and Technology for primary teachers. In R. Ager & C. Benson (Eds.), *International Primary Design and Technology conference 1997* (pp. 2–5). Birmingham, UK: CRIPT at University of Central England.

DATA (The Design and Technology Association). (1995). *Guidance materials for Design & Technology: Key stages 1&2.* Wellesbourne: Author.

DATA (The Design and Technology Association). (2000). *Helpsheets for the national exemplar scheme of work for Design and Technology in primary school.* Wellesbourne: Author.

DATA (The Design and Technology Association). (2000). *Lesson plans for the national exemplar scheme of work for Design and Technology in primary school.* Wellesbourne: Author.

Davies, D., Egan, B., Martin, M., & Rogers, M. (2000). Carrying the torch – Can student teachers contribute to the survival of Design and Technology in the primary curriculum. In R. Kimbell (Ed.), *Design and Technology international millennium conference 2000* (pp. 47–52). Wellesbourne: Author.

Davies, D., & Rogers, M. (2000). Pre-service primary teachers' planning for Science and Technology activities: Influences and constraints. *Research in Science & Technological Education, 18*(2), 213–223.

DES (Department of Education and Science). (1990). *Technology in the national curriculum.* London: HMSO.

DfE (Department for Education). (1995). *Design and Technology in the national curriculum.* London: HMSO.

DfEE (Department for Education and Employment) and QCA (Qualifications and Curriculum Authority). (1999). *Design and Technology: The national curriculum for England.* London: HMSO.

DfES (Department for Education and Skills). (2003). *Excellence and enjoyment - A strategy for primary schools.* London: HMSO.

Dow, W. (2003). Student teachers' perceptions of technology teaching in Scottish primary schools. In C. Benson, M. Martin, & W. Till (Eds.), *Fourth International Primary Design and Technology conference 2003* (pp. 31–34). Birmingham, UK: CRIPT at University of Central England.

Ive, M. (1999). The state of primary Design and Technology education in England – Past success and future developments. In C. Benson & W. Till (Eds.), *Second International Primary Design and Technology conference 1999* (pp. 16–19). Birmingham, UK: CRIPT at University of Central England.

Martin, C. (2001). Using the QCA scheme of work for primary Design and Technology. In C. Benson, M. Martin, & W. Till (Eds.), *Third International Primary Design and Technology conference 2001* (pp. 143–145). Birmingham, UK: CRIPT at University of Central England.

Mitra, J. (1999). Growing a community of good practice through a D&T curriculum development project. In C. Benson & W. Till (Eds.), *Second International Primary Design and Technology conference 1999* (pp. 98–101). Birmingham, UK: CRIPT at University of Central England.

Nuffield Curriculum Project Centre/DATA. (2001). *Primary solutions in Design and Technology.* Wellesbourne: Design and Technology Association/Nuffield Foundation.

Perry, C. (2003). 2 days lessons are….Using a collapsed timetable to teach D&T in primary schools. In C. Benson, M. Martin, & W. Till (Eds.), *Fourth International Primary Design and Technology conference 2003* (pp. 125–127). Birmingham, UK: CRIPT at University of Central England.

QCA (Qualifications and Curriculum Authority). (1998). *Scheme of work for Design and Technology key stages 1 and 2*. London: Author.

QCA (Qualifications and Curriculum Authority). (1999). *Scheme of work for art and design key stages 1 and 2*. London: Author.

Rogers, M. (2004). Is it possible to ensure the survival of primary Design and Technology through ten hour courses? *The Journal of Design and Technology Education*, *9*(1), 14–24.

Rogers, M., & Davies, D. (1999). *What has happened to primary Design and Technology? – Student teachers in search of a foundation subject*. Unpublished conference paper presented at BERA 99, University of Sussex at Brighton (2–5 September).

Rose, J. (2009). *Independent review of the primary curriculum: Final report*. Nottingham: DCSF (Department for Children, Schools and Families) publications.

Rutland, M., Rogers, M., Hope G., Prajapat B., Haffenden, D., Seidel, M., et al. (2006). Student teachers' impressions of primary Design and Technology in English schools: A pilot study. In E. W. L. Norman, D. Spendlove & G. Owen-Jackson (Eds.), *Designing the future, D&T Association International Research conference 2006* (pp. 97–109). Wellesbourne: The D&T Association.

TTA (Teacher Training Agency). (2002). *Qualifying to teach: Professional standards for qualified teacher status and requirements for initial teacher training*. London: Teacher Training Agency/Department for Education and Skills.

Vaughan, S. (1997). Starting from scratch – Planning for Design and Technology. In R. Ager & C. Benson (Eds.), *International Primary Design and Technology conference 1997* (pp. 32–33). Birmingham, UK: CRIPT at University of Central England.

Seminars:

Nuffield Foundation, Design and Technology Association and CRIPT seminar. (2002, February). *Developing and celebrating good practice in primary Design and Technology*. London: Author.

Nuffield Foundation/Design and Technology Association. (2004, February 18). *Report of first meeting of the National Primary Design and Technology Research Group, Enabling research into primary education seminar*. London: Nuffield Foundation.

Nuffield Foundation/Design and Technology Association. (2008, January 16). *Report of second meeting of the National Primary Design and Technology Research Group, Enabling research into Primary Design and Technology*. Retrieved from http://www.primarydandt.org/data//files/itereport-15.pdf

Marion Rutland and Martin Seidel
Roehampton University
London
England

Sally Aston and Dot Jackson
St Mary's University College
London
England

Debbie Haffenden and Bhav Prajapat
University of Brighton
Brighton
England

Gill Hope
Canterbury Christchurch University
Canterbury
England

Maggie Rogers
Centre for Cross Curricular Initiatives
South Bank University
England

SONJA VANDELEUR AND MARC SCHÄFER

24. INDIGENOUS TECHNOLOGY AND CULTURE

INTRODUCTION

At the PATT-18 seminar and conference held in Glasgow, Scotland, in June 2007, it was evident that there is a need in Technology Education to develop an understanding about the basic technological nature of the world. According to Borgmann (2006),

> ... the need is for a penetrating understanding of contemporary life, more particularly of the culture of advanced industrial countries. (p. ix)

Yet there is little in the way of literature or classroom practice that is directly related to a deep understanding of the nature of technology. This needs to include the technologies that are used every day by most of the world's populations in developing countries and not just the culture of advanced industrial countries, as this focus would give only a partial perspective on the nature of technology. The inclusion of 'indigenous technology and culture' in the South African curriculum, for example, is one way of developing learners' sensitivities to the interrelationship between society, the environment, science and technology.

In this chapter we attempt to situate indigenous technology and culture in our technological world and identify some of the inherent problems associated with the inclusion of this aspect in the South African curriculum. The chapter reports on a PhD study completed in 2010 and it gives the context of the study, research method, findings, conclusions and implications for future pedagogical practice.

CURRICULUM REVISION IN SOUTH AFRICA POST-1994

Since 1994, education in South Africa has undergone fundamental transformation. The new curriculum, known as Curriculum 2005 (C2005), was the first single curriculum for all South Africans and it was the pedagogical route out of apartheid education (Chisholm, 2003). The first nine years of schooling, known as the General Education and Training Band (GET), became compulsory and it was in this band that Technology was introduced as a new learning area. The revised National Curriculum Statements (NCS), developed in 2002 for grades R – 9 (ages 5–14), were the result of a process of revision designed to strengthen and streamline the original curriculum statements. The guiding philosophy of C2005 was 'outcomes-based education', a controversial philosophy with links to the 'competency-based' approaches found in the vocational and work-based training areas (Stevens, 2005).

C. Benson and J. Lunt (eds.), International Handbook of Primary Technology Education: Reviewing the Past Twenty Years. 305–317.

Each learning area has its own 'learning outcomes' achieved by attainment of specific 'assessment standards'.

Technology has three Learning Outcomes. 'Technological Processes and Skills' are covered in Learning Outcome 1 (LO1), 'Technological Knowledge and Understanding' in Learning Outcome 2 (LO2) and, of particular interest to this chapter, is Learning Outcome 3 (LO3) which explores the interrelationships between science, technology, society and the environment. The inclusion of this outcome is in line with curriculum revisions undertaken in other countries such as New Zealand (Jones, 2003) and the United States of America (International Technology Education Association, 2002), which acknowledge the interrelationship between science, technology and society. It is noteworthy that the South African curriculum has added the aspect of 'environment' in the exploration of the interrelationships between science, technology and society.

The achievement of LO3 will ensure that learners are aware of indigenous technology and culture, the impacts of technology, and biases created by technology (South Africa. Department of Education, 2002). The 'impacts of technology' and 'biases created by technology' have been consistent throughout the curricula revisions that have taken place in South Africa since 1994. 'Indigenous technology and culture', however, only appeared in the final revised National Curriculum Statements, implemented in 2006. This inclusion is seemingly unique to South Africa as it is explicitly in the National Curriculum for all students, and not only for indigenous students. With this new inclusion, Technology teachers and developers of learning materials have to contend with issues such as:

- what is meant by the interrelationship between science, technology, society and the environment?;
- what is meant by 'indigenous technology'?;
- what is the link between technology and culture?; and
- what does this mean in terms of technological literacy?

The traditional approach of C2005 required students in Technology to explore the positive and negative impacts of technology. This approach addressed the outcomes of technology and cast technology in a perspective of cause and effect relationships which presented a technologically determinist, or at best, an instrumentalist view to students. Technological determinism holds that everything is caused by a sequence of previous conditions and events, operating with regularity and, in principle, with predictability. It presents technological systems as:

> ... ordered accordingly to materials, processes and laws that can be understood from an objective standpoint. (Pannabecker, 1991, p. 44)

Pannabecker (1991) stated that this notion of technological impacts is simple to understand and it has enabled the field to interpret technology in the context of society and culture, something which technology studies have long struggled to do (Russell & Williams, 2002). But one of the problems with the deterministic view is that studying impacts places the emphasis on a restricted point of the sequence of technological development. This view of technology has contributed to a simplistic and inflexible view of the relationship between technology and society and it has reinforced the idea that technology is an autonomous entity developed according to an internal logic which has determinate impacts on society (Williams & Edge, 1996; Russell & Williams, 2002).

The curriculum revision (NCS) implicitly challenged the simplistic, deterministic approach to some extent by suggesting that socio-cultural-ecological patterns are also embedded in the content and processes of technologies. The description for LO3 for Technology in the NCS stated that:

> All technological development takes place in an economic, political, social and environmental context. Values, beliefs and traditions shape the way people view and accept technology, and this may have a major influence on the use of technological products. (South Africa. Department of Education, 2002, p. 9)

The first part of this description implies that technological development is influenced by factors such as economics, politics, society and the environment, suggesting humans' active role in the shaping of technology. The emphasis on context encourages students to explore the challenges and influences faced in specific situations in terms of technological development. The second part of the description emphasises the interpretation of technology in the context of social influences. However, it restricts this social influence to the use of technological products, thereby giving a deterministic view of technology. It neglects the fact that values, beliefs and traditions can also influence the way a technology emerges and develops.

TEACHERS' BELIEFS

A focus of the study was to find out how technology teachers were dealing with indigenous technologies and culture in their pedagogical practices. What teachers know; what they deem to be valuable knowledge; and various other dimensions that influence their choices needed to be identified. Rowell, Gustafson & Guilbert (1999) stated that:

> While curriculum developers set out the orientation of the school subject in documents mandating the goals and content of the instructional program, teachers interpret the program with a focus on aspects congruent with their personal views and interests. (p. 48)

Pedagogical implications for technology education arise from the epistemological debate about the nature of technological knowledge (Rowell et al., 1999). The ways in which technology is conceptualised by teachers will have a direct bearing on the shaping of Technology as a subject. Teachers' conceptualisation of 'technology', 'Technology Education', 'indigenous technology and culture' and 'the interrelationship between science, technology, society and the environment' could influence the way in which they deal with indigenous technology and culture in their classrooms. In other words, teachers' assumptions about the nature of technology will affect how and what they teach. It is important, however, to note that the curriculum, and possibly classroom practice, could also affect the way that teachers view technology. It is necessary therefore to explore the different theories of technology to gain a better understanding about the nature of technology.

The Nature of Technological Knowledge

According to Mitcham (1994), the philosophy of technology has two traditions: the engineering philosophy tradition which emphasises analysing the internal structure of technology, and the humanities philosophy tradition which is more concerned with external relations and the meaning of technology. In other words, most theories of technology distinguish between technology as it relates to artefacts and technology as it relates to social ideas, values and needs. Technology Education curricula in general have focused primarily on the engineering philosophy tradition, perhaps due to the roots of the discipline being in the technical curricula of the industrial arts subjects, such as Technical Drawing and Metalwork. The humanities philosophy tradition is more recent and its influence is evident in curricula revisions in some countries, such as New Zealand (Jones, 2003), the USA (ITEA, 2002), the Netherlands (Eijkelhof, Franssen, & Houtveen, 1998) and South Africa.

Philosophical and sociological perspectives have prompted an extensive debate about the nature of technology (Hansen, 1997). The debate seems to run along two continuums: the extent to which technology is viewed as autonomous or human-controlled, in other words technology's relation to human powers; and the extent to which technology is viewed as neutral or value-laden (Feenberg, 1999). The following diagram illustrates these two continuums and the placement of the more well-known theories of technology on the two continuums:

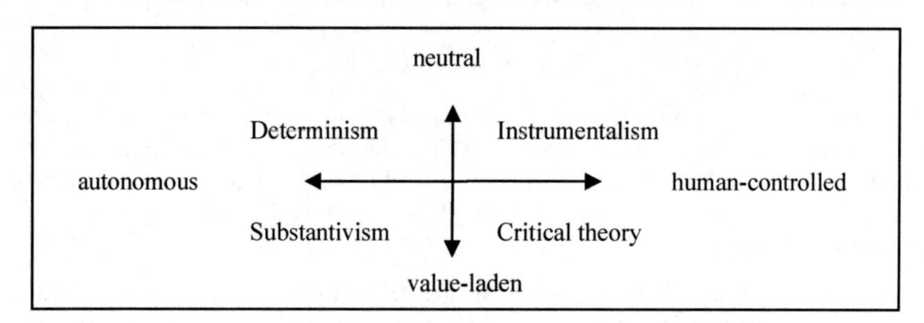

Figure 1. Theories of technology, adapted from Feenberg (1999).

The two theories of determinism and instrumentalism perceive technology as a set of neutral products detached from values. Determinists believe that technology controls humans and in doing so, it shapes society to the requirements of efficiency and progress, as in classical Marxism (Feenberg, 1991). This view of technology has reinforced the idea that technology is an autonomous entity developed according to an internal logic which has determinate impacts on society (Williams & Edge, 1996; Russell & Williams, 2002). Instrumentalism is based on the idea that technologies are tools that are used to provide the means for the realisation of independently chosen ends (Feenberg, 1991). In other words, technology is viewed merely as an instrument of progress. These two theories ignore the influence of contexts, including indigenous knowledge practices.

Theorists such as Feenberg (1999) and Ihde (1990), however, claim that technology can never be removed from a context and therefore can never be neutral.

The non-neutral approaches suggest that attention be given to relationships as well as objects. While the scientific principles used in engineering might be value neutral, the emergence, implementation and impact of a technology are embedded in historical, aesthetic, political and cultural meaning. Substantive theory views means and ends as inseparable: our tools form our environment and therefore who and what we are. Substantivists view technology as a culture of universal control from which there is no escape. Substantivist theories of technology draw attention away from the practical question of what technology *does* to the hermeneutic question of what it *means* (Feenberg, 1999). Substantivists suggest that once society goes down the path of technological development, it will be dedicated to values such as efficiency and power, and traditional values will not survive this challenge. This approach has been criticised for its apocalyptic and dystopian view.

Unlike most other theories of technology, with a critical theory of technology there is no assumption of progress (Feenberg, 1991). Similarly to the substantive view, critical theory argues that the technical order is more than a sum of tools but, like instrumentalism, it rejects the fatalism of the substantivists. In choosing our technology we become what we are which then shapes our future choices. According to Feenberg (1991), a critical theory of technology argues that technology is not a 'thing' in the ordinary sense of the term. Rather, it is an ambivalent process of development suspended between different possibilities. This ambivalence attributes a role to social values in the design and not just the use of technical systems, thereby distinguishing it from the neutral thesis. A critical theory of technology replaces the conventional distinction between artefacts and ideas held by determinism and instrumentalism by a holistic view in which technology reflects the dominant ideologies of the culture in which the technology emerges. It suggests that technology embodies the values of a particular civilisation and can be interpreted as a cultural phenomenon (Hansen, 1997).

INDIGENOUS KNOWLEDGE SYSTEMS

The inclusion of indigenous technology and culture in the South African Technology curriculum is significant and it comes at a time when questions are being asked on the formation of knowledge production, the gap between formal institutions and society, and the vacuum in theorisation (Odora Hoppers, 2002). However, the inclusion of indigenous technology and culture in the Technology curriculum has been problematic. This is discussed further on in this chapter.

Defining Indigenous Knowledge

Various definitions of indigenous knowledge exist and one of the more inclusive definitions is that given by Dei, Hall & Rosenberg (2000):

> Indigenous knowledges are unique to given cultures and societies and they reflect the capabilities and values of the communities that use them. (p. 19)

Woodley (2003) suggested that indigenous knowledge systems should be studied in terms of space and time, emphasising the importance of context. He further stated that the spatial dimension of indigenous knowledge is the embedded, holistic or 'place-based' aspect of knowledge at any one point in time, and that to understand knowledge as embedded in place needs an understanding of the social norms, values, belief systems, institutions and ecological conditions that provide the basis for the place where knowledge is derived. On reading through the various definitions of indigenous knowledge, a question arose that concerned us. The question that needs to be asked is: Is indigenous technology different from any other type of technology? Does this difference not then fragment the way in which we analyse technologies? Horsthemke & Schäfer (2007) stated a similar concern in regard to ethnomathematics:

> ... the term 'indigenous' has, at best, limited applicability. A similar point could be made about the prefix 'ethno'. If ethnomathematics constitutes knowledge in the propositional or factual sense, then it is unclear what purpose the prefix is meant to serve – other than artificially severing ethnomathematics from mathematics as such. (p. 18)

A Renewed Interest

There is a renewed interest in Indigenous Knowledge Systems as is evident in the revised National Curriculum Statements. Up until the 1960s, colonial powers used education as a tool to disseminate the metanarrative form of civilizing culture perceived to be utopian during these times (Boyne & Rattansi, 1990). Metanarratives are stories a culture tells itself about its practices and beliefs. Lyotard (cited in Klages, 1997) suggested that the perceived order maintained in modern societies is through the means of these metanarratives. Colonial administrators viewed western science as the ultimate authority for interpreting reality and indigenous knowledge was perceived to be simplistic and vague. As a result, African and other knowledge systems were marginalised with adverse consequences for the people who were colonised (National Research Foundation, 2002; Pitika Ntuli, 2002). In regard to education, most countries in post-colonial Africa inherited a curriculum that was largely irrelevant to their own circumstances. This is partly due to the fact that students were (and still are) often confronted with sets of world views, knowledge and attitudes that were not their own. Education in most parts of Africa tells children from a traditional culture that their future is rooted not in the knowledge of their parents and grandparents, but in the knowledge imported from a Western pedagogical tradition. This results in views that indigenous knowledge systems are obsolete (Kunnie, 2000; Aikenhead, 2002; International Council for Science, 2002).

However, in the last two decades, these perceptions and views have been changing. There is overwhelming evidence, the result of careful research from many countries and sources, that illustrates the great range, validity and usefulness of indigenous knowledge (Chambers, 1995; McGovern, 1999). Lyotard (cited in Boyne & Rattansi, 1990) suggested that social development in the postmodern era can no longer be seen as fulfilling some metanarrative, but that it should be a pragmatic matter of inventing new rules. The validity of these rules will reside in their effectiveness rather than in their compatibility with some legitimising discourse. It is interesting to note here that the emphasis is on effectiveness and not efficiency. So, in the social development field, less emphasis is being put on the transfer of technology and more on learning from and with indigenous peoples (Massaquoi, 1993; Chambers, 1995).

As people and governments across the world become more concerned with environmental issues and sustainable development, sensitivity to the value of indigenous knowledge grows (Warren, Slikkerveer, & Brokensha, 1995). In the agricultural development field, recent studies about indigenous knowledge are having an effect. Policy makers and planners are starting to recognise the need to understand existing knowledge systems and decision-making processes (Warren & Rajasekaran, 1993). Many people's well being in developing countries depends on using indigenous knowledge practices. It is therefore a valuable resource for the traditional societies from which they originate as well as for scientists, technologists and development agencies. Warren & Rajasekaran (1993) suggested that, in the social development field, it is feasible, efficient and cost-effective to move towards an interactive technology development from the conventional transfer of technology approach.

Indigenous Knowledge and Western Science

The comparison of western scientific knowledge and indigenous knowledge usually creates a dialectical opposition (Shava, 2006). Shiva (2000) brings attention to the dichotomising impact of western scientific research on local knowledges which, through processes of inclusion and exclusion, create boundaries of power. She points out that Western systems of agriculture and medicine were defined as the only scientific systems. Yet the agricultural practices of most farmers in Africa rely heavily on traditional know-how and this sometimes proves to be far more productive than imported techniques (Hountondji, 2002). Indigenous knowledge proponents can also create an oppositional logic of us and them – the subjugated 'us' and the privileged 'them' (Dei cited in Shava, 2006). There is now a growing realisation that these knowledge systems are not necessarily oppositional - they can be complementary.

This has implications for the ways in which these knowledge systems are re-contextualised into learning materials. The historical, political, social, cultural and environmental contexts in which a technology emerges, develops and stabilizes needs to be explored. This needs to be done critically though, and debate and analysis on the internal and external characteristics of all technologies needs to happen. To include indigenous technology and culture in its narrowest sense in pedagogical practices would separate it from 'western' technologies. This would create an unhelpful dichotomy that would only serve to marginalise it from other technologies. To get learners to define what is appropriate by asking questions on the social and environmental influence and effects of all technologies, would encourage a critical disposition.

METHODOLOGY

Goals of the Research

The inclusion of indigenous technology and culture was a new addition to the Technology curriculum (NCS). This study set out to explore what teachers were making of this assessment standard. The broader research goals of this study were to examine and explore pedagogic practice in relation to indigenous technology and culture in the technology curriculum of the revised National Curriculum Statement (NCS).

Research questions. The research attempted to answer the following questions:
- How is the aspect of indigenous technology and culture being proposed for Technology Education processes in policy documents?
- What is the existing pedagogical practice in regards to this aspect of the curriculum?
- Does a process of participatory co-engagement with selected teachers, with reference to indigenous technology and culture in the Technology curriculum, improve teaching practice?

Significance of the study. The purpose of this study was to contribute to a deeper understanding and therefore a more meaningful implementation of indigenous technology and culture by Technology teachers. The results of this study will hopefully impact positively on teachers' practices and contribute to a better quality of teaching and learning in technology education. This study is significant in its own right as it has generated tentative explanations and interpretations around the implementation of indigenous technology and culture as prescribed in the technology curriculum for South Africa. The findings may elicit broader implications for curriculum design and implementation.

Research Methodology

The research method used for this study was a qualitative case study. The case study method was chosen as it allows for depth of investigation into a phenomenon (Merriam, 2001). The case study type was interpretive as the study was concerned with understanding the meaning that participants made of a phenomenon (Ezzy, 2002; Merriam, 2002).

Research design. The research consisted of three interconnected phases. The first phase of the research dealt with how and why indigenous technology and culture has been proposed for Technology Education processes. This phase explored and examined the rationale for the inclusion of this assessment standard. In other words it attempted to answer Lindblad & Popkewitz's (2000) notion of narrative by exploring 'what is the argument put forward?'. Questionnaires sent to curriculum developers, provincial subject advisors and the Department of Education official for Technology were analysed. Texts that referred specifically to indigenous technology and culture in the Technology curriculum were analysed. The issue being explored here was the way the argument was put into a context (Lindblad & Popkewitz, 2000).

Phase 2 of the research analysed existing pedagogical practices in terms of indigenous technology and culture as an assessment standard and explored the issues and problems associated with its implementation. This part of the study probed how teachers were dealing with indigenous technologies and culture in their pedagogical practices. What teachers know; what they deem to be valuable knowledge; and various dimensions that influence their choices, were identified. Teachers' conceptualisation of 'technology' and 'indigenous technology and culture' has a direct bearing on the shaping of Technology as a subject. Feenberg's table (1999) on the varieties of theory provided a useful analytical tool to explore teachers' assumptions on the nature of technology.

The research process for Phase 3 consisted of analysing a process of participatory co-engagement around an area of shared concern. The shared concern in this study was how to implement indigenous technology and culture in the classroom so that it is meaningful.

SOME OF THE FINDINGS AND RECOMMENDATIONS

These findings emanated from the individual interviews and the focus group discussions held with a selected group of teachers.

Unanimous Support for the Inclusion of Indigenous Technology and Culture in the Curriculum

The revisioning of 'nation' and 'citizen' in the National Curriculum Statement was a result of South Africa's historical and political past. The inclusion of indigenous technology and culture in the curriculum emerged from this revisioning, and perhaps due to this, the inclusion had unanimous support from the focus group of teachers. However, a problem extensively discussed in the literature on education in South Africa throughout the three curriculum reforms from 1994 to 2006 concerned the translation of idealistic goals contained in policy texts into transformative practices in the classroom (see Christie, 1999; Chisholm, 2000; Jansen & Sayed, 2001). Some of the findings given below give reasons for this lack of transformation.

Teachers' Interpretation of Policy

The teachers in the focus group did not have a common understanding of the meaning of 'technology' even though a definition was given in the National Curriculum Statement. Their other subject specialities, such as Science, Geography and Art, had a strong influence on their concepts of technology and their classroom practice. Another concept important to the implementation of indigenous technology and culture is, of course, 'indigenous'. The main characteristics of indigenous knowledge derived from various definitions are that it is place-based and therefore makes no claim to universality; it is transmitted (usually orally) from generation to generation; and it is dynamic in nature. However, many issues arise with the use of this term, namely that of inclusion and exclusion. Another issue is that there is no definition for 'indigenous' in any of the curriculum documents for Technology. In some cases, lack of confidence in knowing what to do due to uncertainty about the meaning of the term was the reason why some teachers did not include this aspect in their classroom practice.

Emphasis on Content

According to the Teachers' Guide for the Development of Learning Programmes: Technology (South Africa. Department of Education, 2003), the weighting should be equal for LO2, which focuses on content, and LO3, which deals with the interrelationship between science, technology, society and the environment. A content analysis of six textbooks, all published in 2006, found that there were on average eight times more units given to LO2 than that given to LO3. This finding makes explicit the content-based aspect of these textbooks. The reality is that technology textbooks do not give an appropriate weighting to technological issues. A further analysis was conducted to find out how indigenous technology and culture was being proposed in these textbooks. All the textbooks, except for one, used case studies to propose indigenous technology and culture. The majority of the tasks were instrumental in nature, asking learners to describe the materials used and the manufacturing process used for the product. Very few tasks involved the learners in any questions concerning culture, or how different cultures around the world had adapted technological solutions for optimum usefulness. So these tasks did not assess indigenous technology and culture; instead they assessed the learners' ability to analyse existing products: this means that LO1 was assessed and not LO3. When indigenous technology and culture is presented to learners only as a case study, it promotes a historical stance towards this aspect.

Lack of Meaningful Implementation

One of the curriculum developers stated that, in regards to LO3, 'few teachers are giving enough thought to this aspect', and that there is little evidence of the implementation of this learning outcome in the classroom or at teacher-training workshops, although it must be acknowledged that implementation varies from province to province. This lack of implementation is due to a number of factors: non-qualified Technology teachers teaching the subject; the turn-around time of teachers; and the lack of content knowledge, both conceptual and procedural, that these teachers have. Therefore in teacher workshops, time and resources are based on the first two learning outcomes as these two outcomes are perceived to be the essence of this subject. Lesson plans developed for these workshops often do not assist teachers to implement LO3 effectively.

Another issue on implementation in the classroom was the lack of time given to Technology. Only one of the schools out of five allocated the correct time to the subject. It is my view that Technology teachers are hampered in the implementation of Technology as a subject as there are confusing aspects in policy documents. The lack of implementation of LO3 is detrimental to Technology Education as it places the three Learning Outcomes in a perceived hierarchy. The casting aside of LO3 implies that it less important than the other outcomes. If teachers do not teach LO3 and learners are not enabled to engage with LO3 in a critical way, an instrumental view of technology will prevail. Learners need to start dealing with social and environmental issues surrounding the development and use of technologies so that a critical stance is encouraged.

Recommendations

The following tentative recommendations for Technology Education in general and the implementation of indigenous technology and culture in particular are:

- quality learning materials in which a critical stance of technology is encouraged need to be developed;
- policy documents need to provide more clarity by giving broad descriptions of concepts that are to be taught, especially if these concepts are newly introduced;
- training of Technology teachers is of paramount importance;
- teachers and developers of learning materials need to overcome the dichotomies created between western science and indigenous knowledge by using a heterogeneous, dynamic, plural notion of knowledge and culture;
- schools need to give the correct allocation of classroom time to Technology. Teachers should then be able to allocate time to developing a critical stance in their learners; and
- indigenous technology and culture should be included in all forms of assessment, as well as in design tasks.

IMPLICATIONS FOR FUTURE PEDAGOGICAL PRACTICE

Teachers need more guidance to implement indigenous technology and culture so that it becomes a meaningful and worthwhile aspect of LO3. Broad definitions of terms such as 'technology', 'indigenous' and 'culture', need to be given to teachers so that they have a basis from which to work. These broad definitions could be the start of discussion and debate in the classroom, which in itself would promote a critical disposition. It would get learners to develop a deeper understanding of technology as this sort of discussion would make them engage with aspects such as values, ethics, core beliefs, and environmental and social issues. There are many possible paths that technological development can take. In other words, technologies are not simply instrumental to the goals being set but they also frame a way of life. According to Feenberg (2007), educators need to develop ways in which the basic insights of the philosophy of technology are communicated to children. Getting learners to engage with indigenous technology and culture in the classroom is one possible way of doing this.

In conclusion, I would like to quote O'Riley (1996), who stated:

> I would like to place into question both the adequacy of the selection of technology narratives to represent the study of technology in our current technologized/technocratisized society, and the relevancy of these stories to meet the needs and interests of the diversity of students entering today's technology education classrooms. (p. 28)

REFERENCES

Aikenhead, G. (2002). Whose scientific knowledge? The colonizer and the colonized. In W. Roth & J. Desautels (Eds.), *Science education as/for sociopolitical action* (pp. 151–166). New York: Peter Lang.

Borgmann, A. (2006). Foreword. In J. R. Dakers (Ed.), *Defining technological literacy: towards an epistemological framework* (pp. ix–x). New York: Palgrave Macmillan.

Boyne, R., & Rattansi, A. (1990). The theory and politics of postmodernism: By way of an introduction. In R. Boyne & A. Rattansi (Eds.), *An introduction to post modernism and society* (pp. 1–18). London: Macmillan.
Chambers, R. (1995). Introduction. In D. M. Warren, L. J. Slikkerveer & D. Brokensha (Eds.), *The cultural dimension of development: Indigenous knowledge systems* (pp. xiii–xiv). London: Intermediate Technology Publications.
Chisholm, L. (2000). *A South African curriculum for the twenty first century.* (Report of the Review Committee on Curriculum 2005). Presented to the Minister of Education, Professor Kader Asmal. Pretoria, South Africa.
Chisholm, L. (2003). The politics of curriculum review and revision in South Africa. In *Proceedings of Oxford International Conference on Education and Development* (pp. 1–15). Oxford.
Christie, P. (1999). OBE and unfolding policy trajectories: Lessons to be learned. In J. Jansen & P. Christie (Eds.), *Changing curriculum: Studies on outcomes-based education in South Africa.* Kenwyn: Juta.
Dei, G. J., Hall, B. L., & Rosenberg, D. G. (2000). Situating indigenous knowledges: Definitions and boundaries. In G. J. Dei, B. L. Hall, & D. G. Rosenberg (Eds.), *Indigenous knowledges in global contexts: Multiple readings of our world* (pp. 19–20). Toronto: University of Toronto Press Incorporated.
Eijkelhof, H., Franssen, H., & Houtveen, T. (1998). The changing relation between science and technology in Dutch secondary education. *Journal of Curriculum Studies, 30*(6), 677–690.
Ezzy, D. (2002). *Qualitative analysis.* London: Routledge.
Feenberg, A. (1991). *Critical theory of technology.* New York: Oxford University Press.
Feenberg, A. (1999). *Questioning technology.* London: Routledge.
Feenberg, A. (2007). *What is a philosophy of technology?* Presentation at PATT 18 International Conference on Design and Technology Educational Research, Faculty of Education, University of Glasgow, Glasgow.
Hansen, K. -H. (1997). Science and technology as social relations: Towards a philosophy of technology for liberal education. *International Journal of Technology and Design Education, 7,* 49–63.
Horsthemke, K., & Schäfer, M. (2007). Does 'African' mathematics facilitate access to mathematics? Towards an ongoing critical analysis of ethnomathematics in a South African context. *Pythagoras, 65*(June), 2–9.
Hountondji, P. J. (2002). Knowledge appropriation in a post-colonial context. In C. A. Odora Hoppers (Ed.), *Indigenous knowledge and the integration of knowledge systems: Towards a philosophy of articulation.* Claremont, South Africa: New Africa Books.
Ihde, D. (1990). *Technology and the lifeworld.* Indianapolis, IN: Indiana University Press.
International Council for Science. (2002). *Science, traditional knowledge and sustainable development.* ICSU.
International Technology Education Association (ITEA). (2002). *Standards for technological literacy: Content for the study of technology* (2nd ed.). Reston, VA: Author.
ITEA. (2002). *Standards for technological literacy: content for the study of technology* (second). Reston, VA: Author.
Jansen, J., & Sayed, Y. (2001). *Implementing educational policies: The South African experience.* Cape Town: University of Cape Town Press.
Jones, A. (2003). The development of a national curriculum in technology for New Zealand. *International Journal of Technology and Design Education, 13,* 83–99.
Klages, M. (1997). *Postmodernism.* Retrieved from http://www.colorado.edu/English/ENGL2012Klages/pomo.html
Kunnie, J. (2000). Indigenous African philosophies and socioeducational transformation in 'post-apartheid' Azania. In P. Higgs, N. C. G. Vakalisa, T. V. Mda & N. T. Assie-Lumumba (Eds.), *African voices in education* (pp. 158–178). Landsdowne: Juta.
Lindblad, S., & Popkewitz, T. S. (2000). Research problematics and approaches. In S. Lindblad & T. S. Popkewitz (Eds.), *Public discourses on education governance and social integration and exclusion: Analyses of policy texts in European contexts* (pp. 5–24). Uppsala: Uppsala University.
Massaquoi, J. G. M. (1993). *Indigenous technology for off-farm rural activities.* Retrieved from http://www.nuffic.nl/ciran/ikdm/1-3/articles/massaquoi.html

McGovern, S. (1999). *Education, modern development, and indigenous knowledge: An analysis of academic knowledge production*. New York: Garland.

Merriam, S. B. (2001). *Qualitative research and case study applications in education*. San Francisco: Jossey-Bass.

Merriam, S. B. (2002). *Qualitative research in practice*. San Francisco: Jossey-Bass.

Mitcham, C. (1994). *Thinking through technology: The path between engineering and philosophy*. Chicago: University of Chicago Press.

National Research Foundation. (2002). *Focus area: Indigenous knowledge systems*. Retrieved from http://www.nrf.ac.za/focusareas/iks/

O'Riley, P. (1996). A different storytelling of technology education curriculum re-visions: A storytelling of difference [Electronic version]. *Journal of Technology Education, 7*(2), 28–40.

Odora Hoppers, C. A. (2002). Introduction. In C. A. Odora Hoppers (Ed.), *Indigenous knowledge and the integration of knowledge systems: Towards a philosophy of articulation* (pp. vii–xiv). Claremont: New Africa Books.

Pannabecker, J. R. (1991). Technological impacts and determinism in technology education: Alternate metaphors from social constructivism. (Electronic version). *Journal of Technology Education, 3*(1), 43–54.

Pitika Ntuli, B. (2002). Indigenous knowledge systems and the African Renaissance. In C. A. Odora Hoppers (Ed.), *Indigenous knowledge and the integration of knowledge systems: Towards a philosophy of articulation* (pp. 53–66). Claremont: New Africa Books.

Rowell, P. M., Gustafson, B. J., & Guilbert, S. M. (1999). Characterization of technology within an elementary school science program. *International Journal of Technology and Design Education, 9*, 37–55.

Russell, S., & Williams, R. (2002). Social shaping of technology: Frameworks, findings and implications for policy. In K. H. Sorenson & R. Williams (Eds.), *Shaping technology, guiding principles: Concepts, spaces and tools* (pp. 37–131). Cheltenham: Edward Elgar Publishing.

Shava, S. (2006, March). *Indigenous knowledge research and applications in environmental education processes in Southern Africa: Power/knowledge relationships at community and modern institution interfaces*. PhD research proposal, Grahamstown: Rhodes University.

Shiva, V. (2000). Foreword: Cultural diversity and the politics of knowledge. In G. J. Dei, B. L. Hall & D. G. Rosenberg (Eds.), *Indigenous knowledges in global contexts*. Toronto: University of Toronto Press Incorporated.

South Africa. Department of Education. (2002). *Revised national curriculum statements grades R - 9: Technology*. Pretoria: Government Gazette no 443 of 23406.

South Africa. Department of Education. (2003). *Teacher's guide for the development of learning programmes: Technology*. Pretoria: Department of Education.

Stevens, A. (2005). Technology teacher education in South Africa. In *Proceedings of PATT-15 conference*. Netherlands.

Warren, D., & Rajasekaran, B. (1993). Putting local knowledge to good use. *International Agricultural Development, 13*(4), 8–10. Retrieved April 23, 2004, from http://www4.worldbank.org/afr/ikdb/detail.cfm?id=52

Warren, D., Slikkerveer, L. J., & Brokensha, D. (1995). *The cultural dimension of development: Indigenous knowledge systems*. London: Intermediate Technology Publications.

Williams, R., & Edge, D. (1996). The social shaping of technology. *Research Policy, 25*, 856–899.

Woodley, E. J. (2003). *Local and indigenous knowledge as an emergent property of complexity: A case study in the Solomon Islands*. Retrieved from http://0-wwwlib.umi.com.lochbuie.lib.ac.cowan.edu.au/dissertations/preview_all/NQ75999

Sonja Vandeleur and Marc Schäfer
Faculty of Education
Rhodes University
South Africa

LIST OF AUTHORS

Sally Aston is a lecturer at St. Mary's University College, London specialising in primary Design and Technology. Her previous experiences include primary teaching, consultancy in England and overseas, writing books for children and papers for international journals. Research interests include primary D&T and sustainability.

David Barlex is an acknowledged leader in Design and Technology education, curriculum design and curriculum materials development. He taught Science and Technology in comprehensive schools for 15 years before becoming a teacher educator. He directed the Nuffield Design and Technology Projects from 1990 to 2010, which produced an extensive range of curriculum materials widely used in primary and secondary schools in the UK.

Clare Benson is director for The Centre for Research in Primary Technology (CRIPT) based at Birmingham City University. She taught in primary and secondary schools before moving into Higher Education. She has written extensively about primary Design and Technology and has given numerous presentations in England and overseas. Current research interests include Early Years and children's perceptions of Design and Technology.

Pascale Brandt-Pomares is a senior lecturer at the IUFM Aix-Marseille, Université de Provence and a member of the Gestepro team of the UMR ADEF (learning, teaching, assessment, training). Her research is focused on implementing information and communication technologies (ICT) in education and, in particular, about information retrieval in Technology education, through a range of teaching and learning situations.

Marjolaine Chatoney is a lecturer at the University of Provence-IUFM in Aix-Marseille and director of the Science and Technology department. She is a member of laboratory GESTEPRO - Research Unit ADEF (learning, teaching, assessment, and training). She has undertaken research relating to the teaching and learning process in Science and Technology education. She has published articles in national and international scientific journals and actively contributes to the development of Technology education.

Vicki Compton is a research director with the University of Auckland. She has been involved in research, policy and curriculum development in Technology education for over 17 years. She is currently undertaking research focused on the nature of student technological literacy resulting from the implementation of the 2007 New Zealand Technology curriculum.

Alan Cross has worked as a primary school teacher, deputy headteacher, local authority teacher advisor and university lecturer. He has written books and articles about primary Science and Technology. Alan's classroom research has examined aspects of teaching and learning such as learner autonomy and ways in which teachers and others describe teaching Design and Technology.

John Dakers has published extensively on matters relating to Technology education and the philosophy of technology. His primary research focus is the incorporation of technological literacy into the Technology education curriculum. His most recent book *Analyzing Best Practices In Technology Education* co-edited with Marc De Vries, Rodney Custer and Gene Martin was awarded the Silvius-Wolansky Award for the Outstanding Scholarly Publication in Technology Education.

Wendy Dow is an independent researcher. Previously she was a lecturer in Educational Studies at the University of Glasgow. She has worked collaboratively on a range of projects in, and has carried out research into a number of areas pertaining to, Design and Technology Education. Her main research interests are the effect of implicit theories on pedagogy and learning, and factors affecting motivation.

Wendy Fox Turnbull is a senior lecturer and deputy head of school in the School of Sciences and Physical Education at the University of Canterbury, Christchurch, New Zealand. She predominantly lectures in Technology education and professional studies in primary initial teacher education programmes. Her research interests include the impact of authenticity in Technology education and the nature of conversation in the Technology classroom.

Jacques Ginestié is Professor at the Université de Provence. He is the director of the Teacher Educational Institute IUFM Aix-Marseille and in charge of the research unit Gestepro. His research fields are related to the teaching and learning process in Technology education and about the role of technical languages in the transmission of technological knowledge. He is member of many scientific committees and academic societies.

Keith Good is a senior lecturer at the University of Greenwich, London where he is responsible for primary Design and Technology. He has worked in thirteen countries and has given seminars at the National Exhibition Centre, Birmingham for the D&T/ICT Education show for eight consecutive years. Keith is the author of eight books for children and teachers.

Debbie Haffenden is a senior lecturer for primary Design and Technology Education at University of Brighton, England. She is Year 3 leader for the BA Primary course and subject leader for the Foundation Subjects across the BA and PGCE primary programmes.

Gill Hope is a senior lecturer in Design and Technology at Canterbury Christ Church University, England. Her research interests have mainly focused on young children using drawing for designing. She has published two books on teaching

Design and Technology in the primary school and, more recently, one on the use of drawing across the curriculum. She has worked on the Publications Board of the Design & Technology Association.

Pasi Ikonen is a primary school teacher and secondary school Technology education teacher. He acted as a primary teacher and secondary level Technology teacher between 1992–2003. Since 2003, he has worked as a university lecturer in the Department of Teacher Education in Jyväskylä University, Finland.

Dot Jackson has a primary teaching background and is currently a senior lecturer in primary education at St Mary's University College, London. She teaches Design and Technology and Science on the Primary undergraduate programme and also works on post graduate Masters' programmes. Dot's research interests are in sustainable development, creativity and improving critical reflection in Design and Technology.

Esa-Matti Järvinen is based at the University of Oulu in Finland and is Research Manager for the Technology Education Centre – Teknokas. Järvinen has a teaching and research background in Technology education, with a special interest in students' problem solving processes. During 1998–1999 he worked as an Honorary Research Assistant in the University of Exeter, England.

Steve Keirl has taught Design and Technology in all levels of education, is a D&T curriculum theorist, designer and author. He was a TaskForce member for the National Investigation into the Status of Technology Education in Australian Schools. From January 2011, he is Reader in Design Education at Goldsmiths, University of London.

Julie Lunt is currently Executive Officer – Primary at the Design and Technology Association. She is also associated with The Centre for Research in Primary Technology (CRIPT) based at Birmingham City University. Her research interests include children's perceptions of Design and Technology, the nature of writing tasks, and designing.

Denise MacGregor is the Programme Director for the Bachelor of Education, Design and Technology Education at the University of South Australia. Denise has previously worked as a primary school teacher and a Science and Technology coordinator. She has worked with teachers in South Africa and Papua New Guinea developing Design and Technology curriculum. Her research interests include the development of early career teachers' professional identity, and the link between Design and Technology and place based learning.

Gary O'Sullivan is a Senior Lecturer in Technology education at Massey University College of Education, Palmerston North, New Zealand. He is responsible for all facets of Pre-service Technology Teacher Education within the university. He has published and presented papers on Technology education both nationally and internationally and has been working in this field for over twenty five years.

Bhav Prajapat is currently Senior Lecturer at the University of Brighton and Coordinator for Design and Technology Education across post and undergraduate courses across all age phases. She has worked as a designer and an educator. Ms Prajapat's research interests relate to perceptions of Design and Technology Education.

Aki Rasinen is a teacher of technical subjects. He is working as Senior Lecturer in Pedagogy of Technical Work and Technology at the Department of Teacher Education, University of Jyväskylä, Finland. He has also been involved in developing collaborative projects in Zambia, Namibia and Mozambique. He has visited several countries in Europe, Africa, America and Asia as part of his work.

Timo Rissanen graduated as a primary teacher in 1992. He taught in primary schools before undertaking further studies in Technology education. Since 1996 he has worked as a university teacher in the Department of Teacher Education in the University of Jyväskylä, Finland.

Maggie Rogers set up and coordinated the Design and Technology courses on the Primary BA(Ed) and Postgraduate Certificate of Education at Goldsmiths, University of London from 1987 until 2007, after teaching in primary and secondary schools for nearly twenty years. Her research into Education for Sustainable Development has been disseminated widely. Maggie has been a senior research associate in a variety of nationally funded projects.

Marion Rutland is a principal lecturer for Design and Technology (D&T) Education at Roehampton University, England. She was course tutor for the PGCE Secondary D&T and Curriculum Leader for primary D&T programmes. Currently, she is D&T Leader for school based graduate teachers for the West London Programme, the MA D&T Tutor and a PhD supervisor.

Marc Schäfer is a former head of the Education Department and holds the FRF Mathematics Education Chair at Rhodes University in Grahamstown, South Africa. Although his prime research curiosities lie in Mathematics Education, he has engaged in research and development of Technology education in South Africa, particularly in the rural areas of the Eastern Cape.

Martin Seidel is a senior lecturer at Roehampton University, London and is the subject coordinator for Design & Technology within the Department of Education. Roehampton University offers specialist courses in Design & Technology for both Primary and Secondary Initial Teacher Training. Ongoing research interests are looking at expectations within D&T from both teachers' and pupils' perspectives.

Andrew Stevens is a lecturer in Technology education at Rhodes University, Grahamstown, South Africa. He taught mathematics in high schools and was the principal of one of the first non-racial schools in South Africa before joining the ORT-STEP Institute, a non-government organisation which pioneered the introduction of Technology education in South Africa. He is interested in the development of the field of Technology education, particularly through the expansion of teacher education and research.

Kate Ter Morshuizen is currently retired after 46 years of teaching. She has worked extensively in both Pre-service and In-service teacher training and is particularly interested in working with disadvantaged communities. Currently she is still involved in writing Technology education textbooks and supplementary material. She is also involved in writing children's books.

Tara Treleven has been a primary school teacher in South London since 1999. She is currently both the Science and Design and Technology Co-ordinator in her school. She has always had an interest in Design and Technology in the Early Years and has led workshops at the Design Museum, London and the CRIPT conference on the importance of designerly thinking.

Sonja Vandeleur has taught Technology at Roedean School, South Africa since 1998 and is currently Head of Department. She has undertaken postgraduate study in Technology Education at Rhodes University and her research interests include curriculum development, assessment, thinking skills and technological literacy.

Zanariah Yusef binti Mahyun and **Rama Vengrasalam** both work at the Institut Pendidikan Guru Malaysia, Tun Hussein Onn, which is one of the most highly acclaimed teacher training institutes in Malaysia. Both were members of the collaboration development team, working with Canterbury Christ Church University to upgrade the Malaysian teacher training course from Diploma to Honours Degree bearing course. Rama acted as a consultant to the enquiry that led to the establishment of the initiative. Zanariah was the programme director at Tun Hussein Onn for the collaborative programme and Rama played a major part in the planning and delivery of the course.

CPSIA information can be obtained at www.ICGtesting.com
Printed in the USA
LVOW071053190112

264608LV00001B/34/P